W. J. (William John) Robertson, I. J Birchard

The High School Algebra

Part I.

W. J. (William John) Robertson, I. J Birchard

The High School Algebra
Part I.

ISBN/EAN: 9783337158880

Printed in Europe, USA, Canada, Australia, Japan

Cover: Foto ©berggeist007 / pixelio.de

More available books at **www.hansebooks.com**

THE
HIGH SCHOOL ALGEBRA.

PART I.

BY

W. J. ROBERTSON, B.A., LL.B.,
Mathematical Master, Collegiate Institute, St. Catharines;

AND

I. J. BIRCHARD, M.A., Ph.D.,
Mathematical Master, Collegiate Institute, Brantford.

TORONTO:
WILLIAM BRIGGS, WESLEY BUILDINGS,
29 TO 33 RICHMOND STREET WEST.

Entered, according to the Act of the Parliament of Canada, in the year one thousand eight hundred and eighty-six, by WILLIAM BRIGGS, in the Office of the Minister of Agriculture at Ottawa.

PREFACE.

TEACHERS of mathematics have for some time felt that the Algebras now in use in our High Schools and Collegiate Institutes are not adapted to the wants and requirements of the present day. In these works some of the most important departments of Elementary Algebra, such as Factoring, Symmetry, Theory of Divisors and Theory of Equations, are treated so briefly or so superficially that the pupil has found it impossible to obtain a satisfactory knowledge of these subjects without drawing heavily on the resources of the teacher.

In the following pages an effort has been made to treat with considerable fullness the various departments either deficient or wholly absent in the ordinary text-books. While no branch has been slighted, special attention has been devoted to the Theory of Positive and Negative Numbers, to Factoring, Surds, Symmetry, Theory of Divisors and Theory of Quadratics. Convinced that a large and well-graded selection of problems is a desideratum in any manual intended for class work, we have selected and constructed with great care such as we hope will meet the wants of both teachers and pupils. An effort has been made to secure accuracy, but it is quite possible that errors may have crept in despite our vigilance. We shall be glad to have such pointed out so that they may be removed from subsequent editions.

Some difference of opinion may exist as to the propriety of the order in which the different subjects have been introduced.

PREFACE.

The treatment of Symmetrical Expressions and the Theory of Divisors has been delayed until the pupil has acquired considerable familiarity with algebraic symbols and their manipulation. We trust that this arrangement will commend itself to the experience of the mathematical teachers of the country. Surds and the Theory of Indices have been introduced before Quadratic Equations, as we did not think it possible to deal satisfactorily with the Theory of Quadratics without some knowledge of Surds. An effort has been made in this work to encourage pupils to resort to factoring as much as possible in solving equations. In pursuance of this object at a very early stage equations of an easy character, capable of solution by resolution into factors, have been introduced.

This Algebra is intended for all classes of pupils whose studies do not extend beyond the limits prescribed for Second Class Certificates and Pass Junior Matriculation. Having in view the fact that pupils of very different algebraic attainments study within these limits we have so graded the problems that the judicious teacher can easily select such as are adapted to a junior pupil, and leave those more difficult until he becomes well advanced in his work. It should, however, be understood that many of the problems in Factoring and Symmetry go beyond the requirements for Second Class work. Should this venture prove successful it is the intention of the authors to follow it up with Part II., which will deal with the subjects required for Junior Matriculation with Honors and for First Class "C" Certificates.

W. J. ROBERTSON,
I. J. BIRCHARD.

July, 1886.

CONTENTS.

Chapter I.
	PAGE
Definitions and Explanations of Signs	9
Algebraic Notation	13
Quantity and Number	17
Algebraic Numbers	18

Chapter II.
Addition—Subtraction—Use of Brackets	20
Addition	20
Addition with Literal Coefficients	24
Subtraction	25
Subtraction of Polynomials	28
Brackets	29

Chapter III.
Multiplication	32
Multiplication of Polynomials	34
Powers of a Binomial	42

Chapter IV.
Division	45
Horner's Method of Division	49

Chapter V.
Factoring	54
Monomial Factors	54
Trinomials	55
Complete Squares	56
Difference of two Squares	58
Sum and Difference of Cubes	59
Trinomials	61

Chapter VI.
Simple Equations of One Unknown	65
Problems producing Simple Equations of One Unknown	70

CONTENTS.

Chapter VII.
	PAGE
Common Factors and Common Multiples	76
Highest Common Factor	76
Lowest Common Multiple	84

Chapter VIII.
Fractions	87
Reducing Fractions to their Lowest Terms	88
Changes in the form of Fractions	91
Lowest Common Denominator	93
Addition and Subtraction of Fractions	94
Multiplication of Fractions	101
Division of Fractions	105
Complex Fractions	107
Theorems in Fractions	112

Chapter IX.
Fractional Simple Equations	120
Problems producing Fractional Equations	132

Chapter X.
Simultaneous Equations of the First Degree	140
Simple Simultaneous Equations of Three Unknowns	148
Problems producing Simple Simultaneous Equations	152

Chapter XI.
Square and Cube Root	162
Square Root	162
Cube Root	170

Chapter XII.
Theory of Indices	174

Chapter XIII.
Surds	184
Imaginary Expressions	209

Chapter XIV.
Simple Equations involving Surds	212

Chapter XV.
Symmetrical Expressions	217

Chapter XVI.
Theory of Divisors and Complete Squares	234

CONTENTS.

CHAPTER XVII.
Quadratic Equations of One Unknown 248

CHAPTER XVIII.
Quadratics involving Surds 259

CHAPTER XIX.
Simultaneous Equations involving Quadratics...................... 265

CHAPTER XX.
Problems resulting in Quadratic Equations 272

CHAPTER XXI.
Theory of Quadratics... 277

APPENDIX—Special Forms of Simple Equations 291

ANSWERS 295

ELEMENTS OF ALGEBRA.

CHAPTER I.

DEFINITIONS AND EXPLANATIONS OF SIGNS.

1. Algebra is the science which teaches the use of **Symbols** to denote both numbers and the operations to which numbers may be subjected.

2. The symbols employed in Algebra to denote numbers are **Figures**, as in Arithmetic, and the **Letters** of some Alphabet. Thus $a, b, c, \ldots$; x, y, z; a, β, γ; $a', b', c', \ldots$ (read a dash, b dash, c dash); $a_1, b_1, c_1, \ldots$; $a_2, b_2, c_2, \ldots$ (read a one, b one, c one, $\ldots$; a two, b two, c two, $\ldots$), are used as symbols to denote numbers.

3. The signs $>$, $=$, $<$, stand for the words "is greater than," "is equal to," "is less than," respectively. The signs $\therefore$ and $\because$ stand for the words "therefore" and "because."

4. The sign $+$, *plus*, written between two symbols, signifies that the numbers which these symbols denote are to be added.

Thus $7 + 5 = 12$, read 7 *plus* 5 is equal to 12; $a + b$, read a *plus* b, signifies that the numbers denoted by a and b are to be added. But unless we know the numbers for which they stand there is no other way of expressing their sum. A similar remark applies to subtraction, multiplication, etc.

5. The sign $-$, *minus*, written between two symbols, signifies that the number denoted by the second is to be subtracted from the number denoted by the first.

Thus $7 - 5 = 2$, read 7 *minus* 5 is equal to 2; $a - b$, read *a minus b*, signifies that the number denoted by b is to be subtracted from the number denoted by a.

The sign $\sim$ is sometimes used to denote the difference between two numbers when we do not know which is the greater.

6. The sign $\times$, called the sign of multiplication, written between two symbols, signifies that the numbers which they denote are to be multiplied together; thus $a \times b$, read *a into b*, signifies that the number denoted by a is to be multiplied by that denoted by b. The sign, however, is usually omitted between two letters, or between a figure and a letter; thus ab, $3b$ mean the same as $a \times b$, $3 \times b$. A point is sometimes used instead of the sign $\times$, especially when several numbers expressed in figures are to be multiplied together; thus $3.4.5$ means the same as $3 \times 4 \times 5$.

The numbers multiplied together are called the **Factors** of the product.

7. The sign $\div$, called the sign of division, written between two symbols, signifies that the number denoted by the former is to be divided by that denoted by the latter; thus $a \div b$, read *a by b*, signifies that the number denoted by a is to be divided by that denoted by b. The line between the points is sometimes omitted, thus $a:b$, and sometimes the points are omitted and the symbols written in their places, thus $\dfrac{a}{b}$.

8. When two numbers are multiplied together each is called a <u>Coefficient</u> of the other. When one factor is expressed in figures and the other by letters the former is considered the coefficient. Thus in $7b$, ab; 7 and a are the coefficients of b. A coefficient denoted by a figure is called a *numerical* coefficient; by a letter, a *literal* coefficient. When no numerical coefficient is written 1 is always understood.

NOTE.—It is customary, in order to avoid cumbrous phraseology, to use the word "number" when we mean "symbol denoting number"; also we say "the number a" when we mean "the number denoted by a."

DEFINITIONS AND EXPLANATIONS OF SIGNS. 11

9. A **Power** of a number is the product obtained by multiplying it by itself any number of times.

Thus $a \times a \times a$, or aaa, is called the **Third Power** of a. The Second and Third Powers are also called the square and cube respectively.

10. An **Exponent**, or **Index,** is a small figure placed above and to the right of a number to show the **Power** to which the number is to be raised.

Thus a^4 (read a to the fourth) is a short way of writing $aaaa$. When no exponent is written 1 is understood.

11. A **Root** of a number is one of two or more equal factors whose product equals the given number.

12. The **Index** of a **Root** is a figure which shows how many equal factors are to be found. The roots of numbers are indicated thus: $\sqrt{a}$, $\sqrt[3]{a}$, $\sqrt[4]{a}$, etc., which denote the second, third, fourth roots, etc., of a.

The second and third roots are usually called the square and cube roots respectively. The index 2 is usually omitted from the sign for the square root.

The sign $\sqrt{}$ is a corruption of the letter r in radix, and is called the radical sign.

13. One or more numbers represented by algebraic symbols is called an **Expression.**

Thus $5a$, $3a^2b$, $6x - 7y$, etc., are algebraic expressions.

14. The numbers connected by the signs $+$ and $-$ are called **Terms.**

15. Like Terms are those which differ only in their numerical coefficients; thus $3ab$ and $5ab$, $4x^2y$ and $7x^2y$ are like terms, but $5ax$ and $5ay$, $3a^2b$ and $5ab^2$ are unlike terms.

16. An expression consisting of one term is called a Monomial; an expression consisting of two terms is called a Binomial; an expression consisting of three terms is called a Trinomial;

an expression consisting of four or more terms is called a Multinomial or Polynomial. The word Polynomial, however, is frequently used to denote any algebraic expression except a Monomial.

17. The **Dimensions** of a term are the literal factors in it; the **Degree** is the *number* of such factors.

Thus $4a^2b^3c$ is a term of $2+3+1=6$ dimensions, or of the sixth degree.

18. An expression is **Homogeneous** when all the terms are of the same number of dimensions.

Thus $a^2 + 2ab + b^2$ is homogeneous and of two dimensions.

19. A polynomial is said to be arranged according to the powers of a letter when the exponents of that letter in the various terms are in order of magnitude.

Thus $3a^3 + 4a^2x - 5ax^2 + 6x^3$ is arranged in descending powers of a, but in ascending powers of x.

20. The signs ( ), { }, [ ], called **Brackets**, signify that the terms enclosed form a group, which is to be treated as a single term.

Thus $a - (b + c)$ signifies that the sum of b and c is to be taken from a; $\{a - (b + c)\}^3$ signifies that the former result is to be cubed; $[m - \{a - (b + c)\}]y$ signifies that the sum of b and c is to be taken from a, the remainder subtracted from m, and this last remainder multiplied by y. A line, called a **Vinculum**, drawn over a number of terms, is sometimes used instead of a bracket; thus $a - \overline{b + c}$ signifies the same as $a - (b + c)$.

Examples.—If $a = 12$, $b = 4$,

Then $\quad ab = 12 \times 4 = 48$; $\quad \dfrac{a}{b} = \dfrac{12}{4} = 3$;

$7a^2 = 7 \times 12 \times 12 = 1008$; $\quad a^2 - (a-b)^2 = 144 - 64 = 80$,

$\sqrt{a^2 - 2ab + b^2} = \sqrt{144 - 96 + 16} = \sqrt{64} = 8$.

EXERCISE I.

If $a=1$, $b=2$, $c=3$, $d=4$, $e=5$, $m=0$, find the value of the following expressions:—

1. $2a + 3b - c$.
2. $2c + 3d + 4e - 7$.
3. $a^2 + ab + b^2$.
4. $bc + cd + de$.
5. $c^2 + 2cd + d^2$.
6. $abc + bcd + cde$.
7. $3b^2 - 4a^2 + 7e^2$.
8. $2b^2e^2 - a^3b - me^4$.
9. $3(a+b)^2 - (m+e)^2$.
10. $4e^2 - \{(d-c)^2 + 3b^2\}$.
11. $(a+b)(c+d)$.
12. $e(a+bcd-e^2)^2 + \{e-(d-b)^2\}^2$.

Prove the following equalities:—

13. $a+b+c+d = \frac{1}{2}de$.
14. $\sqrt{c^2 + d^2} = e$.
15. $\sqrt[3]{c^3 + d^3 + e^3} = \frac{1}{2}cd$.
16. $\sqrt{10\sqrt{e^2 - d^2} + 2c} = a+e$.
17. $a^3 + b^3 + c^3 + d^3 + e^3 = (a+b+c+d+e)^2$.
18. $\frac{2}{3}(c+d+e)^3 = c^2(d+e) + d^2(e+c) + e^2(c+d)$.
19. $a^2b^2d^2\left(\dfrac{1}{a^3} + \dfrac{1}{b^3} + \dfrac{1}{d^3}\right) = a^3 + b^3 + d^3$.
20. $2a^2b^2 + 2b^2c^2 + 2c^2a^2 - a^4 - b^4 - c^4 = md$.
21. $\dfrac{8a^2 + 3b^2}{a^2b^2} + \dfrac{4c^2 + 6b^2}{c^2 - b^2} - \dfrac{c^2 + d^2}{e^2} = b^d = d^b$.
22. If $a=16$, $b=10$, $x=5$ and $y=1$, find the value of
$$(b-x)(\sqrt{a}+b) + \sqrt{\{(a-b)(x+y)\}}, \text{ and of}$$
$$(a-y)\{\sqrt{2bx}+x^2\} + \sqrt{\{(a-x)(b+y)\}}.$$

ALGEBRAIC NOTATION.

21. The following examples are designed to furnish additional exercise in Algebraic Notation. It is very important that the student should be able to express the various mathematical relations and the operations to which numbers may be subjected, either in words or in algebraic symbols, and to accurately translate the one form of expression into the other.

ALGEBRAIC NOTATION.

Example 1.—Express the following statement in algebraic symbols :—The difference of the squares of any two numbers is equal to their sum multiplied by their difference.

Let a and b represent the numbers, a being the greater.

Then a^2 and b^2 represent their squares; $a^2 - b^2$ represents the difference of their squares; $a + b$ and $a - b$ represent their sum and their difference. $\therefore a^2 - b^2 = (a + b)(a - b)$ is the expression required.

Example 2.—A pedestrian having agreed to walk a miles in h hours travels the first k hours at the rate of m miles an hour. At what rate must he travel the remainder of the time?

In k hours at m miles an hour he would travel k times m miles, *i.e.*, km miles; $a - km$ is the remaining distance in miles; $h - k$ is the remaining time in hours. $\therefore a - km$ divided by $h - k$, *i.e.*, $\dfrac{a - km}{h - k}$ is the number of miles an hour required.

EXERCISE II.

Express in algebraic symbols the following :—

1. The sum of any two numbers. (Use a and b).
2. The sum of the squares of any two numbers.
3. The square of the sum of two numbers.
4. Six times the product of two numbers.
5. The sum of the cubes of two numbers divided by the sum of the numbers.
6. The square root of the sum of the squares of two numbers.
7. The square of the sum of two numbers is equal to the sum of their squares together with twice their product.
8. The difference of the cubes of two numbers, divided by the difference of the numbers, is equal to the sum of the squares of the numbers, together with their product.
9. How many cents in x dollars? How many dollars in x cents.

ALGEBRAIC NOTATION.

10. How many inches in x feet and y inches? In x yards and y feet?

11. Find the cost of 7 hats at x dollars each.

12. A man gives x dollars in payment for 5 books at a dollars each and 7 books at b dollars each; how much change should he receive?

13. How far will a person travel in x hours at y miles an hour?

14. How long will it take to travel a miles at x miles per hour?

15. A man works q hours a day for n days and p hours a day for m days. He receives x cents per hour; how many dollars does he receive altogether?

16. Find the sum of $x + x + x + \ldots$ where x is written m times.

17. A flower bed is x feet long and y feet wide; how many square feet in the bed?

18. A block is x feet long y feet wide and z feet thick; how many cubic feet in it? How many square feet on all the faces? How many feet in the sum of the lengths of all the edges?

19. A book has x pages; each leaf is y inches long and z inches wide; how many square yards of paper in the book?

20. From a rod x inches long I cut off y inches, and divide the remainder into m equal parts; how many inches in each part?

21. A boy is x years old and his brother y years; find the sum of their ages after five years.

22. What number subtracted from x will leave 10?

23. What number subtracted from x will leave y?

24. The dividend is x and the quotient y; what is the divisor?

25. The divisor is x, the quotient y, and remainder r; what is the dividend?

26. What value of x will make $7x$ equal to 35? $\dfrac{x}{7}$ equal to 14? $x + 2$ equal to 20? $3x + 5$ equal to 26? $4x^2$ greater than 7 by 29?

ALGEBRAIC NOTATION.

27. The number x is to be increased by 3, twice the sum is to be multiplied by $a + b$, and the product, diminished by d, is to be divided by the sum of m and n. Express these operations algebraically.

28. A grocer mixed a pounds of tea worth x cents a pound with b pounds worth y cents a pound; what was the mixture worth a pound?

29. $x + y$ houses have each $a + b$ rooms, and in each room are $p + q$ persons; how many persons in all?

30. There are $x + y$ rows of trees, $x + y$ trees in a row, on each tree $x + y$ bushels of apples worth $x + y$ cents a bushel; how many dollars' worth of apples in all?

31. A man having m dollars buys x pounds sugar at a cents a pound and y pounds of tea at b cents a pound; how many pounds of coffee, at c cents a pound, can he buy with the remainder of his money?

32. A man divided x dollars among m boys and y dollars among n girls. Two boys and three girls put their money together and bought p pounds of candy; how many cents a pound was the candy?

33. A man has a journey of x miles; he travels a hours at b miles per hour and c hours at d miles per hour; how long will it take to finish the journey at y miles per hour?

34. A train, having to make a journey of x miles in h hours, ran for k hours at the rate of r miles an hour and then made a stop of m minutes; how fast must it run during the remainder of its journey to arrive on time?

35. Two cities are m miles distant from each other; two travellers start at the same time, one from each city, and travel towards each other at the rate of x and y miles, respectively, per hour. How long before they will meet? How far will each travel?

QUANTITY AND NUMBER.

22. A Quantity is that which is capable of being divided into parts.

Thus Distance, Time, Weight, etc., are Quantities.

23. A Quantity is measured and its magnitude estimated by selecting some known, definite Quantity of the same kind as a **Standard** or **Unit**, and then finding by trial how many times this unit must be repeated to make up the given Quantity.

24. Number arises from considering the repetitions of the unit necessary to make up a given Quantity.

A number taken in connection with a particular unit is called **Concrete**; without a particular unit, **Abstract**. Thus the magnitude of a Quantity is represented by a **Concrete Number**.

25. When two Quantities are so related to each other that, being taken together, they cancel or destroy each other, either in whole or in part, one of them is called a **Positive Quantity**, the other a **Negative Quantity**.

26. The preceding Arts. may be illustrated as follows:—

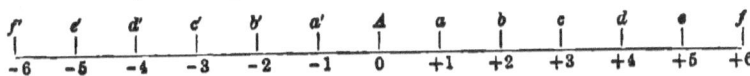

1. Draw any straight line Af; its length will be a Quantity.

2. To measure that Quantity select any length Aa for a unit, and mark off portions each equal to the unit Aa.

3. The numbers 1, 2, 3, 4, 5, 6, in connection with unit Aa, represent the distances of the points a, b, c, d, e, f, from A.

4. Similarly, if the line be produced to f', Af' may be measured in the same way, and the numbers 1, 2, 3, 4, 5, 6, will represent the distances of the points a', b', c', d', e', f', from A.

5. If a point move from A through any number of units of distance to the right, and then through the same number of units to the left, its distance from A will be zero, the two motions having cancelled each other; therefore

6. If distance to the *right* be called a positive Quantity, distance to the *left* is a negative Quantity.

7. Positive Quantities are represented by numbers with the sign $+$ before them, negative Quantities by numbers with the sign $-$ before them; hence the signs $+$ and $-$ are called the positive and the negative sign.

ALGEBRAIC NUMBERS.

27. Numbers taken in connection with the signs $+$ and $-$ are called **Algebraic Numbers**—the former, **Positive Numbers**; the latter, **Negative Numbers**. Without any sign they are called **Absolute** or Arithmetical Numbers.

28. Two algebraic numbers, the one preceded by the sign $+$, the other by the sign $-$, are said to have **unlike signs**. In practice the sign $+$ is usually omitted, and the sign $-$ is used to signify that the number before which it stands represents a Quantity of a nature opposite to some other Quantity previously considered, and which was represented by a number without regard to sign; hence absolute numbers are tacitly considered to be positive, and when no sign is written $+$ may always be understood.

29. Two Quantities, one positive the other negative, containing the same absolute number of units, have the same magnitude, or in other words, a negative Quantity is as *large* as the corresponding positive Quantity: the signs $+$ and $-$ have nothing to do with the magnitude of a Quantity.

30. One number is said to be algebraically greater than another when in a scale of numbers, as in Art. 26, it lies in a *positive* direction from the other.

Thus -2 is said to be algebraically greater than -6, 0 greater than -2, etc.

This is only a convenient way of speaking, and so long as the meaning of such expressions is clearly defined no confusion can arise. It would, of course, be absurd to consider either a number or a Quantity as really less than nothing.

ALGEBRAIC NUMBERS.

31. The following examples show how algebraic numbers are applied to represent positive and negative Quantities. To find the numbers which represent the lines bf, ea, $a'f'$, $e'c$, in magnitude and direction.

From b to f the distance is 4, the *direction* positive.
" e to a " " 4, " " negative.
" a' to f' " " 5, " " negative.
" e' to c " " 8, " " positive.

Hence the lines are represented by the numbers $+4$, -4, -5, $+8$, respectively.

In these examples the signs $+$ and $-$ denote *direction;* in other applications their signification will be readily perceived.

NOTE.—The word "quantity" is frequently used instead of "algebraical expression." When used according to definition, Art. 22, it will be written with a capital, otherwise without.

EXERCISE III.

1. If a line 8 inches long be the unit, what number will represent 22 yards? a quarter of a mile?

2. The number 25 represents half a ton, what is the unit?

3. A certain distance is represented by 36 when the unit is 2 feet 6 inches; what number will represent the same distance if the unit be changed to 10 inches? to 40 inches?

4. The sum of the lengths of the edges of a cube is represented by 36, and the unit of length is 7 inches: find the numbers which represent the area of a face and the volume respectively.

5. If 5 miles to the east of any place be represented by $+5$, what will -5 represent?

6. If a tree 100 feet high be represented by $+20$, what will correctly represent the depth of a well 40 feet deep?

7. If cash in hand be represented by positive numbers, what will negative numbers represent?

8. If we denote a pound *weight* by $+1$, what will denote the force of a balloon lifting 100 pounds?

CHAPTER II.

ADDITION—SUBTRACTION—USE OF BRACKETS.

ADDITION.

32. The **Sum** of two or more algebraic numbers is the single number which correctly represents the Quantity formed by combining the Quantities represented by the several numbers. The process of finding the sum is called **Addition**.

33. The signs + and −, when used to designate positive and negative numbers, are distinguished from the same signs when used to indicate the *operations* of addition and subtraction by enclosing them with the number in a bracket.

Thus $(+4)+(-3)$ indicates the addition of 4 positive and 3 negative units; $(+4)-(-3)$ indicates the subtraction of 3 negative from 4 positive units.

34. Let it be required to perform the following additions:—

1. $(+5)+(+2)$.
2. $(-5)+(-2)$.
3. $(+5)+(-2)$.
4. $(-5)+(+2)$.

Referring to the scale of algebraic numbers, Art. 26, we see that 5 positive units and 2 positive units make 7 positive units, just as 5 inches and 2 inches make 7 inches. Similarly, 5 negative units and 2 negative units make 7 negative units. To add 5 positive and 2 negative units we start from A, the zero point, move 5 units to the *right*, and then 2 units to the *left;* this leaves us at c, whose position is denoted by $+3$. Similarly, moving 5 units to the *left* and two units to the *right* brings us to c', whose position is denoted by -3. We have then the following results:—

1. $(+5)+(+2)=+7$.
2. $(-5)+(-2)=-7$.
3. $(+5)+(-2)=+3$.
4. $(-5)+(+2)=-3$.

ADDITION. 21

35. To add algebraic numbers we have from Art. 34 the following

RULE.—*Take the sum of the absolute values of numbers having like signs and prefix the common sign. Take the difference of the absolute values of numbers having unlike signs and prefix the sign of the greater.*

36. If there are several numbers to be added we may proceed in the same way until all are combined into a single number. It may be assumed, however, that the result will be the same in whatever order the numbers are taken, and it will be found more convenient to add the positive and the negative numbers separately and then to combine the results.

37. If we divide each of the units of length (Art. 26) into m parts the lines there denoted by $2, 3, 4, -2, -3, -4$, etc., will be denoted by $2m, 3m, 4m, -2m, -3m, -4m$, etc.; and by the same reasoning as before we shall get such results as

1. $(+5m)+(+2m) = +7m$. 2. $(-5m)+(-2m) = -7m$.
3. $(+5m)+(-2m) = +3m$. 4. $(-5m)+(+2m) = -3m$.

From which we learn that

Like terms are added by taking the algebraic sum of their coefficients and annexing the common literal factors.

The addition of **unlike terms** can only be indicated by connecting the terms with the proper signs.

38. In practice the brackets used to distinguish the different uses of the positive and negative signs are omitted, and the expressions $(+a)+(+b)$ and $(+a)+(-b)$ are written $a+b$ and $a-b$. The latter expression has already been defined to mean that b is to be subtracted from a. So long as a is greater than b it is evident from Art. 34 that the two interpretations give the same result, and it will be shown hereafter (Art. 44) that this is always true. From these considerations we deduce two conclusions:—

1. An algebraical expression, being the sum of the several

algebraical numbers composing its terms, is itself an algebraical number, and may be treated as such.

2. The terms of an algebraical expression may be arranged in any order. This enables us to simplify expressions by collecting the like terms.

39. The addition of algebraical expressions is *indicated* by writing the terms in succession, each preceded by its proper sign; and the addition is *performed* by collecting like terms, and thus reducing the expression to its simplest form.

Ex.—Add $+3a$, $+2b$, $-5b$, $+7a$, $+b$, $-4a$.

$3a + 2b - 5b + 7a + b - 4a = 3a + 7a - 4a + 2b + b - 5b = 6a - 2b$.

EXERCISE IV.

Add the following:—

1. $+7$, $+8$.
2. -7, -8.
3. $+12$, -3.
4. -17, $+7$.
5. -7, -20, $+3$.
6. $+13$, -20, -8.
7. $+5ab$, $-3ab$, $-7ab$.
8. $-8m^2$, $-3m^2$, $+20m^2$.
9. $+4m^2$, $-3n^2$, $-10n^2$, $+8m^2$, $-2m^2$.
10. $+2a$, $+3b$, $-2b$, $-5a$, $+7c$, $+d$.
11. $4x + (-3y) + (+2x) + (+3y) + (-7z) + (-2z)$.
12. $5x - 7y - 3y - 2x + 4x - 3y + 2z$.
13. $+7a^2 + (-3b^2) + (-4a^2) + (-5ab) + (+4b^2)$.

14. Express in algebraical numbers the dates 44 B.C. and 1885 A.D. What dates are 25 years after each of them?

15. A person travels 20 miles and then returns 15 miles. Express his journeys algebraically and add them: add them arithmetically. What is taken into account algebraically which is neglected arithmetically?

16. A merchant has $1000 cash and goods worth $3000, but he owes $2000 to one man and $2500 to another; find his net capital.

ADDITION.

40. When two or more polynomials are to be added it is convenient to arrange the terms in columns so that like terms shall stand in the same column. The columns are usually added in succession, beginning at the left; but they may be taken in any order, care being taken to prefix the proper sign to the sum of each column as it is written down.

Examples.

1. $4x^2 + 3xy - 2y^2$
   $-3x^2 + 7xy + 3y^2$
   $5x^2 - 2xy + y^2$
   $-2x^2 - 9xy - 2y^2$
   $\overline{4x^2 - xy}$

2. $2a - 3b + 4c$
   $5a - 7b + 2c$
   $-5a + 3b - 6c + x$
   $-2b$
   $\overline{2a - 9b + x}$

Add   EXERCISE V.

1. $2a + 3b + c$, $4a - b + 2c$, $-2a + 5b + 4c$.
2. $4a + 5b - 6c$, $4b + 5c - 6a$, $4c + 5a - 6b$.
3. $7a - 4b - 3c$, $2b - 5c - 4a$, $6a - 11b + 4c$.
4. $2ab - 3bc + 4ac$, $5bc - 3ab$, $4ac - 7bc$.
5. $3x^2 - 5xy - 2y^2$, $2xy + 3y^2$, $7x^2 - 4xy - 5y^2$.
6. $a + b - 2c$, $b + c - 2d$, $c + d - 2a$, $d + a - 2b$.
7. $7(x + y)$, $4(x + y)$, $-5(x + y)$.
8. $5(a^2 + b) + 2c$, $3(a^2 + b) - 7c + d$, $2c - 4d$, $3d - 6(a^2 + b)$, $7(a^2 + b) + 3c$.
9. $3a(x - y)$, $4a(x - y)$, $-9a(x - y)$, $a(x - y)$.
10. $(m + n)^2 + x$, $2(m + n)^2 + y$, $z - 5(m + n)^2$.
11. $4(2a - 3b) + 2c$, $5(2a - 3b) - 3d$, $5c + 4d$, $2a - 3b$, $3(2a - 3b) - 9c - d$.
12. $9(x^2 + y^2) - 3xy$, $x^2 - 7xy + y^2$, $10xy - 10(x^2 + y^2)$.
13. $7a - 3b + 5c - 10d$, $2b - 3c + d - 4e$, $5c - 6a - 4e + 2d$, $-3b - 8c + 7a - e$, $21e - 16c + a - 5d$.
14. $3ab^2 - 4a^2b + a^3$, $-4ac^2 + 5ab^2 - c^3$, $-7b^2 + 2a^2b - 6ac^2$, $5a^3 - 11ab^2 - 12ac^2$.

ADDITION WITH LITERAL COEFFICIENTS.

15. $a^3 + 3a^2b + 3ab^2 + b^3,\quad -5ab^2 + 3a^2b - b^3 + 3a^3,$
    $3ab^2 - 5a^2b + 3b^3 - 3a^3,\quad -5b^3 + 2a^2b - 4a^2 + 3ab^2,$
    $7a^3 + 6b^3 - 5a^2b + 5ab^2,\quad 4a^2b + 3ab^2 - 3b^3 + 4a^3.$

16. $a^5 + 5a^4b + 6a^2b^2c - 7ab,\quad 6a^5 - a^4b - 6a^2b^2c + 10ab,$
    $-2a^5 + 4a^4b + 12a^2b^2c - 10ab,\quad 5a^5 - 16a^4b - 11a^2b^2c + 13ab$
    $-10a^5 + 8a^4b + a^2b^2c - 6ab.$

17. $2(x+y+z),\ 3(x+y-z),\ 5(x+z-y),\ 7(z+y-x).$

18. $3(x - 2y + 3z),\ 4(y - 2z + 3x),\ 5(z - 2x + 3y).$

19. If $x = a + 2b + 3c,\ y = b + 2c + 3a,\ z = c + 2a + 3b$, find the value of $x + y + z$.

20. If $x = a + b - 2c,\ y = b + c - 2a,\ z = c + a - 2b$, find the value of $x + y + z$.

ADDITION WITH LITERAL COEFFICIENTS.

41. When several unlike terms, having one or more literal factors in common, are to be added, it is frequently convenient to collect them into a single term by the aid of brackets, as in the following example.

We reduce such an expression as $6x + 4x - 3x$ to a single term by adding the coefficients; thus $6x + 4x - 3x = (6 + 4 - 3)x = 7x$. Similarly, $ax + bx - cx = (a + b - c)x$. The only difference between the two examples is, in the first case we have a single symbol, 7, to represent the sum of the coefficients 6, 4 and −3, but in the second we have no such symbol, and consequently we must write the coefficients themselves.

Ex. $3ax + 4ay - 2bx + 3by = (3a - 2b)x + (4a + 3b)y,$
$\qquad\qquad\text{or} = (3x + 4y)a + (3y - 2x)b,$

according as we collect the coefficients of x and y, or of a and b.

To find the algebraic sum of the coefficients of any given letter it is convenient to arrange the terms containing that letter in columns, as in the last exercise.

SUBTRACTION. 25

EXERCISE VI.

Collect the coefficients of x and y in the following examples:—

1. $ax + by + mx + ny$.
2. $2ax - 3by - 7x + 10y$.
3. $mx + ny - ay - bx + x - y$.
4. $4ay + 7by - 3ay - 10dx + 5cx$.
5. $ax + by + bx + ay - (a + b)x + (a + b)y$.
6. $(2a - 3b)x + (m - n)y + 3bx + ny + ax - my$.
7. $(m - 3)y + (3 - n)x + 4y + 2nx - 3x - my$.
8. $mx + ny + (m - n)x + (m - n)y - 2mx - my$.
9. $(a - b)x + (b - c)y + (b - c)x + (c - a)y + (c - a)x + (a - b)y$.
10. $(a + b - c)x + (a - b - c)y + (b + c - a)x + (b - c - a)y$
    $+ (c + a - b)x + (c - a - b)y + (a + b + c)y$.

SUBTRACTION.

42. When the sum of two numbers and one of the numbers are given, **Subtraction** is the process by which the other may be found.

The given number is called the **Subtrahend**, the sum of the numbers is called the **Minuend**, and the number to be found is called the **Difference**. The Difference is, therefore, the number to be added to the Subtrahend to make it equal to the Minuend.

43. Let it be required to perform the following subtractions:—

1. $(+2) - (+5)$.
2. $(+2) - (-5)$.
3. $(-2) - (+5)$.
4. $(-2) - (-5)$.

To find what number must be added to the subtrahend to make it equal to the minuend, find on the scale of numbers (Art. 26) the *distance* and *direction* from the former to the latter; the distance will be denoted by a *number of units*, the direction by a sign.

From $+5$ to $+2$ the distance is 3, the direction negative.
" -5 to $+2$ "   "   7, "   "   positive.
" $+5$ to -2 "   "   7, "   "   negative.
" -5 to -2 "   "   3, "   "   positive.

3

We have, therefore, the following results:—

1. $(+2)-(+5) = -3.$ 2. $(+2)-(-5) = +7.$
3. $(-2)-(+5) = -7.$ 4. $(-2)-(-5) = +3.$

44. With the preceding results compare the following examples in addition:—

1. $(+2)+(-5) = -3.$ 2. $(+2)+(+5) = +7.$
3. $(-2)+(-5) = -7.$ 4. $(-2)+(+5) = +3.$

These examples show—

1. To subtract any number of positive units is the same as to add the same number of negative units.

2. To subtract any number of negative units is the same as to add the same number of positive units.

These statements may be briefly expressed in symbols as follows:—

1. $a-(+b) = a+(-b) = a-b,$ (Art. 38.)
2. $a-(-b) = a+(+b) = a+b,$

which show that, when taken in connection with another term,

$$-(+b) = -b \quad \text{and} \quad -(-b) = +b.$$

45. When the sign − stands between two numbers it indicates the operation of subtraction; but when the preceding number is removed that meaning can be no longer retained. The meaning we assign it in such cases is to indicate that the number before which it stands represents a Quantity of a nature opposite to that which the number would represent without the negative sign. Thus if *distance* and *direction* be denoted by $(-b)$, then $-(-b)$ denotes the same distance in the direction opposite to that denoted by $(-b)$, *i.e.*, in the direction denoted by $(+b)$. This meaning being in harmony with Art. 44 enables us to use such combinations of symbols either separately or in connection with other terms, the meaning in each case being the same, viz.:—

$$-(+b) = -b \quad \text{and} \quad -(-b) = +b.$$

SUBTRACTION. 27

46. From Art. 44 we derive, for the subtraction of algebraic numbers, the following

RULE.—*Change the sign of the subtrahend and add it to the minuend.*

47. By reasoning similar to Art. 37 we learn that

Like terms are subtracted by the algebraic subtraction of their coefficients and annexing the common literal factors.

The subtraction of **unlike terms** can only be indicated by connecting the terms with the proper signs.

EXERCISE VII.

1. From $+25$ take $+17$.
2. From $+17$ take $+25$.
3. From -50 take -20.
4. From -20 take -50.
5. From $+16$ take -25.
6. From -35 take $+40$.
7. From $13a$ take $7a$.
8. From $7a$ take $13a$.
9. From $-5m$ take $3m$.
10. From $8x$ take $5x$.
11. From $3ab$ take $11ab$.
12. From $5a^2x$ take $-3a^2x$.
13. From $-6ay$ take $-3ay$.
14. From aby take $4aby$.
15. $13m^2x^2 - (-4m^2x^2)$.
16. $3xy - 15xy$.
17. $9x^2 - 11x^2 + 3x^2 - (-5x^2)$.
18. $8y^2 + 12y^2 - 14y^2 - (-3y^2)$.
19. $5x^3 - (-3x^3) - (+8x^3) + m - n$.
20. $5a - (-3b) + 8b - (-7a) - 12a - 11b$.

21. Rome was founded 753 B.C. and overthrown 476 A.D. Express these dates by algebraical numbers, and find the number of years between them.

22. At the beginning of a year a merchant was $1500 in debt; at the end he had $2000 cash. Find by algebraic subtraction his gain during the year.

SUBTRACTION OF POLYNOMIALS.

48. One polynomial is subtracted from another by subtracting each term in succession, *i.e.*, by changing the sign of each term of the subtrahend and adding it to the minuend. It is convenient to place like terms under each other as in addition.

Ex. 1.—From $3x^3 - 7x^2y + 5xy^2 - y^3$
take $2x^3 - 8x^2y + 7xy^2 + 3y^3$
Result $x^3 + x^2y - 2xy^2 - 4y^3$

The signs should only be changed mentally. For a time the student should test the correctness of each result by adding it to the subtrahend: the sum should equal the minuend.

Ex. 2.—From $(a+b)x + (b+c)y + (c+a)z$
take $(b+c)x + (c+a)y + (a+b)z$
Result $(a-c)x + (b-a)y + (c-b)z$

To perform the above subtraction we subtract the coefficients of x, y and z in succession, *i.e.*, we have to perform the following three easy subtractions:—

From	$a+b$	$b+c$	$c+a$
take	$b+c$	$c+a$	$a+b$
Results	$a-c$	$b-a$	$c-b$

EXERCISE VIII.

1. From $3a - 2b + 5c$ take $2a + b + 7c$.
2. From $5a + 2b - 7c$ take $7a - 5c + 3b$.
3. From $a - b + c$ take $c - a + b + x$.
4. From $4x^3 - 5x^2y - 7xy^2 - y^3$ take $-x^3 + 4xy^2 - y^3$.
5. From $x^3 - 3x^2y + 3xy^2 - y^3$ take $-x^3 + 3x^2y - 3xy^2 + y^3$.
6. From $5x^3 - 7x^2 + 3x - 1$ take $3 + 4x - 5x^2 + 6x^3$.
7. From $2 - 3x + 6x^2 - 5x^3$ take $x^3 - 4x^2 + 7x - 5$.
8. From $a^2 - b^2 - c^2 + 2bc$ take $b^2 - c^2 - a^2 + 2ac$.

BRACKETS.

9. From $x^2 + xy - a$ take $y^2 + 4xy - b$.
10. From $a^4 + b^4$ take $4a^3b - 6a^2b^2 + 4ab^3$, and from the result take $a^4 - 5a^3b + 6a^2b^2 - 5ab^3 + b^4$.
11. From $8x^2 - 9xy + 2y^2 - 3xz + 2yz - z^2$
    take $3x^2 - y^2 - z^2 + 2xy - 3yz + 4xz$.
12. From $a^2bc - ab^2c + abc^2 - abc$ take $abc - a^2bc - ab^2c - abc^2$.
13. From $ax^2 - by^2$ take $cx^2 - dy^2$.
14. From $ax^2 + 2bxy + cy^2$ take $lx^2 + 2mxy + ny^2$.
15. From $(a - b)x^2 + (b - c)xy + (c - a)y^2$
    take $(a - c)x^2 + (b - a)xy + (c - b)y^2$
16. From $4(a - b) + 3(x + y)$ take $3(a - b) - 5(x + y)$.
17. From $3(a + b) - 5(c + d) + 7(x + y) + m$
    take $(a + b) - 3(c + d) - 3(x + y) + n$.
18. From $(a + b - c)x + (b + c - a)y + (c + a - b)z$
    take $(a - b - c)x - (a - b + c)y - (b - c + a)z$.
19. What must be subtracted from the sum of $4x^3 + 3x^2y - y^3$, $4x^2y - 3x^3$, $7x^2y + 9y^3 - 2x^2y$, to leave the remainder $2x^3 - 3x^2y + y^3$.

BRACKETS.

49. The addition and subtraction of polynomials is frequently indicated by enclosing the expression to be added to, or subtracted from another expression, in a bracket, preceded by the sign + or –.

50. To add a polynomial to another expression we add each term in succession; hence *a bracket preceded by the sign + may be removed.*

To subtract a polynomial from another expression we subtract each term in succession, *i.e.*, we change the sign of each term and proceed as in addition; hence *a bracket preceded by the sign – may be removed if we at the same time change the sign of every term within it.*

The rules for introducing brackets follow at once from the rules for their removal.

51. If two or more pairs of brackets be used in the same expression they may be removed, one pair at a time, by the preceding rules. It is easiest to begin with the inside pair, but we may begin with any pair, and a little experience will enable the student to remove several pairs at one operation. At each step of the operation like terms should be combined, to save labor in writing.

Ex. 1. $\quad a - \{b + (c - d)\} = a - \{b + c - d\}$
$\qquad\qquad\qquad\qquad\quad = a - b - c + d.$

Ex. 2. $a - [b - \{a - (b - a) - b\} - a] = a - b + \{a - (b - a) - b\} + a$
$\qquad\qquad\qquad\qquad\qquad\qquad = 2a - b + a - (b - a) - b$
$\qquad\qquad\qquad\qquad\qquad\qquad = 3a - 2b - b + a$
$\qquad\qquad\qquad\qquad\qquad\qquad = 4a - 3b.$

In Ex. 2 the outside brackets were removed each time and like terms combined.

EXERCISE IX.

Remove the brackets from the following expressions and combine like terms:—

1. $(a - b) + (b - c) - (a - c).$
2. $(2a - b - c) - (a - 2b + c) - (a + b - 2c).$
3. $3a - \{b + (2a - b) - (a - b)\}.$
4. $2a + (b - 3c) - \{(3a - 2b) + c\} + 5a - (4b - 3c).$
5. $(3c - 2d) - (2d - 3c) + \{-(c - d) - (3c + 2d)\}.$
6. $x + [x - a - (2a - 2x) + \{a - (a - x)\}].$
7. $-\{(3x - 4y) - (2x - 5y)\} + \{4x - (2y - 3x) - 5y\}.$
8. $a - [b - \{a - (b - \overline{a - b}) - a\} - b].$
9. $3a - [a + b - \{a + b + c - (a + b + c + d)\}].$
10. $a - [2b + \{3c - 3a - (a + b)\} + 2a - (b + 3c)].$
11. $x - [3y - (2z - x) - \{2y - (x + y) - z\} - (y - z) + x].$
12. $x - [y - (x - y) - \{x - (y - x) - y\} - \{x - (y - \overline{x - y} - x)\}].$

13. $a - [5b - \{a - (3c - 3b) + 2c - (a - 2b - c)\}]$.

14. $a - [b - \{a - (b - \overline{a - b} - a) - (b - a) - b\} - (a - b) - a] - b$.

15. $\{(3a - 2b) + (4c - a)\} - \{a - (2b - 3a) - c\} + \{a - (b - 5c - a)\}$.

16. Enclose $a - b + c - d - e + f$ in alphabetical order in brackets, two letters in each: three letters in each.

17. Enclose all but a in an outer bracket, with c, d and e enclosed in an inner bracket.

18. Enclose b and c, d and e, in brackets, and then enclose these groups with f in another bracket.

19. Enclose e and f in brackets, then this group with d in a second pair, this group with c in a third, and this last group with b in a fourth pair.

20. Add $1 - \{1 - (1 - x)\}$, $1 + \{1 - (1 + x)\}$, $x - \{x - (x - 1)\}$, $c + \{x - (x + 1)\}$, and from their sum subtract $a - b + c$.

21. If $a = 1$, $b = 2$, $c = \dfrac{1}{6}$, $d = \dfrac{1}{9\frac{1}{2}}$, find the value of

$a - [3a - 5b - \{7a - 9b - 11c - (13a - 15b - 17c - 19d)\}]$.

CHAPTER III.

MULTIPLICATION.

52. The **Product** of two algebraical numbers is formed by substituting one of the numbers for the unit in the other, and reducing the resulting expression to its simplest form.

The former number is called the **Multiplicand**; the latter the **Multiplier**; the process of finding the product is called **Multiplication**.

53. Let it be required to perform the following multiplications:
Multiply

    1. $+7$ by $+3$.   2. -7 by $+3$.
    3. $+7$ by -3.   4. -7 by -3.

Since $+3 = 1+1+1$ and $-3 = -1-1-1$.

Substituting the multiplicand for the unit in the multiplier in the several cases we get

1. $(+7) \times (+3) = (+7) + (+7) + (+7) = +21.$ (Art. 35.)
2. $(-7) \times (+3) = (-7) + (-7) + (-7) = -21.$ (Art. 35.)
3. $(+7) \times (-3) = -(+7) - (+7) - (+7)$
                 $= -7-7-7 = -21.$ (Art. 45.)
4. $(-7) \times (-3) = -(-7) - (-7) - (-7)$
                 $= 7+7+7 = 21.$ (Art. 45.)

Similarly, if a and b are any absolute numbers,

    $(+a) \times (+b) = +ab.$ $(-a) \times (+b) = -ab.$
    $(+a) \times (-b) = -ab.$ $(-a) \times (-b) = +ab.$

54. From Art. 53 we have for the multiplication of algebraic numbers the following

RULE.—*Take the product of the absolute values of the numbers and prefix the sign + or −, according as the two factors have like or unlike signs.* This rule is often abbreviated thus: "*Like signs give plus, unlike signs give minus,*" and is called "*The Rule of Signs.*"

55. The following points should be carefully noted :—
1. The product of any number of **positive** factors is **positive**.
2. The product of any *even* number of **negative** factors is **positive**.
3. The product of any *odd* number of **negative** factors is **negative**.
4. If the sign of one factor be changed the sign of the product is changed.

56. The absolute value of the product of any number of numerical factors does not depend on the order in which the factors are taken, and the sign depends only on the *number* of *negative factors;* therefore the product is the same whatever be the arrangement of factors.

57. The product of two or more powers of any number is that number with an exponent equal to the sum of the exponents of the factors. For example, $a^2 \times a^3 = aa \times aaa = aaaaa = a^5$.

Similarly, if m and n are any positive integers, $x^m \times x^n = x^{m+n}$, which is the rule.

58. The product of numbers represented by different letters can only be indicated by writing the letters side by side. The product of the *same* powers of different letters may be represented by a single exponent. Thus

$a^2 b^2 = (ab)^2$, for $(ab)^2 = ab \times ab = a \times a \times b \times b = a^2 \times b^2 = a^2 b^2$.

Similarly, $a^3 b^3 c^3 = (abc)^3$, and so on for any number of factors.

59. The product of any monomial factors may be formed by the aid of the preceding Arts.

Exs. $\qquad 3ax \times -7a^2 bx = -21a^3 bx^2.$
$\qquad -a^2 b \times 3b^2 c \times -4ac^2 \times -abc = -12a^4 b^4 c^4.$

EXERCISE X.

1. $+3 \times +5.$
2. $+3 \times -5.$
3. $-7 \times +6.$
4. $-9 \times -11.$
5. $(-14)^2.$
6. $(-1)^2 \times 15.$
7. $(-3)^3 \times -5.$
8. $(+3)^3 + (-3)^3.$
9. $(+3)^4 + (-3)^4.$
10. $(5-6)(-4+2) - 2^2.$
11. $(25-40)(40-25) - (-15)^2.$
12. $(-1)^3(40-16) - 3(-1).$
13. $(9^3 - 11^3)(9 - 11)^3.$
14. $7a^2 \times -5a^3.$
15. $-5ab \times 16ac.$
16. $3a^2x^3 \times -12a^3y^2.$
17. $-8m^3n^2 \times -3mn^2p.$
18. $2x^2yz \times 3xy^2z \times -5xyz^2 \times xyz.$
19. $3ab \times -4bx \times 5xy \times -ay \times abxy.$
20. Find the value of $a^3 + b^3 + c^3 - 3abc$ when

    1. $a = 2$ 2. $a = 4$ 3. $a = -5$
        $b = 3$ $b = -10$ $b = -5$
        $c = -5$ $c = 6$ $c = 10$

21. $(-1)^2(b+c) - [(-1)^3(c+a) - (-1)^7\{a+b-(3b-3c)\}].$

MULTIPLICATION OF POLYNOMIALS.

60. Let a, b, c, d, represent any absolute whole numbers, and

1. Let it be required to multiply $a - b$ by c.

$c(a-b) = (a-b) + (a-b) \ldots$ with $(a-b)$ written c times.
$\quad\quad\;\; = a + a \ldots c$ times, $-b - b - \ldots c$ times
$\quad\quad\;\; = ca - cb.$

2. Let it be required to multiply $a - b$ by $c - d$.

$(c-d)(a-b) = \quad (a-b) + (a-b) + \ldots$ with $(a-b)$ written c times
$\quad\quad\quad\quad\quad\; - \{(a-b) + (a-b) + \ldots$ with $(a-b)$ written d times$\}$
$\quad\quad\quad\quad\; = ca - cb - (da - db)$
$\quad\quad\quad\quad\; = ca - cb - da + db.$

61. The multiplication of algebraic numbers has been shown to differ from the multiplication of absolute numbers only in determining the sign of the product; therefore, by taking signs

MULTIPLICATION OF POLYNOMIALS.

into consideration in accordance with the rules already given, the preceding results are true for all integral values of a, b, c, d.

62. From Arts. 60 and 61 we learn that

1. To multiply a polynomial by a monomial we multiply each term in succession, and connect the partial products by the proper signs.

2. To multiply a polynomial by a polynomial we multiply each term of the multiplicand by each term in the multiplier, and connect the partial products by the proper sign.

EXERCISE XI.

Multiply

1. $2x^2 - 3x + 4$ by $3x$.
2. $5x^3 - 2xy^2$ by $3xy$.
3. $a^2 - ab + b^2$ by $-2ab^2$.
4. $2m^2n - 5mn^2$ by $-4mnp$.
5. $7a^2x - 5aby + 11b^2y - 6ay^2$ by $-5aby^2$.
6. $3x^3 - 5y^2 - 7xy - 4x - 3y + 2$ by $-8xy$.

Simplify

7. $3x(x^2 - 4x + 3) + 5x(2x^2 - 3x + 7)$.
8. $4a(a-b) - 5b(2a-3b) + 3ab$.
9. $4x\{2y - 3(2x-y)\} - 2y\{5x - 2(2y-x)\}$.
10. $20x^4 - 2x[3x^3 - 4x^2 - 3x\{4x^2 - 5x - 2(3x^2 - 4x + 1)\}]$.
11. $a(a+b-c) + b(b+c-a) + c(c+a-b)$.
12. $a(b-c) + b(c-a) + c(a-b)$.
13. $(a+b)(c+d) - (a-b)(c-d)$.
14. $(x+y)(m-n) - (x-y)(m+n)$.
15. $(a-b)(a+b-c) + (b-c)(b+c-a) + (c-a)(c+a-b)$.
16. $3(a-b+c) - 5(a-2b+3c) + 4(a-3b+2c) - 2(a-7b-2c)$.
17. $(ax+by) + (x+y) + (a-1)x - (b+1)y$.
18. $(a+b)x + (b+c)y - \{(a-b)x - (b-c)y\}$.
19. $\{(a+b)x + (a+c)y\} + \{(b-c)x + (b-a)y\} - \{(a-c)x - (c-b)y\}$.
20. $(a-b)x + (b-c)y + (c-a)z - \{a(x-y) + b(y-z) + c(z-x)\}$
    $- (ax+by+cz) + (a+b+c)(x+y+z) - 2(cx+ay+bz)$.

MULTIPLICATION OF POLYNOMIALS.

63. In the multiplication of polynomials like terms in the several partial products should be collected so as to give the result in the simplest form. The following examples show the best methods of proceeding:—

Ex. 1.

$$2x^3 - x^2 + 3x - 1$$
$$3x^2 - x + 2$$
$$\overline{6x^5 - 3x^4 + 9x^3 - 3x^2}$$
$$- 2x^4 + x^3 - 3x^2 + x$$
$$+ 4x^3 - 2x^2 + 6x - 2$$
$$\overline{6x^5 - 5x^4 + 14x^3 - 8x^2 + 7x - 2}$$

Ex. 2.

$$x^2 - (a+b)x + ab$$
$$x - c$$
$$\overline{x^3 - (a+b)x^2 + abx}$$
$$- cx^2 + (ac+bc)x - abc$$
$$\overline{x^3 - (a+b+c)x^2 + (ab+bc+ac)x - abc}$$

64. The multiplicand and multiplier should be arranged in powers of the leading letter, both descending, as in the examples, or both ascending. In working long examples it is frequently convenient to omit the letters and work with the coefficients only. The student should work the following example in full and compare with the work given:—

Multiply $x^5 - 3x^4 + 2x^2 - 5x + 7$ by $x^3 - 2x + 3$.

$$1 - 3 + 0 + 2 - 5 + 7$$
$$1 + 0 - 2 + 3$$
$$\overline{1 - 3 + 0 + 2 - 5 + 7}$$
$$- 2 + 6 - 0 - 4 + 10 - 14$$
$$+ 3 - 9 + 0 + 6 - 15 + 21$$
$$\overline{1 - 3 - 2 + 11 - 14 + 3 + 16 - 29 + 21}$$

Result: $x^8 - 3x^7 - 2x^6 + 11x^5 - 14x^4 + 3x^3 + 16x^2 - 29x + 21$.

A cipher is introduced, both in the multiplicand and multiplier, in place of a regular term in the series which is wanting, to keep the other terms in their proper columns. When the terms are written in full this is unnecessary. The above process is called "multiplying by detached coefficients."

Multiply

EXERCISE XII.

1. $x^2 - 3x + 4$ by $x + 2$. 2. $2x^2 - 4x + 5$ by $x - 2$.
3. $4x^2 + 3x - 7$ by $2x - 3$. 4. $4x^2 - 3x - 5$ by $-2x + 3$.
5. $x^2 + x + 1$ by $x^2 - x + 1$. 6. $x^2 + 2x + 2$ by $x^2 - 2x + 2$.
7. $a^2 + ab + b^2$ by $a - b$. 8. $a^2 - ab + b^2$ by $a + b$.
9. $a^2 - 2a + 3$ by $a^2 + 2a - 3$. 10. $2a^2 - 5ab + 3b^2$ by $2a^2 + 5ab + 3b^2$.
11. $3x^3 - 7x^2 + 2x - 5$ by $2x - 3$.
12. $x^4 - x^2 + x + 1$ by $2x^2 - 4x + 7$.
13. $1 - 2x + 3x^2 + 4x^3$ by $1 + 2x - 3x^2$.
14. $3 - x^2 + 5x^3 + x^4$ by $x^3 - 2x + 1$.
15. $4x - 3 + 2x^3 - x^2$ by $-3x + x^2 - 6$.
16. $x^6 + x^4 - x^2 + 1$ by $x^3 - x - 1$.
17. $x^4 + 2x^3 + 3x^2 + 2x + 1$ by $x^4 - 2x^3 + 3x^2 - 2x + 1$.
18. $x^3 + 4x^2 + 5x - 24$ by $x^2 - 4x + 11$.
19. $x^3 - 4x^2 + 11x - 24$ by $x^2 + 4x + 5$.
20. $a^2 + b^2 + c^2 - ab - bc - ac$ by $a + b + c$.
21. $a^2 + 2ab + b^2 - c^2$ by $c^2 - a^2 + 2ab - b^2$.
22. $x^2 - xy + y^2 + x + y + 1$ by $x + y - 1$.
23. $x^4 + x^2 - 4x - 11 + 2x^3$ by $x^2 - 2x + 3$.
24. $49x^6 + 56x^3y + 64y^2$ by $7x^3 - 8y$.
25. $x^4 + 4x^3y + 6x^2y^2 + 4xy^3 + y^4$ by $x^4 - 4x^3y + 6x^2y^2 - 4xy^3 + y^4$.
26. $x^2 + 4y^2 + z^2 + 2xy + 2yz - xz$ by $x - 2y + z$.

Find the continued product of

27. $x + a$, $x + b$ and $x + c$.
28. $x - a$, $x - b$ and $x - c$.
29. $x - 2$, $x - 3$ and $x - 4$.
30. $x - 3$, $x - 1$, $x + 1$ and $x + 3$.

31. x^2+ax+a^2, x^2-ax+a^2, $x+a$ and $x-a$.
32. $x^4-x^2y^2+y^4$, x^2-xy+y^2 and x^2+xy+y^2.
33. $9m^2+3am+a^2$, $9m^2-3am+a^2$, $3m+a$ and $3m-a$.

65. The following examples are of great importance and should be carefully remembered:—

| 1. $a + b$ | 2. $a - b$ | 3. $a + b$ |
$a + b$	$a - b$	$a - b$
$a^2 + ab$	$a^2 - ab$	$a^2 + ab$
$\quad + ab + b^2$	$\quad - ab + b^2$	$\quad - ab - b^2$
$a^2 + 2ab + b^2$	$a^2 - 2ab + b^2$	$a^2 \quad\quad - b^2$

These results should also be remembered in words thus:—

1. The square of the sum of two quantities is equal to the sum of their squares together with twice their product.

2. The square of the difference between two quantities is less than the sum of their squares by twice their product.

3. The product of the sum and the difference of two quantities is equal to the difference of their squares.

66. A truth expressed by algebraic symbols is called a formula. Thus $(a+b)^2 = a^2 + 2ab + b^2$ is a formula.

67. The formulæ of Art. 65 may sometimes be used to obtain results in the multiplication of numbers thus:—

$$51 \times 51 = (50+1)^2 = 2500 + 2 \times 50 + 1 = 2601.$$
$$48 \times 48 = (50-2)^2 = 2500 - 4 \times 50 + 4 = 2304.$$
$$67 \times 73 = (70-3)(70+3) = 4900 - 9 = 4891.$$

EXERCISE XIII.

Perform the operations indicated.

1. $(x+y)^2$.
2. $(x-y)^2$.
3. $(x+y)(x-y)$.
4. $(2x+y)^2$.
5. $(x-2y)^2$.
6. $(x+2y)(x-2y)$.
7. $(2x+3y)^2$.
8. $(2x-3y)^2$.
9. $(2x+3y)(2x-3y)$.
10. $(x^2+y^2)^2$.
11. $(x^2-y^2)^2$.
12. $(x^2+y^2)(x^2-y^2)$.

MULTIPLICATION OF POLYNOMIALS. 39

13. $(a^2+bc)^2$.   14. $(a^2-bc)^2$.   15. $(a^2+bc)(a^2-bc)$.
16. $(ax+by)^2$.   17. $(ax-by)^2$.   18. $(ax+by)(ax-by)$.
19. $(1+x^2)^2$.   20. $(3x^2-4)^2$.   21. $(4x^5+1)(4x^5-1)$.
22. $(5m^2n+6mn^2)^2$.   23. $(8m^3-5mp^2)^2$.   24. $(9x+7y)(7y-9x)$.
25. $(x-y)(x+y)(x^2+y^2)(x^4+y^4)$.   26. 81×79.
27. 97×97.   28. 88×92.   29. $257^2 - 243^2$.
30. $9 \times 11 \times 101 \times 10001$.   (Use Ex. 25.)
31. $4(a-3b)(a+3b) - 2(a-6b)^2 - 2(a^2+6b^2)$.
32. $x^2(x^2+y^2)^2 - 2x^2y^2(x+y)(x-y) - (x^3-y^3)^2$.
33. $16(a^2+b^2)(a^2-b^2) - (2a-3)(2a+3)(4a^2+9)$
$\quad\quad + (2b-3)(2b+3)(4b^2+9)$.
34. $(a-2b)(a+2b) + (2b-3c)(2b+3c) + (3c-d)(3c+d)$.
35. Show that
$\{(ac+bd)^2+(ad-bc)^2\}\{(ac+bd)^2-(ad+bc)^2\} = (a^4-b^4)(c^4-d^4)$

68. To form the square of a trinomial.

$\quad a + b + c \quad\quad\quad\quad\quad a + b - c$
$\quad a + b + c \quad\quad\quad\quad\quad a + b - c$
$\quad\overline{a^2 + ab + ac}\quad\quad\quad\overline{a^2 + ab - ac}$
$\quad\quad + ab + b^2 + bc \quad\quad\quad + ab + b^2 - bc$
$\quad\quad\quad + ac + bc + c^2 \quad\quad\quad\quad - ac - bc + c^2$
$\overline{a^2 + 2ab + 2ac + b^2 + 2bc + c^2}\quad\overline{a^2 + 2ab - 2ac + b^2 - 2bc + c^2}$
$= a^2 + b^2 + c^2 + 2ab + 2ac + 2bc.\quad = a^2 + b^2 + c^2 + 2ab - 2ac - 2bc.$

These results consist, in each case, of two sets of terms:—
1. The sum of the squares of each term of the trinomial.
2. Twice the product of each pair of terms.

The sign of each of the square terms is positive. The sign of any product is positive or negative according as the signs of the terms from which it is formed are alike or different. It is worthy of note that the signs of the products in the square of any trinomial are either all positive, or two are negative and one positive.

69. A little consideration will show that the square of any polynomial is formed in the same way as that of a trinomial. The most convenient arrangement of terms is according to the following

RULE.—*To the sum of the squares of each term add twice the product of each term into each of the terms that follow it.*

EXERCISE XIV.

Perform the operations indicated.

1. $(x+y+z)^2$.   2. $(x+y-z)^2$.   3. $(x-y+z)^2$.
4. $(x-y-z)^2$.   5. $(-x-y+z)^2$.   6. $(-x+y+z)^2$.
7. $(a+b+2c)^2$.   8. $(a-2b+3c)^2$.   9. $(2a-b+3c)^2$.
10. $(1+x+x^2)^2$.   11. $(1-x+x^2)^2$.   12. $(x^2+xy+y^2)^2$
13. $(2x^2-3x+4)^2$.   14. $(x-x^2+2)^2$.   15. $(4-2x^2+x)^2$.
16. $(a-b+c-d)^2$.   17. $(a^2+ab-ac-bc)^2$.   18. $(x^3-x^2+x-1)^2$.
19. $(x+y+z)^2+(x+y-z)^2+(x-y+z)^2+(-x+y+z)^2$.

70. To form the product of two expressions which differ only in the sign of one or more terms, we first arrange the terms of each expression in two groups, placing those which have the same sign in the two expressions in the first group in each case, and those which have different signs in the second group. The terms of the second group in the two expressions, having different signs as they stand, will have the same sign after being enclosed in brackets with a positive sign before one group and a negative sign before the other. We have now to find the product of the sum and the difference of two quantities, which, by Art. 65, is the difference of their squares. The work may be arranged as in the following example:—

$Ex.$—Multiply $a-b+c-d$ by $a+b-c-d$.

$$(a-b+c-d)(a+b-c-d) = \{(a-d)-(b-c)\}\{(a-d)+(b-c)\}$$
$$= (a-d)^2-(b-c)^2$$
$$= a^2-2ad+d^2-(b^2-2bc+c^2)$$
$$= a^2-b^2-c^2+d^2-2ad+2bc.$$

EXERCISE XV.

Perform the operations indicated.

1. $(a+b+c)(a+b-c)$.
2. $(a+b-c)(a-b+c)$.
3. $(2x-y-3z)(2x-y+3z)$.
4. $(x-2y+3z)(3z-x+2y)$.
5. $(x^2-x+1)(x^2+x+1)$.
6. $(a^2+ab+b^2)(a^2-ab+b^2)$.
7. $(3a^2-ab+2b^2)(3a^2+ab+2b^2)$.
8. $(x^2+2ax+2a^2)(x^2-2ax+2a^2)$.
9. $(a+b+c-d)(a+b-c+d)$.
10. $(a-b+c+d)(-a+b+c+d)$.
11. $(y^2-2yz-z^2)(y^2+2yz-z^2)-(y^2-yz-2z^2)(y^2+yz-2z^2)$.
12. $(a^2+b^2-c^2+2ab)(c^2-a^2-b^2+2ab)$
$\qquad -(a^2-b^2+c^2+2ac)(b^2-c^2-a^2+2ac)$.
13. $(a+b+c)(a+b-c)(a-b+c)(-a+b+c)$.

71. To form the product of two binomials which differ only in their second terms, the coefficients of each of the first terms being unity and positive.

$$
\begin{array}{ll}
\text{1.} \quad x+2 & \text{2.} \quad x-2 \\
\quad x+3 & \quad x-3 \\
\hline
\quad x^2+2x & \quad x^2-2x \\
\qquad +3x+6 & \qquad -3x+6 \\
\hline
\quad x^2+5x+6 & \quad x^2-5x+6
\end{array}
$$

$$
\begin{array}{ll}
\text{3.} \quad x+2 & \text{4.} \quad x-2 \\
\quad x-3 & \quad x+3 \\
\hline
\quad x^2+2x & \quad x^2-2x \\
\qquad -3x-6 & \qquad +3x-6 \\
\hline
\quad x^2-x-6 & \quad x^2+x-6
\end{array}
$$

In these examples observe—

1. The first term of the product is the square of the common term of the binomials.

2. The coefficient of the second term is the algebraic sum of the different terms of the binomials,

3. The third term is the product of the different terms of the binomials.

The substance of this Art. is briefly expressed by the single formula—
$$(x+a)(x+b) = x^2 + (a+b)x + ab,$$
which also includes the case in which the first coefficient is any quantity whatever.

EXERCISE XVI.

Perform the operations indicated.

1. $(x+3)(x+4)$.
2. $(x+2)(x+5)$.
3. $(x+1)(x+6)$.
4. $(x+9)(x+12)$.
5. $(x+6)(x+15)$.
6. $(x+9)(x+11)$.
7. $(x-2)(x-7)$.
8. $(x-3)(x-10)$.
9. $(x-5)(x-15)$.
10. $(x+3)(x-2)$.
11. $(x+8)(x-1)$.
12. $(x+20)(x-5)$.
13. $(x-20)(x+5)$.
14. $(x-30)(x+10)$.
15. $(x-21)(x+1)$.
16. $(x^2-10)(x^2+2)$.
17. $(x^3+14)(x^3-4)$.
18. $(x^4+25)(x^4-5)$.
19. $(x-5y)(x+y)$.
20. $(x^2-2yz)(x^2-7yz)$.
21. $(x+a)(x-b)$.
22. $(2x-7y)(2x-10y)$.
23. $(5x-2)(5x-10)$.
24. $(3a^2-b)(3a^2+c)$.
25. $(ax+b)(ax+c)$.
26. Show that $x(x+1)(x+2)(x+3)+1 = (x^2+3x+1)^2$.

POWERS OF A BINOMIAL.

72. From Art. 65, $(a+b)^2 = a^2 + 2ab + b^2$
Multiplying by $a+b$

$$\begin{array}{r} a^2 + 2ab + b^2 \\ a + b \\ \hline a^3 + 2a^2b + ab^2 \\ a^2b + 2ab^2 + b^3 \\ \hline \end{array}$$

we get $(a+b)^3 = a^3 + 3a^2b + 3ab^2 + b^3$

$$\begin{array}{r} a + b \\ \hline a^4 + 3a^3b + 3a^2b^2 + ab^3 \\ a^3b + 3a^2b^2 + 3ab^3 + b^4 \\ \hline \end{array}$$

and $(a+b)^4 = a^4 + 4a^3b + 6a^2b^2 + 4ab^3 + b^4$

POWERS OF A BINOMIAL. 43

In the above example observe—

1. The first term in each result is a raised to the same power as the binomial; in each succeeding term its exponent is reduced by unity.

2. The second term contains the first power of b, and in each succeeding term its exponent is increased by unity.

3. Each result is homogeneous, and of the number of dimensions indicated by the exponent of the binomial.

73. The powers of $a-b$ should be written out in the same way as those of $a+b$ in the preceding Art. The results will be the same except that the terms containing *odd* powers of b will have the sign $-$ prefixed. (See Art. 55, 3.)

74. It is sometimes convenient to write the cube of a binomial in the following form, which may easily be verified:—

$$(a+b)^3 = a^3 + b^3 + 3ab(a+b);$$
$$(a-b)^3 = a^3 - b^3 - 3ab(a-b).$$

75. The coefficients of the terms of the successive powers of a binomial, up to the fifth, should be committed to memory. They may be arranged thus:—

```
1st power,  1,  1
2nd   "     1,  2,  1
3rd   "     1,  3,  3,  1
4th   "     1,  4,  6,  4,  1
5th   "     1,  5,  10, 10, 5,  1
```

This list may be continued to any extent by carefully studying Art. 72.

EXERCISE XVII.

Perform the operations indicated.

1. $(x+1)^3$.
2. $(x-1)^3$.
3. $(x+1)^4$.
4. $(x-1)^4$.
5. $(x+1)^5$.
6. $(x-1)^5$.
7. $(x+2y)^3$.
8. $(2x-y)^3$.
9. $(2a-3b)^3$.
10. $(a-2b)^4$.
11. $(2a+b)^4$.
12. $(x-y)^5$.

13. $(a+x)^3+(a-x)^3$. 14. $(a+x)^3-(a-x)^3$.
15. $(x+y)^5+(x-y)^5$. 16. $(x+y)^4-(x-y)^4$.

17. Find the value of $a^3+b^3+c^3-3abc$ in terms of a and b (1) if $c=(a+b)$; (2) if $c=-(a+b)$.

18. Find the value of $a^2+b^2+c^2+ab+bc+ca$ in terms of x, y and z if $a=x-y$, $b=y-z$, $c=z-x$.

19. Find the value of $a^3+b^3+c^3-3abc$ in terms of x, y and z if

(1) $a=y+z$, $b=z+x$, $c=x+y$;
(2) $a=y-z$, $b=z-x$, $c=x-y$.

20. Find the value of $a^2+b^2+c^2-ab-bc-ca$ in terms of x, y and z if $a=x+y$, $b=y+z$, $c=z+x$.

CHAPTER IV.

DIVISION.

76. When a **Product** and one of its factors are given, **Division** is the process by which the other factor may be found.

77. With regard to this process the given product is called the **Dividend**, the given factor the **Divisor**, and the factor to be found the **Quotient**.

Since $(+a) \times (+b) = +ab$, $\therefore \dfrac{+ab}{+b} = +a$, (Art. 53)

$(-a) \times (+b) = -ab$, $\therefore \dfrac{-ab}{+b} = -a$,

$(+a) \times (-b) = -ab$, $\therefore \dfrac{-ab}{-b} = +a$,

$(-a) \times (-b) = +ab$, $\therefore \dfrac{+ab}{-b} = -a$,

we have for the division of algebraic numbers the following

RULE.—*Divide as in absolute numbers and prefix the sign + or −, according as the divisor and the dividend have like or unlike signs.*

78. Since the dividend contains all the factors of both divisor and quotient, we have for the division of monomials the following

RULE.—*Remove from the dividend all the factors of the divisor, the remaining factors will be the quotient.*

DIVISION.

79. Should the divisor contain any factor not found in the dividend, the division can be *indicated* but cannot really be performed.

Thus the quotient of a by b is $\dfrac{a}{b}$; the quotient of $3ab$ by $5bc$ is $\dfrac{3a}{5c}$, etc.

80. Since all factors found in the divisor are removed from the dividend to obtain the quotient, it is evident that the quotient will not be changed by multiplying or dividing both divisor and dividend by the same factor.

81. A Power of a number is divided by a lower power of the same number by subtracting the index of the latter from that of the former; thus $x^5 \div x^2 = x^3$, since two factors removed from five factors leave three factors, etc.

82. A polynomial is divided by a monomial by dividing each of its terms in succession and connecting the partial quotients by the proper signs. This follows at once from multiplication (Art. 62).

Examples:—

1. $-12 \div 4 = -3.$ 2. $20 \div -5 = -4.$
3. $30abc \div 3ab = 10c.$ 4. $15a^3b^3 \div 3a^2b = 5ab^2.$
5. $(6a^3b - 9a^2b^2 + 12ab^3) \div 3ab = 2a^2 - 3ab + 4b^2.$

EXERCISE XVIII.

Divide

1. $+20$ by $+5.$ 2. -20 by $+5.$ 3. -750 by $-15.$
4. $+850$ by $-17.$ 5. $(-8)^3$ by $(-2)^2.$ 6. -250 by $(-1)^3.$
7. a^3b^2 by $a^2b^2.$ 8. $8a^4b^3c^2$ by $4ab^3.$ 9. $72a^5b^3$ by $9a^2b^2.$
10. $256a^3xy^2$ by $16axy.$ 11. $96a^7b^5$ by $-4ab.$
12. $-45x^6y^7z^2$ by $-5x^2y^2z^2.$
13. $750a^{10}m^6x^5$ by $-30a^5m^5x^5.$
14. $-1728a^{12}b^{10}c^8d^6$ by $-864a^6b^5c^4d^3.$

15. $(4a^2xy \times 15ax^2y^2)$ by $30a^3x^3$.
16. $(-3ax \times -5xy \times -12yz)$ by $(5ay \times -3xz)$.
17. $x^3 - 2x^2 - 7x$ by x.   18. $y^5 - ay^4 + by^3$ by $-y^2$.
19. $8a^3 - 16a^2b - 24ab^2$ by $8a$. 20. $25a^5 - 30a^4b - 40a^3b^2$ by $-5a^3$.
21. $mpx^4 - m^2p^2x^2 + mp^3$ by mp.
22. $16x^4y - 28x^3y^2 + 36x^2y^3$ by $-4x^2y$.
23. $-49x^5yz^3 + 63x^4y^2z$ by $-7x^3yz$.
24. $52a^8b^3 - 65a^4b^7 + 78a^3b^8$ by $13a^3b^3$.
25. $34a^5b + 51a^4b^2 - 68a^2b^2$ by $17a^2b$.
26. $-144x^3y + 132x^2y^2 - 120xy^3$ by $-12xy$.
27. $-46a^3b^2c + 69a^3b^3c^2 - 115a^4b^2cd^2$ by $-23\ a^3b^2c$

83. When the divisor consists of more than one term we proceed according to the following

RULE.—*Arrange the terms of both divisor and dividend according to the powers of some common letter, both in descending or both in ascending order.*

Divide the first term of the dividend by the first term of the divisor; the result will be the first term of the quotient.

Multiply each term of the divisor by the first term of the quotient and subtract the product from the dividend.

If there be a remainder consider it a new dividend, and proceed as before.

If a remainder occurs of lower dimensions, with regard to the letter of reference, than the divisor, the division cannot be exactly performed. Such examples will be considered in the chapter on Fractions.

84. The reasons for the preceding rule are the following:—
1. The term containing the highest exponent of the letter of reference in the dividend must be the product of the terms containing the highest exponents of that letter in the factors of the dividend, *i.e.*, the divisor and the quotient.

DIVISION.

2. Since the dividend is the sum of the products of the divisor by each term of the quotient, if we subtract the product of the divisor by one term of the quotient the remainder must be the sum of the products of the divisor and the other terms of the quotient.

85. The following examples show the proper arrangement of the work in the division of polynomials by polynomials:—

Divide

1. $2x^3 + 7x^2 + 5x + 100$ by $x + 5$.
2. $a^3 + b^3$ by $a + b$.
3. $a^3 - b^3$ by $a^2 + ab + b^2$.
4. $16a^4 + 4a^2x^2 + x^4$ by $4a^2 - 2ax + x^2$.

$$
1. \ x+5 \overline{\smash{\big)}\, 2x^3 + 7x^2 + 5x + 100} \ (2x^2 - 3x + 20 \\
\underline{2x^3 + 10x^2} \\
-3x^2 + 5x \\
\underline{-3x^2 - 15x} \\
20x + 100 \\
\underline{20x + 100}
$$

$$
2. \ a+b \overline{\smash{\big)}\, a^3 + b^3} \ (a^2 - ab + b^2 \\
\underline{a^3 + a^2b} \\
-a^2b + b^3 \\
\underline{-a^2b - ab^2} \\
ab^2 + b^3 \\
\underline{ab^2 + b^3}
$$

$$
3. \ a^2 + ab + b^2 \overline{\smash{\big)}\, a^3 - b^3} \ \underline{\big\lfloor a - b} \\
\underline{a^3 + a^2b + ab^2} \\
-a^2b - ab^2 - b^3 \\
\underline{-a^2b - ab^2 - b^3}
$$

$$
4. \ 4a^2 - 2ax + x^2 \overline{\smash{\big)}\, 16a^4 + 4a^2x^2 + x^4} \ \underline{\big\lfloor 4a^2 + 2ax + x^2} \\
\underline{16a^4 - 8a^3x + 4a^2x^2} \\
8a^3x + x^4 \\
\underline{8a^3x - 4a^2x^2 + 2ax^3} \\
4a^2x^2 - 2ax^3 + x^4 \\
\underline{4a^2x^2 - 2ax^3 + x^4}
$$

HORNER'S METHOD OF DIVISION.

86. Long examples in division, in which the exponents of the terms in both dividend and divisor follow each other in regular order, may be conveniently worked by the following

RULE.—Arrange both divisor and dividend in descending powers of a common letter, and place a cipher in place of any term of the regular series which is wanting. Write the coefficients of the dividend in a horizontal line, and those of the divisor in a vertical line to the left of the dividend, changing the sign of every term in the divisor except the first.

Divide the first term of the dividend by the first term of the divisor; the result will be the first term of the quotient.

Multiply all the terms of the divisor, except the first, by the first term of the quotient, and arrange the partial products diagonally under the second and following terms of the dividend.

Add the terms in the second column; divide the sum by the first term of the divisor; the result will be the second term of the quotient.

Multiply as before and arrange the partial products under the third and following terms of the dividend. Continue the process until the number of terms in the quotient is greater by one than the difference of the exponents of the first terms of the dividend and the divisor.

Add up the remaining columns; their sums will be the coefficients of the remainder.

Attach the proper literal factors to the coefficients of the quotient, the exponent of the first term being the difference of the exponents of the first terms of the dividend and the divisor, and the others following in regular order.

The literal factors of the terms of the remainder will be the same as those of the terms of the dividend under which they stand.

HORNER'S METHOD OF DIVISION.

87. The following examples show the arrangement of the work:—

Ex. 1.—Divide

$6x^7 - x^6 - 11x^5 + 16x^4 + x^3 + 8x^2 - 19x + 20$ by $2x^3 + x^2 - 3x + 4$.

```
  2 | 6 - 1 - 11 + 16 + 1 + 8 - 19 + 20
 -1 |    - 3 +  2 +  0 + 1 -  5
 +3 |       +  9 -  6 + 0 - 3 + 15
 -4 |            - 12 + 8 + 0 +  4 - 20
    ─────────────────────────────────────
      3 - 2 +  0 -  1 + 5
```

Quotient, $3x^4 - 2x^3 - x + 5$. Remainder, 0, since the sum of each of the last three columns is zero.

Ex. 2.—Divide

$2x^7 + x^6y - 6x^5y^2 - 11x^4y^3 + 2x^3y^4 + 6x^2y^5 + 16xy^6 - 12y^7$

by $x^3 + 2x^2y - 3y^3$.

```
  1 | 2 + 1 - 6 - 11 +  2 + 6 + 16 - 12
 -2 |    - 4 + 6 +  0 + 10 - 6
 +0 |         + 0 +  0 +  0 + 0 +  0
 +3 |             +  6 -  9 + 0 - 15 +  9
    ───────────────────────────────────────
      2 - 3 + 0 -  5 +  3 |  + 1 -  3
```

Quotient, $2x^4 - 3x^3y - 5xy^3 + 3y^4$. Remainder, $xy^6 - 3y^7$.

In the above example the coefficient of the first term being -1 no division of the sums of the columns is necessary.

From its brevity this method is also known as "Synthetic Division."

EXERCISE XIX.

Divide

1. $x^2 + 15x + 50$ by $x + 10$.
2. $x^2 - 11x + 28$ by $x - 7$.
3. $x^2 - x - 56$ by $x - 8$.
4. $x^2 + x - 90$ by $x + 10$.
5. $x^3 + 13x^2 + 54x + 72$ by $x + 6$.
6. $x^3 + 2x^2 + 2x + 1$ by $x + 1$.
7. $a^2 - b^2$ by $a - b$.
8. $a^3 - b^3$ by $a - b$.
9. $a^3 + b^3$ by $a^2 - ab + b^2$.
10. $a^5 + b^5$ by $a + b$.

HORNER'S METHOD OF DIVISION. 51

11. $x^3 - 7x - 6$ by $x - 3$.   12. $4x^3 + 5x + 21$ by $2x + 3$.
13. $8x^3 + 27y^3$ by $4x^2 - 6xy + 9y^2$.   14. $64x^3 - 1$ by $1 + 4x + 16x^2$.
15. $x^5 - 5x^3 + 7x^2 + 6x + 1$ by $x^2 + 3x + 1$.
16. $6x^4 - x^3y + 2x^2y^2 + 13xy^3 + 4y^4$ by $2x^2 + 4y^2 - 3xy$.
17. $a^5 - 4a^4b + 4a^3b^2 + 4a^2b^3 - 17ab^4 - 12b^5$ by $a^2 - 3b^2 - 2ab$.
18. $a^6 + 13a^4x^2 - 6a^5x + 4a^2x^4 - 12a^3x^3$ by $a^3 - 3a^2x + 2ax^2$.
19. $a^4 + a^2b^2 + b^4$ by $a^2 + ab + b^2$.   20. $x^3 + x^4 + 1$ by $x^4 - x^2 + 1$.
21. $x^{15} + y^{15}$ by $x^{10} - x^5y^5 + y^{10}$.   22. $x^{15} - y^{15}$ by $x^3 - y^3$.
23. $x^6 - 2x^3 + 1$ by $x^2 - 2x + 1$.   24. $x^6 + 2x^3 + 1$ by $x^2 - x + 1$.
25. $x^{10} - x^5y^5 + y^{10}$ by $x^2 - xy + y^2$.   26. $4x^5 - x^3 + 4x$ by $2 + 2x^2 + 3x$.
27. $2x^4 + 2x^2y^2 - 2xy^3 - 7x^3y - y^4$ by $2x^2 + y^2 - xy$.
28. $1 - 51a^3b^3 - 52a^4b^4$ by $4a^2b^2 + 3ab - 1$.
29. $x^7y - xy^7$ by $x^3y + 2xy^3 - 2x^2y^2 - y^4$.
30. $8y^6 - x^6 + 21x^2y^3 - 24xy^5$ by $3xy - x^2 - y^2$.
31. $81x^6y + 18x^2y^6 - 54x^5y^2 - 18x^3y^4 - 18xy^6 - 9y^7$ by $3x^4 + x^2y^2 + y^4$.
32. $2a^7b - 5a^6b^2 - 11a^5b^3 + 5a^4b^4 - 26a^3b^5 + 7a^2b^6 - 12ab^7$
     by $a^4 - 4a^3b + a^2b^2 - 3ab^3$.
33. $x^9 - 3x^8 - 31x^7 + 25x^6 + 3x^5 - 22x^4 + 44x^3 - 2x^2 - 15x + 10$
     by $x^4 - 7x^3 + 3x - 2$.
34. $x^{12} - x^9y^3 + x^6y^6 - x^3y^9 + y^{12}$ by $x^4 - x^3y + x^2y^2 - xy^3 + y^4$.

88. When several terms contain the same power of the letter of reference it is usually best to collect them into one term by the use of brackets. The following are examples:—

1. Divide $x^3 - (a + b + c)x^2 + (ab + bc + ca)x - abc$ by $x - a$.
2. Divide $a^3 + b^3 + c^3 - 3abc$ by $a + b + c$.

1. $x - a\,)\,x^3 - (a+b+c)x^2 + (ab+bc+ca)x - abc\ \underline{|\ x^2 - (b+c)x + bc}$
   $x^3 - ax^2$
   $\overline{}$
   $ -\ (b+c)x^2 + (ab+bc+ca)x$
   $ -\ (b+c)x^2 + (ab+ca)x$
   $ bcx - abc$
   $ bcx - abc$

HORNER'S METHOD OF DIVISION.

2. $a+b+c) a^3+b^3 +c^3 - 3abc \lfloor a^2 - a(b+c) + b^2 - bc + c^2$
$\underline{a^3 + a^2 b + a^2 c}$
$\underline{-a^2(b+c) - 3abc + b^3 + c^3}$
$-a^2(b+c) - a(b^2 + 2bc + c^2)$
$\underline{a(b^2 - bc + c^2) + b^3 + c^3}$
$a(b^2 - bc + c^2) + b^3 + c^3$

Beginners usually find considerable difficulty in examples of this kind, especially in the subtraction. In difficult cases the brackets might be removed, the subtraction performed, and the terms again arranged in order, before proceeding with the division.

EXERCISE XX.

Divide

1. $x^3 - (a+b+c)x^2 + (ab+bc+ca)x - abc$ by $x - b$ and by $x - c$.
2. $x^3 + (a+b+c)x^2 + (ab+bc+ca)x + abc$ by $x^2 + (b+c)x + bc$.
3. $x^4 - (a+b+c+d)x^3 + (ab+bc+cd+ac+ad+bd)x^2$
$ - (abc+bcd+acd+abd)x + abcd$
by each of the divisors $x^2 - (a+c)x + ac$ and $x^2 - (b+c)x + bc$.
4. $x^4 + (5+a)x^3 - (4-5a+b)x^2 - (4a+5b)x + 4b$ by $x^2 + 5x - 4$.
5. $(1+m)x^3 - (m+n)x^2 y + (m+n)xy^2 - (n-1)y^3$ by $x^2 - xy + y^2$.
6. $a^3 + b^3 - c^3 + 3abc$ by $a+b-c$. 7. $a^2 - b^2 - c^2 + 2bc$ by $a-b+c$.
8. $8a^3 - b^3 + c^3 + 6abc$ by $2a - b + c$.
9. $x^2 + xy + 2xz - 2y^2 + 7yz - 3z^2$ by $x - y + 3z$.
10. $2x^2 - 6y^2 - 12z^2 + xy + 17yz - 2xz$ by $x + 2y - 3z$.
11. $a^2(b+c) + b^2(c+a) + c^2(a+b) + 3abc$ by $a+b+c$.
12. $a^2(b+c) + b^2(a-c) + c^2(a-b) + abc$ by $a+b+c$.
13. $x^3 - 8y^3 + 27z^3 + 18xyz$ by $x^2 + 4y^2 + 9z^2 + 2xy + 6yz - 3xz$.
14. $(x+y)^3 - 3(x+y)^2 z + 3(x+y)z^2 - z^3$ by $x+y-z$.
15. $b(x^3 + a^3) + ax(x^2 - a^2) + a^3(x+a)$ by $(a+b)(x+a)$.
16. $b(x^3 - a^3) + ax(x^2 - a^2) + a^3(x-a)$ by $(a+b)(x-a)$.
17. $(x^3 - 1)a^3 - (x^3 + x^2 - 2)a^2 + (4x^2 + 3x + 2)a - 3(x+1)$
 by $(x-1)a^2 - (x-1)a + 3$.

HORNER'S METHOD OF DIVISION.

89. The following cases of division are of frequent occurrence and should be carefully remembered:—

If n stands for any positive integer,

$x^n - y^n$ is divisible by $x - y$ always;
$x^n - y^n$ " " " $x + y$ when n is even;
$x^n + y^n$ " " " $x + y$ when n is odd;
$x^n + y^n$ " " " $x - y$ never.

By actual division we obtain

$$\frac{x^3 - y^3}{x - y} = x^2 + xy + y^2; \qquad \frac{x^4 - y^4}{x - y} = x^3 + x^2y + xy^2 + y^3;$$

$$\frac{x^3 + y^3}{x + y} = x^2 - xy + y^2; \qquad \frac{x^5 + y^5}{x + y} = x^4 - x^3y + x^2y^2 - xy^3 + y^4.$$

In these examples observe that

1. The first term of the quotient is obtained by dividing the first term of the dividend by the first term of the divisor.

2. The exponents of the first letter decrease, and those of the second increase, by unity in each succeeding term.

3. When the connecting sign of the divisor is $-$, the signs of the quotient are all $+$; otherwise they are $+$ and $-$ alternately.

EXERCISE XXI.

Write down the quotients by inspection in the following examples:—

1. $x^4 - y^4$ by $x + y$.
2. $x^5 - y^5$ by $x - y$.
3. $x^6 - y^6$ by $x - y$.
4. $x^7 + y^7$ by $x + y$.
5. $x^3 + 1$ by $x + 1$.
6. $x^6 - 1$ by $x + 1$.
7. $1 - x^3y^3$ by $1 - xy$.
8. $x^5 + y^5z^5$ by $x + yz$.
9. $x^5 + 32$ by $x + 2$.
10. $27 - x^3$ by $3 - x$.
11. $243 + x^5$ by $3 + x$.
12. $8x^3 - 27y^3$ by $2x - 3y$.
13. $x^3 + 8$ by $x + 2$.
14. $x^5 - 32$ by $x - 2$.
15. $16x^4 - 1$ by $2x + 1$.
16. $x^4 - 81$ by $x - 3$.
17. $216 - x^3$ by $6 - x$.
18. $8x^3 - 343y^3$ by $2x - 7y$.
19. $27x^3 + 1000y^3$ by $3x + 10y$.
20. $64x^3 - 343y^3$ by $4x - 7y$.
21. $125x^3 + 512y^3$ by $5x + 8y$.
22. Divide $(x^2 + xy + y^2)^3 + (x^2 - xy + y^2)^3$ by $2x^2 + 2y^2$.

CHAPTER V.

FACTORING.

90. In multiplication two factors are given, and their product is required; in division the product and one factor are given from which to find the other factor. We have now to consider how to find both factors from the product alone.

91. The only rule which can be given is to examine carefully the process of the multiplication of various kinds of factors, to note the results, and then learn to retrace the steps from the product to the factors which produced it.

MONOMIAL FACTORS.

92. The factors of a **Monomial** are evident by inspection. If each term of a **Polynomial** contains a monomial factor it may be discovered by inspection, and then the other factor may be obtained by division.

$Ex.\ 1.$ $10ac + 15bc = 5c(2a + 3b)$.

$Ex.\ 2.$ $3x^3 + 15x^2 - 9x = 3x(x^2 + 5x - 3)$.

$Ex.\ 3.$ $4a^3b + 6a^2b^2 + 2ab^3 = 2ab(2a^2 + 3ab + b^2)$.

93. This principle may frequently be extended to groups of terms, thus:—

$Ex.\ 1.$ $ax + ay + bx + by = a(x+y) + b(x+y) = (x+y)(a+b)$.

$Ex.\ 2.$ $ac - bc - ad + bd = c(a-b) - d(a-b) = (a-b)(c-d)$.

EXERCISE XXII.

Resolve into factors

1. $3x^2 - 15$.
2. $10x^2 - 15xy + 20y^2$.
3. $7p^3 - 63p^2$.
4. $22m^2 - 33mn - 110n^2$.

TRINOMIALS.

5. $ax^3 - abx^2 + ax.$
6. $54a^3b^6 + 108a^6b^8 - 243a^8b^9.$
7. $35x^2yz + 70xy^2z - 105xyz^2.$
8. $(a+b)x + (a+b)y.$
9. $(a-b)x - (a-b)y.$
10. $2ac + 2ad + 3bc + 3bd.$
11. $3ax - ay - 3bx + by.$
12. $ac + bc + a + b.$
13. $x^2 + ax + bx + ab.$
14. $x^2 - ax - bx + ab.$
15. $x^2 - ax + bx - ab.$
16. $x^2 + ax - bx - ab.$
17. $ax - bx + ab - x^2.$
18. $bx - ax + ab - x^2.$
19. $a^2cx - abdx - abcy + b^2dy.$
20. $acx - acy + bdx - bdy.$
21. $adx + ady - bcx - bcy.$
22. $x^3 + x^2 + x + 1.$
23. $x^5 + x^4 - x^3 - x^2 + x + 1.$
24. $2ax - 3by + cy - 2ay + 3bx - cx.$
25. $6a^2b^2 - 2a^3c - 9b^3c + 3abc^2.$
26. $abx^2 + acxy - abxy - acy^2.$

TRINOMIALS.

COEFFICIENT OF FIRST TERM UNITY AND POSITIVE.

94. From Art. 71 we learn that a **Trinomial** can be resolved into the product of two **Binomials**—

1. If the exponent of the first term be double that of the second, and
2. If two numbers can be found whose algebraic sum is the coefficient of the second term and whose product is the third term.

Consider the following examples:—

1. $x^2 + 7x + 12.$
2. $x^2 - 7x + 12.$
3. $x^2 + x - 12.$
4. $x^2 - x - 12.$

We have to find two numbers in

Ex. 1 whose sum is $+7$ and product $+12$, viz., $+3$ and $+4$;
" 2 " -7 " $+12$, " -3 " -4;
" 3 " $+1$ " -12, " -3 " $+4$;
" 4 " -1 " -12, " $+3$ " $-4.$

Therefore

$x^2 + 7x + 12 = (x+3)(x+4);$ $\quad x^2 - 7x + 12 = (x-3)(x-4);$
$x^2 + x - 12 = (x-3)(x+4);$ $\quad x^2 - x - 12 = (x+3)(x-4).$

COMPLETE SQUARES.

EXERCISE XXIII.
Resolve into factors

1. $x^2+7x+12$.
2. $x^2+7x+10$.
3. x^2+7x+6.
4. $x^2+13x+30$.
5. $x^2+11x+30$.
6. $x^2-17x+30$.
7. $x^2-18x+72$.
8. $x^2-22x+40$.
9. $x^2-20xy+51y^2$.
10. x^4-22x^2+72.
11. x^2+x-30.
12. x^2+x-42.
13. $x^2+2x-35$.
14. $x^2+3x-88$.
15. $x^2+9x-10$.
16. $x^2-4x-77$.
17. $x^2-7x-18$.
18. x^4-11x^2-12.
19. x^6-13x^3-68.
20. x^6-x^3-12.
21. $x^{10}+19x^5-20$.
22. $x^2y^2+29xy-390$.
23. $x^4y^2+15x^2y-100$.
24. $x^2y^2z^2-15xyz-100$.
25. $3x^2-3x-216$.
26. $4x^2+4x-24$.
27. $5x^4-10x^2y^2-400y^4$.
28. $2ax^6-10a^2x^3-28a^3$.
29. $11x^{10}+55x^5-3300$.
30. $x-5x^2-6x^3$.
31. $a^2b+18a^2bx-19a^2bx^2$.
32. $-x^2-5x+6$.
33. $29-28x-x^2$.
34. $x^2-(a+b)x+ab$.
35. $ab-ac+bc-b^2$.

COMPLETE SQUARES.

95. Since $(a+b)^2=a^2+2ab+b^2$ and $(a-b)^2=a^2-2ab+b^2$, we see that a **Trinomial** is the exact square of a **Binomial** when two of its terms are exact squares, and the other term equal to plus or minus twice the square root of their product.

96. When the terms are in order and the coefficient of the first term is unity, the coefficient of the last term must equal the square of half the coefficient of the middle term. This is a special case of the preceding Art. which deserves careful attention.

Ex. 1. $4x^2+12xy+9y^2=(2x+3y)^2$.
Ex. 2. $x^4-12x^2+36=(x^2-6)^2$.

COMPLETE SQUARES.

97. A careful study of Art. 68 will enable the student to recognize an expression which is the exact square of a trinomial.

Ex. $a^2 + 4b^2 + 9c^2 - 4ab + 6ac - 12bc = (a - 2b + 3c)^2$.

The terms of the trinomial are found by taking the square roots of the square terms of the expression. To determine their signs we notice that the product $6ac$ is positive, and that consequently the terms containing a and c must have the same sign; the other products being negative the term containing b must have the contrary sign.

EXERCISE XXIV.

Express as complete squares

1. $x^2 + 10x + 25$.
2. $x^4 + 18x^2 + 81$.
3. $x^2 + 20xy + 100y^2$.
4. $a^2x^2 + 4ax + 4$.
5. $a^4x^2 + 8a^2xy^3 + 16y^6$.
6. $m^4 - 16m^2n^2 + 64n^4$.
7. $x^6 - 38x^3 + 361$.
8. $l^2x^2 - 2lmxy + m^2y^2$.
9. $a^{10} - 14a^5b + 49b^2$.
10. $81x^4 - 18x^2 + 1$.
11. $4x^2 + 12xy + 9y^2$.
12. $9x^2 - 30xy + 25y^2$.
13. $16a^4 - 24a^2b + 9b^2$.
14. $a^2 + 9b^2 + 6ab$.
15. $4a^2 + 25b^2 - 20ab$.
16. $25a^2x^2 - 70abxy + 49b^2y^2$.
17. $9a^2x^2 + 49b^2x^2 - 42abx^2$.
18. $4a^2x^2 + 25a^2y^2 - 20a^2xy$.
19. $36a^2 + 16b^2 - 48ab$.
20. $24x^2 + 54y^2 + 72xy$.
21. $3a^2x^3 - 18ax^2y + 27xy^2$.
22. $3x^2y^2 - 18axy^3 + 27a^2y^4$.
23. $(a+b)^2 + 2(a+b)c + c^2$.
24. $(a-b)^2 + 8(a-b) + 16$.
25. $(x-y)^2 + 2(x-y)y + y^2$.
26. $(x-y)^2 + 2(x^2-y^2) + (x+y)^2$.
27. $(x+y)^2 - 2(x+y)y + y^2$.
28. $(x-y)^2 + 4y(x-y) + 4y^2$.
29. $4(a+b)^2 + 12(a+b)(c+d) + 9(c+d)^2$.
30. $4a^2 + b^2 + c^2 - 4ab + 4ac - 2bc$.
31. $a^4 + b^4 + c^4 - 2a^2b^2 - 2b^2c^2 + 2a^2c^2$.

DIFFERENCE OF TWO SQUARES.

98. The formula $a^2 - b^2 = (a+b)(a-b)$ enables us to resolve the difference of the squares of any two quantities into two factors, one of which is the sum of these quantities and the other their difference.

Ex. 1. $\quad 9a^2 - 16b^2 = (3a)^2 - (4b)^2$
$\qquad\qquad = (3a + 4b)(3a - 4b)$

Ex. 2. $\quad x^4 - y^4 = (x^2)^2 - (y^2)^2$
$\qquad\qquad = (x^2 + y^2)(x^2 - y^2)$
$\qquad\qquad = (x^2 + y^2)(x + y)(x - y).$

Ex. 3. $\quad a^2 - (b-c)^2 = (a + \overline{b-c})(a - \overline{b-c})$
$\qquad\qquad = (a + b - c)(a - b + c).$

Ex. 4. $\quad a^2 - 4b^2 - 9c^2 + 12bc = a^2 - (4b^2 - 12bc + 9c^2)$
$\qquad\qquad = a^2 - (2b - 3c)^2$
$\qquad\qquad = (a + 2b - 3c)(a - 2b + 3c).$

EXERCISE XXV.

Resolve into factors

1. $x^2 - y^2$.
2. $x^2 - 16$.
3. $25 - y^2$.
4. $4x^2 - 9y^2$.
5. $16x^4 - 49y^2$.
6. $x^4 - 16$.
7. $16x^4 - 81y^4$.
8. $256x^8 - 1$.
9. $x^2y^4 - 100z^6$.
10. $5x^4 - 20y^2$.
11. $3a^4 - 27b^4$.
12. $162x^2y^2 - 242z^2$.
13. $(a+b)^2 - c^2$.
14. $(a-b)^2 - c^2$.
15. $a^2 - (b+c)^2$.
16. $a^2 - (b-c)^2$.
17. $a^2 - (a-b)^2$.
18. $(a+2b)^2 - b^2$.
19. $(a+b)^2 - (c+d)^2$.
20. $(a-b)^2 - (c-d)^2$.
21. $(a+b)^2 - (a-b)^2$.
22. $x^2 - 2xy + y^2 - z^2$.
23. $2bc + b^2 + c^2 - a^2$.
24. $2bc - b^2 - c^2 + a^2$.
25. $a^2x^2 + b^2y^2 + 2abxy - 1$.
26. $a^2 - 4b^2 - 9c^2 + 12bc$.
27. $1 - 16a^2 - 25b^2 + 40ab$.
28. $a^2 + b^2 - c^2 - d^2 + 2ab + 2cd$.
29. $a^2 + b^2 - c^2 - d^2 - 2ab - 2cd$.
30. $a^2 - b^2 + c^2 - d^2 - 2ac + 2bd$.
31. $a^2 - b^2 + c^2 - d^2 + 2ac - 2bd$.
32. $(5x - 2)^2 - (x - 4)^2$.

SUM AND DIFFERENCE OF CUBES.

33. $(x^2-5x+5)^2-1.$
34. $(x^2-10)^2-36.$
35. $(x^2-25)^2-(2x+10)^2.$
36. $(2x^2+3x-5)^2-(x^2-9x-40)^2.$
37. $a^2b^2+c^2d^2-a^2c^2-b^2d^2.$
38. $(x^2-n^2y^2)^2-(y^2-n^2x^2)^2.$

99. A trinomial having two of its terms exact squares can sometimes be expressed as the difference of two squares, and thus be resolved into factors as in the following example:—

$$x^4+a^2x^2+a^4 = x^4+2a^2x^2+a^4-a^2x^2$$
$$= (x^2+a^2)^2-a^2x^2$$
$$= (x^2+a^2+ax)(x^2+a^2-ax)$$
$$= (x^2+ax+a^2)(x^2-ax+a^2).$$

EXERCISE XXVI.

Resolve into factors

1. $x^4+x^2+1.$
2. $x^4+4x^2+16.$
3. $x^4+9x^2+81.$
4. $x^4+4y^4.$
5. $x^4+5x^2y^2+9y^4.$
6. $x^4-3x^2y^2+y^4.$
7. $x^4-5x^2y^2+4y^4.$
8. $4x^4+1.$
9. $4x^4+3x^2y^2+9y^4.$
10. $x^4+x^2+25.$
11. $9x^4-10x^2y^2+y^4.$
12. $x^8+x^4+1.$
13. $4x^4-37x^2y^2+9y^4.$
14. $4x^4-13x^2y^2+9y^4.$
15. $(a+b)^4+(a^2-b^2)^2+(a-b)^4.$
16. $(a+b)^4+4(a-b)^4.$
17. $a^4+b^4+c^4-2a^2b^2-2b^2c^2-2c^2a^2.$

SUM AND DIFFERENCE OF CUBES.

100. Since
$$\frac{a^3+b^3}{a+b} = a^2-ab+b^2,$$

and
$$\frac{a^3-b^3}{a-b} = a^2+ab+b^2,$$

we can always resolve the sum or the difference of the cubes of any two quantities into two factors.

Ex. 1. $8a^3+27b^6 = (2a)^3+(3b^2)^3$
$$= (2a+3b^2)(4a^2-6ab^2+9b^4)$$

SUM AND DIFFERENCE OF CUBES.

Ex. 2. $a^6 + b^6 = (a^2)^3 + (b^2)^3$
$\qquad = (a^2 + b^2)(a^4 - a^2b^2 + b^4).$

Ex. 3. $a^6 - b^6 = (a^3 + b^3)(a^3 - b^3)$
$\qquad = (a + b)(a^2 - ab + b^2)(a - b)(a^2 + ab + b^2).$

101. Since $x + y$ is a factor of $x^n + y^n$ when n is any odd whole number, and $x - y$ is a factor of $x^n - y^n$ when n is any whole number, we can always resolve the sum of two equal odd powers, or the difference of any equal powers of any two quantities, into factors.

Ex. $a^{10} + b^{10} = (a^2)^5 + (b^2)^5$
$\qquad = (a^2 + b^2)(a^8 - a^6b^2 + a^4b^4 - a^2b^6 + b^8).$

Similarly the sum of any even powers can be resolved if both exponents contain the *same odd factor*.

EXERCISE XXVII.

Resolve into factors

1. $a^3 + b^3$.
2. $a^3 + 8$.
3. $27 + b^3$.
4. $8x^3 + 27y^3$.
5. $27x^3 + 64y^3$.
6. $x^3 - y^3$.
7. $x^3 - 1000$.
8. $729x^3 - 512y^3$.
9. $125x^3y^3 - 343z^3$.
10. $64a^3b^3 - 1000c^3$.
11. $x^5 + y^5$.
12. $x^5 + 243$.
13. $243 + y^5$.
14. $x^5 - y^5$.
15. $x^5 - 32y^5z^{10}$.
16. $1 - x^5y^{10}z^{10}$.
17. $a^6 - b^6$.
18. $a^6 + b^6$.
19. $a^9 + b^9$.
20. $a^9 - b^9$.
21. $a^{30} - b^{30}$.
22. $a^{10} - b^{10}$.
23. $a^{10} + b^{10}$.
24. $a^{14} + b^{14}$.
25. $a^{16} - b^{16}$.
26. $(a + b)^3 - a^3$.
27. $(a + b)^3 + a^3$.
28. $(a - b)^3 + a^3$.
29. $(a - b)^3 + 8b^3$.
30. $(a + b)^3 - 8b^3$.
31. $(a + b)^3 + (a - b)^3$.
32. $(a + b)^3 - (a - b)^3$.
33. $(a - 2b)^3 + (2a - b)^3$.
34. $(2a - b)^3 - (a - 2b)^3$.
35. $a^3 + b^3 + 3ab(a + b)$.
36. $a^3 - b^3 - 3ab(a - b)$.
37. $a^3 - 3a^2b + 3ab^2 - b^3 - c^3$.
38. $x^6 - 3x^5 + 3x^4 - x^3 - 8$.

TRINOMIALS.

FIRST COEFFICIENT NOT UNITY.

102. Trinomials whose first coefficient is not unity may sometimes be resolved into binomial factors. The method will be understood from the following example:—

Ex.—Resolve $6x^2 - 13xy + 6y^2$ into factors.

1. Each factor must evidently contain both x and y.
2. The product of the first terms of the two factors must be $6x^2$, and of the last terms $6y^2$.
3. The algebraic sum of the products of the first term of each factor into the last term of the other must be $-13xy$.
4. Since the last term is positive the connecting signs of the factors must be alike, and since the middle term is negative they must also be negative.
5. It is useless to try such a factor as $2x - 2y$ since 2 is not a factor of the given expression.

The above considerations lead us to reject all but two sets of factors, viz.: $(x - 6y)(6x - y)$ and $(2x - 3y)(3x - 2y)$, and it is only by trial that we determine that the latter set is correct.

Thus $\quad (6x^2 - 13xy + 6y^2) = (2x - 3y)(3x - 2y)$.
Similarly $\quad (6x^2 - 37xy + 6y^2) = (\ x - 6y)(6x -\ y)$.

EXERCISE XXVIII.

Resolve into factors

1. $2x^2 + 5xy + 2y^2$.
2. $2x^2 + 5xy + 3y^2$.
3. $2x^2 + 7xy + 3y^2$.
4. $15x^2 - 26xy + 8y^2$.
5. $12x^2 - 5xy - 3y^2$.
6. $14x^2 + 83xy - 6y^2$.
7. $34x^4 + 21x^2 + 2$.
8. $6x^4 - 13x^2y^2 + 2y^4$.
9. $8x^2 + 22x + 9$.
10. $12x^2 - 2xy - 30y^2$.
11. $6ax^2 - 35axy - 6ay^2$.
12. $10a^2x^2 - 7a^2xy - 33a^2y^2$.
13. $ax^2 + (a^2 + 1)xy + ay^2$.
14. $4x^4 - 17x^2y^2 + 4y^4$.

103. To find the factors of $6x^2 + 11xy + 3y^2 + 8xz - 2yz - 8z^2$.

Reject the terms containing z and factor the remaining ones as a trinomial, thus: $6x^2 + 11xy + 3y^2 = (2x + 3y)(3x + y)$.

Take two terms whose product is $-8z^2$, attach one to each factor, and then find by trial whether the factors thus formed give the terms $8xz - 2yz$ in the product. Take $4z$ and $-2z$, thus: $(2x + 3y + 4z)(3x + y - 2z)$. In the product of these factors the term containing xz will be $(2x)(-2z) + (3x)(4z) = 8xz$; the term containing yz will be $(3y)(-2z) + y(4z) = -2yz$, which proves the factors correct.

104. By rejecting the terms containing each letter in succession, and factoring the remaining trinomials, we can determine the complete factors without further trial, thus:—

$$6x^2 + 8xz - 8z^2 = (2x + 4z)(3x - 2z);$$
$$3y^2 - 2yz - 8z^2 = (y - 2z)(3y + 4z).$$

In these factors $4z$ occurs with $2x$ and also with $3y$; place it, then, with the factor $2x + 3y$ first obtained; similarly with regard to $-2z$. We thus get the same factors as before; the first method, however, is generally the better.

105. To find the factors of $a^3 + b^3 + c^3 - 3abc$.

$$a^3 + b^3 + c^3 - 3abc = a^3 + b^3 + 3ab(a+b) + c^3 - 3ab(a+b) - 3abc$$
$$= (a+b)^3 + c^3 - 3ab(a+b+c)$$
$$= (a+b+c)\{(a+b)^2 - (a+b)c + c^2 - 3ab\}$$
$$= (a+b+c)(a^2 + b^2 + c^2 - ab - bc - ca).$$

From this result the factors of Exs. 10–13 may be written down by inspection.

EXERCISE XXIX.

Resolve into factors

1. $2x^2 + 5xy + 2y^2 - 7xz - 8yz + 6z^2$.
2. $2x^2 - 5xy + 2y^2 + 7xz - 5yz + 3z^2$.
3. $6x^2 - 37xy + 6y^2 - 5xz - 5yz - z^2$.

TRINOMIALS.

4. $6x^2 - 13xy + 6y^2 + 12x - 13y + 6$.
5. $x^2 + 2xy + y^2 + 5x + 5y + 6$.
6. $4x^2 - 12xy + 9y^2 - 6x + 9y + 2$.
7. $x^2 - 12y^2 + 15z^2 - 4xy + 8yz - 8xz$.
8. $72x^2 - 8y^2 + 55xy + 12y - 169x + 20$.
9. $15x^4 - 16y^4 - 22x^2y^2 + 15z^4 + 14y^2z^2 + 50x^2z^2$.
10. $a^3 + b^3 - c^3 + 3abc$.   11. $a^3 - b^3 - c^3 - 3abc$.
12. $a^3 + b^3 + 8c^3 - 6abc$.   13. $8a^3 + b^3 - 27c^3 + 18abc$.

106. The following exercise consists of a miscellaneous set of examples arranged by combining the elementary forms explained in this chapter. The student should not try to work them all consecutively, but one or two each day will prove a valuable exercise while proceeding with the following chapters.

EXERCISE XXX.

Resolve into factors

1. $4b^2c^2 - (a^2 - b^2 - c^2)^2$.
2. $4(ab + cd)^2 - (a^2 + b^2 - c^2 - d^2)^2$.
3. $(a^2 - 5a)^2 + 10(a^2 - 5a) + 24$.
4. $(a^2 - 9a)^2 + 4(a^2 - 9a) - 140$.
5. $13x^2y^2 - 9x^4 - 4y^4$.
6. $(a^2 - b^2)(x^2 - y^2) - 4abxy$.
7. $a^6 - b^6 + a^4 + b^4 + a^2b^2$.
8. $a^6 + b^6 + a^4 + b^4 - a^2b^2$.
9. $a^5 + b^5 - a^3b^3 - a^2b^3$.
10. $a^5 - b^5 - a^3b^2 + a^2b^3$.
11. $(x + y)^4 - 7(x + y)^2z^2 + z^4$.
12. $(a + b)^4 + 2(a^2 - b^2)^2 - 3(a - b)^4$.
13. $a^2b^2x^4 - (a^4 + b^4)x^2y^2 + a^2b^2y^4$.
14. $x^4 - 2xy(x^2 - y^2) - y^4$.
15. $(x^2 + xy - y^2)^2 - (x^2 - xy - y^2)^2$.
16. $(a + b)^2 + a^2 + b^2 + 2(a^3 - b^3)$.
17. $x^5 + a^5 - ax(x^3 + a^3)$.
18. $(a^2 - b^2)(x^2 - y^2) + 4abxy$.
19. $81x^4 + 90x^3 - 10x - 1$.
20. $ax^3 - (a^2 + b)x^2 + b^2$.
21. $x^3 - 2ax^2 - a^2x + 2a^3$.
22. $(ac + bd)^2 + (ad - bc)^2$.
23. $x^4 - (p^2 + 1)x^2 + p^2$.
24. $(a^2 + b^2)^2 - (a^2 - b^2)^2 - (a^2 + b^2 - c^2)^2$

TRINOMIALS.

25. $a^3 + b^3 + c^3 + 3ab^2 + 3a^2b$.
26. $a^3 + b^3 + a^2b + ab^2 + a + b$.
27. $x^3 - y^3 - x^2z - xyz - y^2z$.
28. $x^4 + x^2y^2 + y^4 + x^2 + xy + y^2$.
29. $x^3 - y^3 + x^4 + y^4 + x^2y^2$.
30. $a^3 + b^3 + a^2 + b^2 - ab$.
31. $(a^2 - b^2 - c^2 + d^2)^2 - 4(ad - bc)^2$.
32. $x^5 + x^4y + x^3y^2 + x^2y^3 + xy^4 + y^5$.
33. $c^2(a-b) + ab(a-b) - c(a^2 - b^2)$.
34. $a^3 - b^3 - a(a^2 - b^2) + b(a-b)$.
35. $(x-y)(y^2 - z^2) - (x^2 - y^2)(y - z)$.
36. $(x-y)(y^3 - z^3) - (x^3 - y^3)(y - z)$.
37. $(ax + by)^2 + (ay - bx)^2 + c^2x^2 + c^2y^2$.
38. $b^3 + b^2c + bc^2 + c^3 - (a^3 + a^2b + ab^2 + b^3)$.
39. $(a - b)(b^4 - c^4) - (a^4 - b^4)(b - c)$.
40. $(x^2 + xy + y^2)^4 + (x^4 + x^2y^2 + y^4)^2 + (x^2 - xy + y^2)^4$.
41. $a^4 + 16b^4 + c^4 - 8a^2b^2 - 8b^2c^2 - 2c^2a^2$.
42. $8a^2b^2 + 32b^2c^2 + 8c^2a^2 - a^4 - 16b^4 - 16c^4$.
43. $(a^2 - b^2)^2(x^4 + y^4) - \{(a+b)^4 + (a-b)^4\}x^2y^2$.
44. $(m^2 - n^2)^2\{(x+y)^4 + (x-y)^4\} - (x^2 - y^2)^2\{(m+n)^4 + (m-n)^4\}$.
45. $4\{2(x^2 + x)^2 + 3\}^2 + 28(x^2 + x)\{2(x^2 + x)^2 + 3\} - 275(x^2 + x)^2$.

CHAPTER VI.

SIMPLE EQUATIONS OF ONE UNKNOWN.

107. An **Equation** is a statement of the equality of two algebraic expressions.

Thus $3x - 4 = 2x - 6$ is an equation.

108. An **Identical Equation**, or, as it is sometimes called, an **Identity**, is one in which the two expressions are equal for *all* values of the letters involved.

Thus $(a+b)^2 + (a-b)^2 = 2(a^2 + b^2)$ is an identity.

109. A **Conditional Equation** is one in which the two expressions are not equal for all values of the letters involved, but for *one or more* particular values.

Thus $4x + 8 = 2x + 2$ is true only when $x = -3$.

The term "Equation" is generally used for "Conditional Equation."

110. A letter to which a particular value or particular values must be given to make the two expressions equal is called an **Unknown Quantity**.

111. That value of the unknown quantity which will make the two sides of an equation identical is called a **Root** of the equation, and is said to *satisfy* the equation.

NOTE.—The two expressions connected by the sign = are called the sides of the equation.

112. To **solve** an equation is to find the value or values which will satisfy the equation.

113. A **Simple Equation** is one in which the *first* power only of the unknown quantity occurs, and is also called an equation of the *first degree*.

114. The following axioms are used in solving equations:—

1. If equals be added to equals the wholes are equal.

Thus, if $\qquad x - 3 = 4,$
add 3 to each side $\qquad x - 3 + 3 = 4 + 3,$
that is $\qquad x = 7.$

2. If equals be taken from equals the remainders are **equal**.

Let $\qquad 2x + 5 = 3x + 7.$
Take $2x + 5$ from each side,
then $\qquad 2x + 5 - (2x + 5) = 3x + 7 - (2x + 5),$
or $\qquad 0 = x + 2.$

3. If equals be multiplied by equals the products are equal.

Let $\qquad 2x + 1 = 3.$
Multiply both sides by 3,
then $\qquad 6x + 3 = 9.$

4. If equals be divided by equals the quotients are equal.

Let $\qquad 4x + 8 = 6x + 12.$
Divide both sides by 2,
then $\qquad 2x + 4 = 3x + 6.$

5. If the product of two or more factors be equal to *zero*, then *one* or more of the factors must be equal to *zero*.

Thus, if $(x-2)(x-4) = 0$, then either $x - 2 = 0$ or $x - 4 = 0$

115. From axioms 1 and 2 we derive the following

Rule.—*Any term may be transposed from one side of an equation to the other, provided its sign be changed.*

Thus, if $\qquad 5x - 4 = 2x + 5,$
take $2x$ from and add 4 to both sides,
then $\qquad 5x - 2x = 4 + 5,$
and $\qquad 3x = 9.$

116. Transferring a *positive* quantity from one side of an equation to the other is the same as *subtracting* that quantity from both sides.

SIMPLE EQUATIONS OF ONE UNKNOWN. 67

117. Transferring a *negative* quantity from one side to the other is the same as *adding* to both sides the quantity with its sign changed.

118. To *solve* a simple equation, bring all the terms containing the unknown quantity to one side of the equation and all the remaining terms to the other side. Combine the like terms, and divide both sides by the coefficient of the unknown quantity.

Examples.

1. Solve the equation $\quad 3x - 4 = 7x + 12.$
Transposing, $\quad 7x - 3x = -4 - 12.$
Combining, $\quad 4x = -16.$
Dividing by coefficient of x, $\quad x = -4.$

2. Solve $\quad 2x - 3 = 3(x - 2) + 4.$
Simplifying, $\quad 2x - 3 = 3x - 6 + 4.$
Transposing, $\quad 3x - 2x = 6 - 4 - 3.$
Combining, $\quad x = -1.$

3. Solve $\quad 2x - a = b + c - 3x.$
Transposing, $\quad 2x + 3x = a + b + c.$
Therefore $\quad 5x = a + b + c$
and $\quad x = \dfrac{a+b+c}{5}.$

4. Solve $\quad 3(x-a) + 4(x-b) = 2(x+a) - x + b.$
Simplifying, $\quad 3x - 3a + 4x - 4b = 2x + 2a - x + b.$
Transposing, $\quad 3x + 4x - 2x + x = 3a + 4b + 2a + b.$
Combining, $\quad 6x = 5a + 5b.$
Therefore $\quad x = \dfrac{5a + 5b}{6}.$

5. Solve $\quad (x-1)^2 + (x-2)^2 = 2(x-3)^2.$
Expanding, $\quad x^2 - 2x + 1 + x^2 - 4x + 4 = 2x^2 - 12x + 18.$
Therefore $\quad x^2 + x^2 - 2x^2 - 2x - 4x + 12x = -1 - 4 + 18.$
Combining, $\quad 6x = 13,$
and $\quad x = \dfrac{13}{6} = 2\tfrac{1}{6}.$

SIMPLE EQUATIONS OF ONE UNKNOWN.

EXERCISE XXXI.

Find the value of x in

1. $3x + 6 = 12$.
2. $24x - 17x = 30 + 4 - 10x$.
3. $12 - 5x = 19 - 12x$.
4. $5x + 50 = 4x + 56$.
5. $3x + 16 = 31 - 2x$.
6. $5x - 9 = 10x + 11$.
7. $13 - 3x = 5x - 3$.
8. $5x - (3x - 7) = 4x - (6x - 35)$.
9. $9x - 3(5x - 6) = -30$.
10. $6x - 2(9 - 4x) + 3(5x - 7) = 10x - (4 + 16x + 35)$.
11. $x - 7(4x - 11) = 14(x - 5) - 19(8 - x) - 61$.
12. $(x + 7)(x - 3) = (x - 5)(x - 15)$.
13. $(x - 8)(x + 12) = (x + 1)(x - 6)$.
14. $(x + 10)(x - 2) = (x + 9)(x - 3)$.
15. $(x - 2)(x - 3)(x - 4) = x(x - 1)(x - 8)$.
16. $(x + 7)(x + 2)(x - 1) = (x + 4)(x - 2)(x + 6)$.
17. $(x - 2)(7 - x) + (x - 5)(x + 3) - 2(x - 1) + 12 = 0$.
18. $(2x - 7)(x + 5) = (9 - 2x)(4 - x) + 229$.
19. $(x + 5)^2 - (4 - x)^2 = 21x$.
20. $(x - 1)^2 + (x - 2)^2 = 2(x - 4)^2$.
21. $(x - 1)^2 + (x - 2)^2 + (x - 3)^2 = 3(x - 5)^2$.
22. $(x - 1)^3 + (x - 2)^3 + (x - 3)^3 = 3(x - 1)(x - 2)(x - 3)$.
23. $(x - a)(x + a) = (x^2 - b^2) + x(b - a)$.
24. $2x - (a - b) = x + (a + b)$.
25. $5(a + x) - 2x = 3(a - 5x)$.
26. $4(a - 2x) + 5(a + x) = 9(a - x)$.
27. $ax + a^2 = bx + 2a^2 - b^2$.
28. $x(a + b) + a^2 + b^2 = c^2 - 2ab - cx$.
29. $2x - \{1 - (2x - 3)\} = 0$.
30. $x = 4 + 3x - \{3 - (x + 2)\}$.
31. $ax - bx = a + bx - b$.
32. $(a + x)(b + x) = (c + x)(d + x)$.
33. $(x - a)(x - b) - (x + a)(x + b) + ab = 0$.

SIMPLE EQUATIONS OF ONE UNKNOWN. 69

34. $(a+x)^3 + (b+x)^3 + (c+x)^3 = 3(a+x)(b+x)(c+x)$.

35. $a(b^2 - ax) = ab^2 - b(a^2 - bx)$.

36. $(x-a)(x-b) + (x-b)(x-c) - 2(x-c)(x-a) = 0$.

37. $x^2(a-b) + x^2(b-c) + x^2(c-a) + x(ab-bc) + x(bc-ca) + ab - ac = 0$.

38. $a(b-c)x^2 + b(c-a)x^2 + cx^2(a-b) + a^2x(b-c) + b^2x(c-a)$
    $+ c^2x(a-b) + ab(a-b) + bc(b-c) + ca(c-a) = 0$.

119. We propose introducing here a few examples of easy equations of a higher degree than the first, but limited to one unknown quantity.

120. All equations simply express conditions or a condition to be satisfied by the letters involved. If $x^2 - 2x + 1 = 0$, then x must have a numerical value such that when it is substituted for x the two sides of the equation will be identical.

So also if $(x-b)(x-d) = 0$, the value of x which will make either factor $=0$ will make both sides identical, and therefore satisfies the equation.

121. Hence, if an equation can be put into the form of the product of a number of factors, each of *one* dimension, its roots can be found by putting each factor in turn $= 0$; for if any one factor $= 0$ the whole quantity must $= 0$.

Ex. 1.—Solve $\quad x^2 - m^2 = 0$.

Factoring, $\quad (x-m)(x+m) = 0$.

Therefore, if $x - m = 0$ or $x + m = 0$ the equation will be satisfied; therefore $\quad x = m$ or $x = -m$.

Ex. 2.—Solve $\quad (x-1)(x-2)(x-3) = 0$.

If $\quad x - 1 = 0$ the equation is satisfied; therefore 1 is a root of the equation. Similarly 2 and 3 are roots of the equation.

Ex. 3.—Solve $\quad 2x^2 - 4x - 6 = 0$.

Factoring, $\quad 2(x-3)(x+1) = 0$.

Therefore $\quad x - 3 = 0$ or $x + 1 = 0$;

therefore $\quad x = 3$ or $x = -1$.

SIMPLE EQUATIONS OF ONE UNKNOWN.

EXERCISE XXXII.

What values of x will satisfy

1. $(x-a)(x-b) = 0$.
2. $(x+a)(x-b) = 0$.
3. $(x-a)(x+b) = 0$.
4. $(x+a)(x+b) = 0$.
5. $x^2 - (b-c)x - bc = 0$.
6. $x^2 + (b+c)x + bc = 0$.
7. $x^2 - 8x + 12 = 0$.
8. $2x^2 - 8x + 8 = 0$.
9. $x^3 - x^2(a+b+c) + x(ab+bc+ca) - abc = 0$.
10. $x^3 - 6x^2 + 11x - 6 = 0$.
11. $3x^2 - 10x + 3 = 0$.
12. $(a+b)x^2 - (a^2-b^2)(a-b) = 0$.
13. $x^2 - ax - 2a^2 = 0$.
14. $x^3 - 2bx^2 - 3b^2x = 0$.
15. $6x^2 + x - 12 = 0$.
16. $9x^2 - 9x - 28 = 0$.
17. $4x^2 - 36 = 0$.
18. $25x^2 - 49 = 0$.
19. $x^3 - 3x^2 + 3x - 1 = 0$.
20. $x^3 - 9x^2 + 23x - 15 = 0$.
21. $x^3 + x^2 - x - 1 = 0$.

PROBLEMS PRODUCING SIMPLE EQUATIONS OF ONE UNKNOWN.

122. Algebra is extensively used in the solution of problems of practical value. In proceeding to solve a problem the first thing to do is to observe carefully the facts given, or *data*, and to exclude all that have no bearing on the solution. The next thing in order is to state in algebraic language the facts given, which, if correctly done, will result in an equation whose solution will give the required result. It must be noticed that before we can have an equation we must have *two different* algebraic expressions of the *same value*. Thus, if we say that John has 5 apples more than James we have but one expression; for let x represent the number of apples James has, then John will have $x+5$. To obtain an equation we must be able to say that $x+5$ is equal to a *different expression*. If we add to the fact already given that John has 10 apples we have the statement: $x+5=10$, from which x can be found.

SIMPLE EQUATIONS OF ONE UNKNOWN. 71

The art of expressing in algebraic language the *facts* of a problem must be acquired by practice. The following examples will illustrate some of the simpler methods of solving problems:—

Ex. 1.—What number is that to which, if 7 be added, twice the sum will be equal to 32 ?

Let $x =$ the number;
then $x + 7 =$ the number when 7 is added,
and $2(x+7) =$ twice the number when 7 is added;
also $32 =$ twice the number when 7 is added;
therefore $2(x+7) = 32$, or $x+7 = 16$, or $x = 9$.

Ex. 2.—What two numbers are those whose sum is 48 and difference 22 ?

Let $x =$ the number; then, since the sum of the two numbers is 48, the second number must be $48 - x$.

But the difference of the numbers is 22; therefore, if x is the greater number,

$$x - (48 - x) = 22,$$
or $\quad x - 48 + x = 22,$
or $\quad 2x - 48 = 22.$
Transposing, $\quad 2x = 22 + 48 = 70$
$\quad x = 35 =$ first number;
therefore $\quad 48 - x = 48 - 35 = 13 =$ second number.

We might have assumed in this problem that the second number is $x + 22$; then the statement would be·

$$x + x + 22 = 48$$
$$2x + 22 = 48$$
$$2x = 26$$
$$x = 13$$
$$\therefore x + 22 = 35$$

Here the second number is the greater.

SIMPLE EQUATIONS OF ONE UNKNOWN.

Ex. 3.—Thirty yards of cloth and 40 yards of silk together cost £66, and the silk is twice as valuable as the cloth. Find the cost of a yard of each.

Let $x =$ cost in £'s of a yard of cloth;
then $2x =$ cost in £'s of a yard of silk;
therefore $30x =$ cost of 30 yards of cloth,
and $80x =$ cost of 40 yards of silk.

But the whole cost is £66,
therefore $30x + 80x = 66$,
or $110x = 66$;
therefore $x = \tfrac{66}{110} = \tfrac{3}{5}$

Hence value of cloth is £$\tfrac{3}{5}$ or 12s.,
and value of silk is £$\tfrac{6}{5}$ or 24s.

EXERCISE XXXIII.

1. Find two numbers whose sum shall be 100 and whose difference 10.

2. What value of x will make the difference between $(2x+4)(3x+6)$ and $(3x-2)(2x-8)$ equal to 96?

3. Add 24 to a certain sum and the amount will be as much above 80 as the sum is below 80. What is the sum?

4. The sum of the ages of two brothers is 49, and one of them is 13 years older than the other. Find their ages.

5. I bought 20 yards of cloth for 10 guineas, part at 11s. 6d. a yard and the rest at 7s. 6d. a yard. How many yards of each did I buy?

NOTE.—Be careful to express the various sums of money in the *same* denomination.

6. At an election where 979 votes were given, the successful candidate had a majority of 97. What were the numbers for each?

7. Divide 87 into three parts such that the first may exceed the second by 7 and the third by 17.

8. Find a number which, being multiplied by 6, and having 12 added to the product, the sum shall be 66.

SIMPLE EQUATIONS OF ONE UNKNOWN.

9. A person left $700 to be divided among three persons in such a way that the first was to receive double of what the second received, and the second double of what the third received. Find each person's share.

10. Two persons, A and B, put equal sums of money in business. A gains $126 and B loses $87, and A's money is now double that of B's. What did each lay out?

11. What two numbers are those whose difference is 10, and if 15 be added to their sum the whole will be 43?

12. A father's age is twice as great as that of his son, but 10 years ago it was three times as great. Find the age of each.

13. The ages of two men differ by 10 years, and 15 years ago the elder was twice as old as the younger. Find the ages of the men.

14. A had 7 times as many apples, and B three times as many, as C had. A parted with 16 to B, and then these two persons had equal quantities. How many apples had C?

15. Find that number of which it can be said that four times the number is as much less than 25, as five times the number is greater than 11.

16. Two trains start at the same time from Toronto and Montreal, a distance of 333 miles, travelling, the former at 25 miles an hour, the latter at 22 miles. Where will they meet, and in what time from starting?

17. A man bought a number of cows for $40 apiece, and 5 more horses for $120 apiece, and paid altogether $2200. How many did he buy of each kind?

18. How much tea at 65 cents a pound must be mixed with 50 pounds at 75 cents a pound that the mixture may be worth 70 cents a pound?

19. A cistern is filled in 20 minutes by 3 pipes, one of which conveys 10 gallons more, and another 10 gallons less, than the third per minute. The cistern holds 900 gallons. How much flows through each pipe per minute?

20. The sum of $500 is divided among A, B, C and D. A and B have together $280; A and C, $260; A and D, $220. How much does each receive?

21. Divide $1520 among A, B and C, so that A has $100 less than B, and B $270 less than C.

22. Find two numbers differing by 8 such that four times the less may exceed twice the greater by 10.

23. A vessel containing some water was filled up by pouring in 42 gallons, and there was then in the vessel seven times as much as at first. How many gallons did the vessel hold?

24. In a company of 266 persons, composed of men, women and children, there are twice as many men as there are women, and twice as many women as there are children. How many are there of each?

25. Divide 90 into two such parts that four times one part may be equal to five times the other.

26. Divide 60 into two such parts that one part is greater than the other by 24.

27. A bill of $100 was paid in dollar and fifty-cent pieces, and 80 more fifty-cent pieces were used than dollars. How many of each were used?

28. A bill of £100 was paid with guineas and half-crowns, and 48 more half-crowns were used than guineas. How many of each were used?

29. Thirty yards of cloth and 20 yards of silk cost $120, and the silk cost twice as much as the cloth. How much did each cost per yard?

30. A man bought 30 pounds of sugar of two different kinds, and paid for the whole $2.94. The better kind cost 10 cents a pound, and the poorer kind 7 cents a pound. How many pounds were there of each kind?

31. A man has three times as many quarters as half-dollars, four times as many dimes as quarters, and twice as many half-dimes as dimes. The whole sum is $7.30. How many coins has he altogether?

SIMPLE EQUATIONS OF ONE UNKNOWN.   75

32. A gentleman gave some children 10 cents each, and had a dollar left. He found that he would have required one dollar more to enable him to give them 15 cents each. How many children were there?

33. A wine merchant has two kinds of wine, one worth 50 cents a quart and the other 75 cents a quart. From these he wishes to make a mixture of 100 gallons worth $2.40 a gallon. How many gallons must he take of each kind?

34. Two casks contain equal quantities of vinegar. From the first cask 34 quarts are drawn; from the second, 20 gallons. The quantity remaining in one vessel is now twice that in the other. How much did each cask contain at first?

35. A man leaves his property, amounting to $7500, to be divided among his wife, two sons and three daughters. A son is to have twice as much as a daughter, and the wife $500 more than all the children together. How much was the share of each?

36. A has $80 and B has $32, and each loses a certain sum; then A has five times as much as B. What is the sum lost by each?

37. A and B have equal sums of money. A gains $300 and B loses $100, and then three times A's is equal to five times B's. How much had each at first?

38. A bankrupt owes B twice as much as he owes A, and C as much as he owes A and B together. Out of $3000 which is to be divided among them what should each receive?

39. A workman was employed for 60 days on condition that for every day he worked he should receive 15 pence, and for every day he was absent he should forfeit 5 pence. At the end of the time he had 20 shillings to receive. Find the number of days he worked.

40. The length of a field is twice its breadth; another field, which is 50 yards longer and 10 yards broader, contains 6800 square yards more than the former. Find the size of each.

CHAPTER VII.

COMMON FACTORS AND COMMON MULTIPLES.

HIGHEST COMMON FACTOR.

123. The methods of resolving a single algebraical expression into factors have already been given. We proceed now to explain how to determine the factors, if any, which are contained in each of two or more given expressions.

124. A **Common Factor** of two or more algebraical expressions is any expression which is a factor of each of them. Two expressions having no common factor except unity are said to be **prime** to each other.

125. The **Highest Common Factor** of two or more algebraical expressions is the factor of highest dimensions and greatest numerical coefficients which is a factor of each of them.

The Highest Common Factor is usually denoted by the letters H. C. F.

126. A **Multiple** of an algebraical expression is any expression of which the given expression is a factor.

The terms "Measure" and "Multiple," used in arithmetic, are not very appropriate for algebraical expressions. The former is, therefore, replaced by the correct term, "Factor;" the latter is retained because we have no other word to supply its place.

127. The H. C. F. of two or more expressions will evidently be the product of all the factors common to each, and may therefore be found by the following

RULE.—*Resolve each expression into elementary factors, and the product of all the factors common to each expression will be the H. C. F. required.*

HIGHEST COMMON FACTOR.

128. If only one of two given expressions can be readily factored its factors may be tried in succession as divisors of the other expressions, and thus all the common factors will be readily found.

Ex. 1.—Find the H. C. F. of $45a^2b^3c^4$, $60a^3bc^3$, $35a^2b^2c^4$.

The factors 5, a^2, b, c^3, are all those which are found in each expression; therefore $5a^2bc^3$ is the H. C. F. required.

Ex. 2.—Find the H. C. F. of $a^2 - b^2$, $ab + b^2$, $a^3 + b^3$.

$$a^2 - b^2 = (a+b)(a-b)$$
$$ab + b^2 = b(a+b)$$
$$a^3 + b^3 = (a+b)(a^2 - ab + b^2).$$

The only factor found in each is $a+b$, which is therefore the H. C. F. required.

Ex. 3.—Find the H. C. F. of

$$(2x-3)(x-5)(3x-4) \text{ and } 15x^3 - 128x^2 + 309x - 220.$$

In this example the second expression cannot readily be factored, but the factors of the first are given. The *common* factors, if any, must be found among the three given factors; but $2x-3$ is *not* a factor of the second expression, because 2 and 3 are not factors of 15 and 220. Trying $3x-4$ we find the quotient is $5x^2 - 36x + 55$, and by further trial we find $5x^2 - 36x + 55$ is exactly divisible by $x-5$; therefore $(3x-4)(x-5)$ is the H. C. F. required.

EXERCISE XXXIV.

Find the H. C. F. of

1. a^3b^2c and a^2b^3d.
2. $18ab^3m$ and $24a^2b^2m^3$.
3. $7a^3b^2$, $14a^2b$ and $35ab^3$.
4. $17x^2y^2$, $34x^3y^4$ and $51xy^3$.
5. $8x^2y^2z^4$, $12x^3y^2z^3$ and $20x^4y^3z^2$.
6. $57x^7y^9z^{11}$, $76x^5y^7z^9$ and $95x^6y^5z^6$.
7. $5x^3y^3$ and $15x^3y^2 + 20x^2y^3$.
8. $14m^2x$ and $21m^3p - 7mx$.
9. $3mx + 5nx^2$ and $3my + 5nxy$.
10. $3x^4 + 3x^2y^2$ and $5x^4 + 5x^2y^2$.
11. $ay + y^2$ and $abc + bcy$.
12. $a^3b^2 + a^2b^3$ and $a^2bcd + ab^2cd$.

13. x^2-5x+6 and $x^2-11x+24$.   14. $x^2-5x-14$ and x^2+x-56.
15. x^2-1 and $x^2+20x+19$.   16. $x^2-14x-51$ and $x^2-7x-170$.
17. x^4-y^4 and x^6+y^6.   18. $x^4+x^2y^2+y^4$ and x^3-y^3.
19. x^9-y^9 and x^6-y^6.   20. x^6+y^6 and $x^{10}+y^{10}$.
21. $x^8+x^4y^4+y^8$ and $x^{12}-y^{12}$.   22. $x^2-4x-77$ and x^3+343.
23. $8x^3+27y^3$ and $6x^2+5xy-6y^2$.   24. $a^3+b^3+2ab(a+b)$ and a^3-b^3.
25. x^4+4x^2+16 and x^3-8.   26. x^3+1 and x^3+mx^2+mx+1.
27. $125x^3-64y^3$ and $75x^3+60x^2y+48xy^2$.
28. $x(x+1)^2$, $x^2(x^2-1)$ and $2x(x^2-x-2)$.
29. $6(a-b)^4$, $8(a^2-b^2)^2$ and $10(a^4-b^4)$.
30. $ac(a-b)(a-c)$ and $bc(b-a)(b-c)$.
31. $(a+b)^2-(c+d)^2$, $(a+c)^2-(b+d)^2$ and $(a+d)^2-(b+c)^2$.
32. $6x^3-6xy^2+2x^2y-2y^3$ and $12x^2-15xy+3y^2$.
33. $3a^3-3a^2b+ab^2-b^3$ and $4a^2-5ab+b^2$.
34. $3x^3-3x^2y+xy^2-y^3$ and $4x^3-x^2y-3xy^2$.
35. $20x^4+x^2-1$ and $75x^4+15x^3-3x-3$.

129. The following principles enable us to find the H. C. F. of any two algebraical expressions. For their demonstration see Arts. 136–139.

1. If one expression is a factor of another it will also be a factor of any multiple of the other.

2. If one expression is a factor of two others it will also be a factor of the sum or the difference of any multiples of the others.

3. If one expression be divided by another the remainder, if any, will be the difference between the dividend and a multiple of the divisor, and will therefore contain all the factors common to the two expressions.

130. To find the H. C. F. of any two algebraical expressions we have from Arts. 127 and 129 the following

RULE.—*If the expressions contain any common factors which can be discovered by inspection strike them out and reserve them*

HIGHEST COMMON FACTOR. 79

as factors of the H. C. F. From each of the remaining expressions strike out and reject any other elementary factors which can be discovered.

Arrange the resulting expressions according to the power of some common letter, select that expression which is not of lower dimensions than the other, and divide the former by the latter; take the remainder as divisor and the former divisor as dividend and repeat the process until there is no remainder. The last divisor, multiplied by the reserved factors, if any, will be the H. C. F. required.

131. When the first term of any dividend is not exactly divisible by the first term of the corresponding divisor, we multiply each term of the dividend by such a factor as will render it exactly divisible. This cannot affect the H. C. F., since all monomial factors are supposed to have been removed from the given expressions, and therefore multiplying *one* expression cannot introduce any *common* factor. Similarly, any factor which is clearly not a part of the H. C. F. may be removed from any remainder before using it as a divisor.

132. Since each remainder contains all the common factors, it follows that if any remainder can be resolved into elementary factors the H. C. F. may be found by trying these factors in succession as divisors of *one* of the given expressions; for any factor of a remainder will divide *both* of the given expressions or *neither* of them. This follows from Art. 129. Also, any of the preceding remainders may be tested instead of one of the given expressions.

133. The following examples show the application of the preceding rules:—

Ex. 1.—Find the H. C. F. of

$18a^5b - 3a^4b - 12a^3b - 3a^2b$ and $12a^5c - 6a^4c - 9a^3c + 3a^2c$.

$18a^5b - 3a^4b - 12a^3b - 3a^2b = 3a^2b(6a^3 - a^2 - 4a - 1)$
$12a^5c - 6a^4c - 9a^3c + 3a^2c = 3a^2c(4a^3 - 2a^2 - 3a + 1)$

Now, $3a^2$ is evidently a factor of the H. C. F., but b and c are not. Proceeding with the remaining expressions,

$$4a^3 - 2a^2 - 3a + 1 \overline{\smash{\big)}\, \begin{matrix} 6a^3 - a^2 - 4a - 1 \\ 2 \\ \hline 12a^3 - 2a^2 - 8a - 2 \\ 12a^3 - 6a^2 - 9a + 3 \\ \hline 4a^2 + a - 5 \\ = (4a+5)(a-1) \end{matrix}} \, (3$$

Now, if these expressions have any common factor it must be one of the factors of this remainder. $4a+5$ is evidently not a factor of either, and by trial we find $a-1$ is a factor of the former; therefore $3a^2(a-1)$ is the H. C. F. required.

Ex. 2.—Find the H. C. F. of
$4x^4 + 2x^3 - 18x^2 + 3x - 5$ and $6x^5 - 4x^4 - 11x^3 - 3x^2 - 3x - 1$.

$$4x^4+2x^3-18x^2+3x-5 \overline{\smash{\big)}\, \begin{matrix} 6x^5 - 4x^4 - 11x^3 - 3x^2 - 3x - 1 \\ 2 \\ \hline 12x^5 - 8x^4 - 22x^3 - 6x^2 - 6x - 2 \\ 12x^5 + 6x^4 - 54x^3 + 9x^2 - 15x \\ \hline -14x^4 + 32x^3 - 15x^2 + 9x - 2 \\ 2 \\ \hline -28x^4 + 64x^3 - 30x^2 + 18x - 4 \\ -28x^4 - 14x^3 + 126x^2 - 21x + 35 \\ \hline 78x^3 - 156x^2 + 39x - 39 \\ = 39(2x^3 - 4x^2 + x - 1). \end{matrix}} \, (3x$$

$$2x^3 - 4x^2 + x - 1 \overline{\smash{\big)}\, \begin{matrix} 4x^4 + 2x^3 - 18x^2 + 3x - 5 \\ 4x^4 - 8x^3 + 2x^2 - 2x \\ \hline 10x^3 - 20x^2 + 5x - 5 \\ 10x^3 - 20x^2 + 5x - 5 \end{matrix}} \, (2x+5$$

Therefore $2x^3 - 4x^2 + x - 1$ is the H. C. F. required.

EXPLANATION.—We first multiply the dividend by 2 to make the first term exactly divisible by the first term of the divisor. This will not increase the H. C. F., because 2 is not a factor of the divisor. Again, we reject the factor 39 from a remainder. This cannot decrease the H. C. F., since 39 is not a factor of either of the given expressions.

Sometimes the work can be abbreviated by reversing the order of terms in the given expressions. If this be done in Ex. 1 the first multiplication will be avoided. It is a general principle that either the highest or the lowest power of the leading letter must be eliminated by division, but only experience can determine the best method in any particular example.

134. The following example illustrates a difference between algebraical and arithmetical operations, and is worthy of careful attention:—

The H. C. F. of x^2+4x+3 and x^2+8x+7 is $x+1$.

Now, if for x we substitute any whole number we may be certain that the value of $x+1$ will be a common measure of the values of x^2+4x+3 and x^2+8x+7, but we cannot be certain that it will be their greatest common measure. If $x=2$ the expressions become 15 and 27, and $x+1=3$, and this *is* their greatest common measure; but if $x=3$ the expressions become 24 and 40, and $x+1=4$, which is *not* their greatest common measure. The explanation lies in the fact that the other factors, $x+3$ and $x+7$, have no common measure for *all* values of x, but they have for *particular* values of x, viz., when x is any *odd number*.

135. The following is the substance of the theory usually given in connection with this subject. The reasoning, however, is not suitable for beginners, and should, therefore, be omitted until considerable progress has been made. The same remark applies to the corresponding Arts. in L. C. M.

136. If P is a factor of A then it is a factor of mA; for, since P is a factor of A, we may suppose $A=aP$; then $mA=maP$; thus P is a factor of mA.

137. If P is a factor of A and B then it is a factor of $mA \pm nB$; for, suppose $A=aP$ and $B=bP$, then $mA \pm nB = maP \pm nbP = P(ma \pm nb)$.

The double sign $\pm$ signifies both the sum and the difference of the quantities between which it is written.

138. Let A and B denote any two expressions whose H. C. F. is required. Let them be arranged and the divisions performed according to Art. 130, and let the various quotients and remainders be denoted by the letters in the margin; then we have the following results:—

$$B) \overline{A} \,(p$$
$$pB$$
$$\overline{C) B} \,(q$$
$$qC$$
$$\overline{D) C} \,(r$$
$$rD$$

$$A = pB+C, \quad B = qC+D, \quad C = rD;$$

$$\therefore A = pB+C \qquad\qquad B = qC+D$$
$$\quad = p(qC+D)+rD \qquad = rD+D$$
$$\quad = p(qrD+D)+rD \qquad = D(r+1)$$
$$\quad = D(pqr+p+r)$$

Therefore D is a factor of both A and B.

139. Again, since every expression which divides A and B divides $A-pB$, that is, C, therefore C contains all the factors common to A and B. Similarly D contains all the factors common to B and C, that is, all the factors common to A and B; therefore D is the H. C. F. required. Also, every factor of A and B is a factor of D.

140. If we are required to find the H. C. F. of three expressions, A, B and C, find the H. C. F. of two of them, A and B; let it be D. Then the H. C. F. of D and C will be the H. C. F. required; for every factor of D and C is a factor of A, B and C, and every factor of A, B and C is a factor of D and C; therefore D is the H. C. F. of A, B and C.

EXERCISE XXXV.

Find the H. C. F. of

1. $x^3+3x^2+4x+12$ and x^3+4x^2+4x+3.
2. x^3+x^2-2x-8 and x^3+2x^2+x-4.
3. x^3+x^2+x-3 and x^3+3x^2+5x+3.
4. $x^3+x^2-13x-4$ and $3x^3+10x^2-13x-20$.

5. $3x^3 - 17x^2 - 5x + 10$ and $3x^3 - 23x^2 + 23x - 6$.
6. $2x^3 - x^2 + x + 4$ and $2x^4 - x^3 - x^2 + 7x - 4$.
7. $2x^3 - 19x^2 + 38x - 21$ and $3x^3 - 20x^2 - 12x + 35$.
8. $3x^3 - 13x^2 + 23x - 21$ and $6x^3 + x^2 - 44x + 21$.
9. $x^4 - 3x^3 + 2x^2 + x - 1$ and $x^3 - x^2 - 2x + 2$.
10. $x^5 + 5x^4 - x^2 - 5x$ and $x^4 + 3x^3 - x - 3$.
11. $2x^4 - 12x^3 + 19x^2 - 6x + 9$ and $4x^3 - 18x^2 + 19x - 3$.
12. $2x^5 - 11x^3 - 9$ and $4x^5 + 11x^4 + 81$.
13. $x^5 + 3x^4 - 8x^2 - 9x - 3$ and $x^5 - 2x^4 - 6x^3 + 4x^2 + 13x + 6$.
14. $2x^5 - 5x^4 + x^3 - 9x^2 - 11x + 6$ and $3x^5 - 8x^4 - 4x^3 + 5x^2 - 5x - 3$.
15. $9x^5 + 11x^3 - 2$ and $81x^5 + 11x + 4$.
16. $x^5 - 209x + 56$ and $56x^5 - 209x^4 + 1$.
17. $x^6 + 1$ and $x^5 + x^4 - x^3 - x^2 + x + 1$.
18. $2a^4 + 3a^3x - 9a^2x^2$ and $6a^4x - 17a^3x^2 + 14a^2x^3 - 3ax^4$.
19. $a^3x^3 - a^2bx^2y + ab^2xy^2 - b^3y^3$ and $2a^2bx^2y - ab^2xy^2 - b^3y^3$.
20. $x^3 - 9x^2 + 26x - 24$, $x^3 - 10x^2 + 31x - 30$ and $x^3 - 11x^2 + 38x - 40$.
21. $x^4 - 10x^2 + 9$, $x^4 + 10x^3 + 20x^2 - 10x - 21$
    and $x^4 + 4x^3 - 22x^2 - 4x + 21$.
22. $x^4 - px^3 + px^2 - p^2x$ and $x^3 - p^3$.
23. $x^3 + (p+1)x^2 + (p+1)x + p$ and $x^3 + (p-1)x^2 - (p-1)x + p$.
24. $ax^3 - bx^2 + ax - b$ and $ax^3 + (a-b)x^2 + (a-b)x - b$.
25. $x^4 + 2mx^3 + m^2x^2 - n^2$ and $x^3 + (m-1)x^2 + (n-m)x - n$.
26. $x^3 + (5m-3)x^2 + (6m^2 - 15m)x - 18m^2$
    and $x^3 + (m-3)x^2 - (2m^2 + 3m)x + 6m^2$.
27. $px^3 - (p-q)x^2 + (p-q)x + q$ and $px^3 - (p+q)x^2 + (p+q)x - q$.
28. $ax^3 - (a-b)x^2 - (b-c)x - c$ and $2ax^3 + (a+2b)x^2 + (b+2c)x + c$.
29. $a^3 + b^3 + c^3 - 3abc$ and $a(a+2b) + b(b \pm 2c) + c(c+2a)$.
30. $ay(x^3 + b^3) + bx(by^2 + a^2x)$ and $ax(y^3 + b^3) + by(bx^2 + a^2y)$.

LOWEST COMMON MULTIPLE.

141. A **Common Multiple** of two or more algebraical expressions is any expression of which each of the given expressions is a factor.

142. The **Lowest Common Multiple** of two or more algebraical expressions is the expression of lowest dimensions and smallest numerical coefficients of which each of the given expressions is a factor.

The Lowest Common Multiple is usually denoted by the letters L. C. M.

143. The L. C. M. must evidently contain all the elementary factors which any one of the given expressions contains, but no other factor; therefore for finding the L. C. M. of two or more algebraical expressions we have the following

RULE.—*Resolve each of the given expressions into elementary factors, and the product of the highest powers of all the different factors which occur will be the L. C. M. required.*

EXERCISE XXXVI.

Find the L. C. M. of

1. $2a^2b$, $3ab^2$, $6abc$.
2. $6a^3b^2c$, $10b^3cd$, $15abcd^2$.
3. $18p^4q^2$, $40q^4r^2$, $75r^4p^2$.
4. $21l^5m^2$, $35m^4n^3$, $70m^2n^2p$.
5. a^2-b^2, $ab+b^2$, a^2+ab.
6. a^3-b^3, $a-b$, a^2+ab+b^2.
7. x^3-1, x^2-1, x^2+x+1.
8. x^3-xy^2, xy^2+y^3, x^3-x^2y.
9. $4(x^2-y^2)$, $6(x-y)^2$, $10(x+y)^2$. 10. $3x(x^3-1)$, $5y(x+1)$, x^2-x+1
11. a^2-b^2, a^3-b^3, $a^4+a^2b^2+b^4$, a^6-b^6.
12. a^2+b^2, $a^4-a^2b^2+b^4$, a^6-b^6, a^3-b^3, a^4-b^4.
13. x^2-5x+6, x^2-4, x^2-9, $x^2-7x+12$, x^2-16.

14. $x^2-(a+b)x+ab$, $x^2-(b+c)x+bc$, $x^2-(c+a)x+ca$.

15. $(x^2-4a^2)^2$, $(x+2a)^3$, $(x-2a)^3$.

16. $8(a^2-b^2)(a-b)^2$, $10(a^4-b^4)(a+b)^2$, $15(a^2+b^2)^3$.

17. $12(a^3-b^3)(a+b)^3$, $9(a^4-b^4)(a-b)^2$, $24(a^2-b^2)^3$.

18. x^4-10x^2+9, x^3-7x+6, x^3-7x-6.

19. $a^2-b^2+c^2+2ac$, $a^2+b^2-c^2+2ab$, $b^2+c^2-a^2+2bc$.

20. $4(x^3-xy^2)$, $20(x^3+x^2y-xy^2-y^3)$, $12(xy^2+y^3)$, $8(x^3-x^2y)$.

21. x^2-1, x^3+1, x^3-x, x^7-x, x^5+x^3+x.

22. x^4-a^4, $x^3-a^2x-ax^2+a^3$, $ax^3+a^3x-a^2x^2-a^4$.

144. If the expressions cannot readily be resolved into factors we must proceed as follows:—

The L. C. M. of two algebraical expressions is their product divided by their H. C. F.

Let A and B denote the two expressions, D their H. C. F.; let a and b denote the other factors of A and B, so that $A = aD$, $B = bD$.

The expression of lowest dimensions which contains either A or B as a factor is evidently abD.

Then $$abD = \frac{aD \cdot bD}{D} = \frac{AB}{D}.$$

In practice it is easier to divide one of the expressions by the H. C. F., and multiply the quotient by the other expression.

145. Every common multiple of two algebraical expressions is a multiple of their lowest common multiple; for, using the notation of the last Art., if M be any common multiple of A and B it must contain the factors abD; if it contains no others, then $M = abD$; if it contains another factor, m, then $M = mabD$, which proves the proposition.

146. To find the L. C. M. of any number of algebraical expressions, A, B, C, etc., we have the following

RULE.—*Find the L. C. M. of A and B, denote it by M; find the L. C. M. of M and C. and proceed as before until all the expressions have been used. The last multiple thus found will be the L. C. M. required;*

For the L. C. M. of M and C contains, without excess or defect, the factors of M and C, that is, the factors of A, B and C. Similarly for any number of expressions.

EXERCISE XXXVII.

Find the L. C. M. of

1. $x^3 - 6x^2 + 11x - 6$ and $x^3 - 9x^2 + 26x - 24$.
2. $x^3 - 9x^2 + 19x + 4$ and $x^3 - 6x^2 + 4x + 1$.
3. $2x^3 + 9x^2 + 7x - 3$ and $3x^3 + 5x^2 - 15x + 4$.
4. $x^4 - x^3 - x + 1$ and $x^4 - 2x^3 - x^2 - 2x + 1$.
5. $6x^3 + x^2y - 11xy^2 - 6y^3$ and $6x^3 + 11x^2y - xy^2 - 6y^3$.
6. $x^4 + ax^3 - 9a^2x^2 + 11a^3x - 4a^4$ and $x^4 - ax^3 - 3a^2x^2 + 5a^3x - 2a^4$.
7. $x^3 - 7x - 6$ and $x^3 - 4x^2 + 4x - 3$.
8. $x^4 - 10x^2 + 9$, $x^4 + 10x^3 + 20x^2 - 10x - 21$
   and $x^4 + 4x^3 - 22x^2 - 4x + 21$.
9. $x^4 + 3x^3 - x - 3$, $2x^4 + 5x^3 - 5x^2 - 5x + 3$ and $2x^4 + 3x^3 + 2x^2 - 1$.
10. $x^4 + 4x^2 + 16$, $x^5 + 2x^4 + 4x^3 + 8x^2 + 16x + 32$
    and $x^5 - 2x^4 + 4x^3 - 8x^2 + 16x - 32$.
11. $x^3 + 6x^2 + 11x + 6$, $x^3 + 7x^2 + 14x + 8$ and $x^3 + 8x^2 + 19x + 12$.
12. $a^3x + a^2x^2 - 2ax^3$, $a^3x - a^2x^2 - 6ax^3$ and $a^3 - 2a^2x - 5ax^2 + 6x^3$.
13. $x^3 - 3x^2 + 3x - 1$, $x^3 - x^2 - x + 1$, $x^4 - 2x^3 + 2x - 1$
    and $x^4 - 2x^3 + 2x^2 - 2x + 1$

CHAPTER VIII.

FRACTIONS.

147. A **Fraction** is the quotient of one expression by another when the division is indicated but cannot really be performed. Thus $\frac{3}{4}$, $\frac{a}{b}$, are fractions; but $\frac{6}{3}$, $\frac{a^2-b^2}{a-b}$ are fractions in form but not in reality.

148. The expression to be divided is called the **Numerator**, and that by which it is to be divided is called the **Denominator**.

The Numerator and the Denominator are called the **Terms** of a fraction.

149. Any expression may be written in the form of a fraction by considering it the quotient of itself by unity.

150. As division is the reverse of multiplication any expression may be written as a fraction with a given denominator by first multiplying it by that denominator and writing the product as numerator over the given denominator.

151. Since the numerator is a dividend and the denominator its divisor we learn, from Arts. 78 and 80, the following facts:—

1. If the numerator of a fraction be multiplied or divided by any factor the fraction itself will be multiplied or divided by that factor.

2. If the denominator of a fraction be multiplied or divided by any factor the fraction itself will be divided or multiplied by that factor.

3. If both numerator and denominator be either multiplied or divided by the same factor the value of the fraction will not be changed.

REDUCING FRACTIONS TO THEIR LOWEST TERMS.

4. If the signs of all the terms in both numerator and denominator be changed the value of the fraction will remain the same. This is equivalent to multiplying both by -1.

5. If the sign of either numerator or denominator be changed the sign of the fraction will be changed.

Thus $\dfrac{a}{b} = \dfrac{-a}{-b} = -\dfrac{-a}{b} = -\dfrac{a}{-b}$, $\dfrac{a}{b-c} = -\dfrac{a}{c-b}$, $\dfrac{x-y}{a-b} = \dfrac{y-x}{b-a}$, etc.

These examples should be carefully observed.

REDUCING FRACTIONS TO THEIR LOWEST TERMS.

152. When the numerator and denominator contain no common factor except unity the fraction is said to be in its **lowest terms**.

A fraction is reduced to its lowest terms by dividing both numerator and denominator by their H. C. F.

153. In the following examples the factors of the various numerators and denominators are evident by inspection, or may be determined by methods already given:—

Ex. 1. $\dfrac{2a^2 - 2ab}{3ab - 3b^2} = \dfrac{2a(a-b)}{3b(a-b)} = \dfrac{2a}{3b}$.

Ex. 2. $\dfrac{a^4 - a^3b - a^2b^2 + ab^3}{a^5 - a^4b - ab^4 + b^5} = \dfrac{a^3(a-b) - ab^2(a-b)}{a^4(a-b) - b^4(a-b)} = \dfrac{a^3 - ab^2}{a^4 - b^4}$

$= \dfrac{a(a^2 - b^2)}{(a^2 + b^2)(a^2 - b^2)} = \dfrac{a}{a^2 + b^2}$.

EXERCISE XXXVIII.

Reduce to their lowest terms the following fractions:—

1. $\dfrac{3a^2b}{6ab^2}$.

2. $\dfrac{8b^3d^2}{12b^2cd^3}$.

3. $\dfrac{14l^2mn^3}{21lmn}$.

4. $\dfrac{18a^3b^5c^2}{27a^2b^2c^3}$.

5. $\dfrac{125axy}{150abc}$.

6. $\dfrac{13a^{10}b^5}{65a^{10}b^{10}}$.

REDUCING FRACTIONS TO THEIR LOWEST TERMS.

7. $\dfrac{360 x^3 y^2 z^3}{72 x^3 y^2 z^3}$.

8. $\dfrac{240 p^4 q^5 r^6}{180 p^6 q^5 r^5}$.

9. $\dfrac{a^3 b^2}{a^4 b^3}$.

10. $\dfrac{375 x^3 y^3}{125 x^2}$.

11. $\dfrac{a^2}{a^2 + ab}$.

12. $\dfrac{16 a^2 b}{24 a^3 b - 32 a b^3}$.

13. $\dfrac{xy}{3 x^2 y - 5 x y^2}$.

14. $\dfrac{lm + m^2}{abl + abm}$.

15. $\dfrac{x^2 y - xy^2}{abx - aby}$.

16. $\dfrac{6 a^5 x y^3 - 6 b^2 x y^3}{12 a^5 x^2 y - 12 b^2 x^2 y}$.

17. $\dfrac{3 x^2 y - 3 x y^2}{6 x^2 - 6 y^2}$.

18. $\dfrac{a^2 - 4 b^2}{2 a^2 + 4 a b}$.

19. $\dfrac{12 a^2 + 18 b^2}{8 a^4 - 18 b^4}$.

20. $\dfrac{a^2 b c x - a^2 b c}{a b c^2 x^2 - a b c^2}$.

21. $\dfrac{4 x - 4 y}{8 x^2 - 16 x y + 8 y^2}$.

22. $\dfrac{x^4 - a^4}{x^5 - a^2 x^3}$.

23. $\dfrac{x^6 - a^6}{x^4 - a^4}$.

24. $\dfrac{x^4 - a^2 x^2 + a^4}{x^6 + a^6}$.

25. $\dfrac{a^2 - ab - 2 b^2}{a^2 - 3 a b + 2 b^2}$.

26. $\dfrac{x^4 + 4 a^4}{x^2 + 2 a x + 2 a^2}$.

27. $\dfrac{x^3 - 1}{x^3 + 2 x^2 + 2 x + 1}$.

28. $\dfrac{a^2 + b^2 - c^2 + 2 a b}{a^2 - b^2 + c^2 + 2 a c}$.

29. $\dfrac{c^2 - a^2 - b^2 + 2 a b}{b^2 - c^2 - a^2 + 2 a c}$.

30. $\dfrac{x^2 - a x - b x + a b}{x^2 - a x + b x - a b}$.

31. $\dfrac{a c x^2 + (a d - b c) x - b d}{a^2 x^2 - b^2}$.

32. $\dfrac{(3 x^2 - 1)(2 x^2 - 1) - x^2 (5 x^2 - 7)}{(3 x^2 - 1)^2 + (x^3 - 3 x)^2}$.

33. $\dfrac{x^5 - x^4 - x + 1}{x^4 - x^3 - x^2 + x}$.

34. $\dfrac{(a - b + c)\{(a + b)^2 - c^2\}}{4 b^2 c^2 - (a^2 - b^2 - c^2)^2}$.

35. $\dfrac{(a^2 - b c)^2 - (b^2 - a c)(c^2 - a b)}{(b^2 - c a)^2 - (c^2 - a b)(a^2 - b c)}$.

154. When the factors of neither numerator nor denominator can be easily obtained the H. C. F. must be found by the method of division.

Ex.—Reduce $\dfrac{x^3 - 4 x^2 - 19 x - 14}{2 x^3 - 9 x^2 - 38 x + 21}$ to its lowest terms.

The H. C. F. of the numerator and denominator found by the usual process is $x - 7$. Dividing both terms of the fraction by $x - 7$ we get $\dfrac{x^2 + 3 x + 2}{2 x^2 + 5 x - 3}$, the result required.

EXERCISE XXXIX.

Reduce to their lowest terms the following fractions:—

1. $\dfrac{x^2 - 11x + 28}{x^2 - 4x - 21}$.

2. $\dfrac{4x^2 + 12x + 9}{2x^2 - 5x - 12}$.

3. $\dfrac{x^4 + x^2 + 1}{x^6 - 1}$.

4. $\dfrac{x^3 - x}{x^4 - 5x^2 + 4}$.

5. $\dfrac{x^4 + 2x^2 + 9}{x^4 - x^2 + 6x}$.

6. $\dfrac{3a^3 - 22a - 15}{5a^4 - 17a^3 + 18a}$.

7. $\dfrac{x^3 - 4x^2 + 9x - 10}{x^3 + 2x^2 - 3x + 20}$.

8. $\dfrac{x^3 - 7x + 6}{x^3 + 3x^2 - 4x}$.

9. $\dfrac{x^3 - 5x^2 + 10x - 8}{2x^3 - 9x^2 + 17x - 12}$.

10. $\dfrac{3x^3 - 16x^2 + 23x - 6}{2x^3 - 11x^2 + 17x - 6}$.

11. $\dfrac{2x^3 - 13x^2 + 19x - 20}{2x^3 + 7x^2 - 11x + 20}$.

12. $\dfrac{2x^3 + 9x^2 + 7x - 3}{3x^3 + 5x^2 - 15x + 4}$.

13. $\dfrac{4x^2 - xy - 3y^2}{12x^3 - 12x^2y + 4xy^2 - 4y^3}$.

14. $\dfrac{x^4 - x^3 - x + 1}{x^4 + x^3 - x - 1}$.

15. $\dfrac{x^3 + 2x^2 + 2x}{x^5 + 4x}$.

16. $\dfrac{3a^2x^4 - 2ax^3 - 1}{4a^3x^6 - 2a^2x^4 - 3ax^2 + 1}$.

17. $\dfrac{(a+b)^3 + (a-b)^3}{a^4 + 5a^2b^2 + 6b^4}$.

18. $\dfrac{(x^2 + 4x)^2 - 2(x^2 + 4x) - 15}{x^4 - 10x^2 + 9}$.

19. $\dfrac{x^5 + x^4y + x^3y^2 + x^2y^3 + xy^4 + y^5}{x^5 - x^4y + x^3y^2 - x^2y^3 + xy^4 - y^5}$.

20. $\dfrac{ab(x^2 + y^2) + xy(a^2 + b^2)}{ab(x^2 - y^2) + xy(a^2 - b^2)}$.

21. $\dfrac{x^2 - y^2 - z^2 + 2yz + x + y - z}{z^2 - x^2 - y^2 + 2xy + y + z - x}$.

22. $\dfrac{a^3 + b^3 + c^3 - 3abc}{(a-b)^2 + (b-c)^2 + (c-a)^2}$.

23. $\dfrac{x^4 + (2b^2 - a^2)x^2 + b^4}{x^4 + 2ax^3 + a^2x^2 - b^4}$.

24. $\dfrac{x^2 + (a+b+c)x + (a+b)c}{a^2 + 2ab + b^2 - x^2}$.

25. $\dfrac{ab(ab - 3c^2) - c(3a^3 - b^3)}{ac(ac - 4b^2) - b(4a^3 - c^3)}$.

26. $\dfrac{2x^3z^2 - 3x^2y^2 - 2y^2z^2 + 3y^4}{3x^2z^2 + 2y^4 - 2x^2y^2 - 3y^2z^2}$.

27. $\dfrac{ax^2 + (bc - ab - ac)x + abc - b^2c}{b^2c^2 - 2abc^2 + a^2c^2 - a^2x^2}$.

28. $\dfrac{(yz - a)^2 - (ca - y^2)(ab - z^2)}{(bc - 1)(yz - a) - (z - by)(y - cz)}$.

CHANGES IN THE FORM OF FRACTIONS. 91

29. $\dfrac{(a^3-1)x^3 + (2a^2+a)x^2 + 2ax + 1}{(a-1)^3x^3 - 2(a-1)x^2 - 2ax - 1}$.

30. $\dfrac{(1+a)^2(1+c^2) - (1+c)^2(1+a^2)}{a(1-c)(1+c) - c(1-a)(1+a)}$.

31. $\dfrac{m^4 - n^4 + 2n(m^3+n^3) - (m^2-n^2)^2}{(m+n)^3 - m^3 - n^3 - mn(m+n)}$.

32. $\dfrac{(a-b)^6 + 6ab(a-b)^4 + 8a^2b^2(a-b)^2}{(a+b)^6 - 6ab(a+b)^4 + 8a^2b^2(a+b)^2}$.

33. $\dfrac{a^2b + c^2d + a^2d + bc^2}{ab^2 + cd^2 + abd + bcd + acd + abc + b^2d + bd^2}$.

34. $\dfrac{(ax+by)^2 - (a-b)(x+z)(ax+by) + (a-b)^2xz}{(ax-by)^2 - (a+b)(x+z)(ax-by) + (a+b)^2xz}$.

35. $\dfrac{\{(a+b)(a+c) + 2a(b+c)\}^2 - (a-b)^2(a-c)^2}{\{(b+c)(b+a) + 2b(c+a)\}^2 - (b-c)^2(a-b)^2}$.

CHANGES IN THE FORM OF FRACTIONS.

155. If the degree of the numerator of a fraction equals or exceeds that of the denominator the fraction may be changed to the form of a mixed or integral expression by dividing the numerator by the denominator. The quotient will be the integral part of the result; the remainder, if any, will be the numerator, and the divisor the denominator, of the fractional part.

Ex. $\dfrac{x^4+y^4}{x+y} = x^3 - x^2y + xy^2 - y^3 + \dfrac{2y^4}{x+y}$ or $y^3 - y^2x + yx^2 - x^3 + \dfrac{2x^4}{x+y}$,

according as we consider x or y the leading letter. In arithmetic the fractional part is written beside the quotient, with no sign between; but this is not admissible in algebra, since when no sign is written between two parts of an expression multiplication is understood. Thus $3\dfrac{4}{5}$ means $3 + \dfrac{4}{5}$, but $a\dfrac{b}{c}$ means $a \times \dfrac{b}{c} = \dfrac{ab}{c}$. The dividing line between numerator and denominator has the same effect as a bracket; hence when a minus sign is placed before a fraction the sign of every term in either numerator or denominator must be changed.

156. A mixed expression may be reduced to a complete fraction as follows:—

Multiply the integral expression by the denominator, to the product annex the numerator, and under the result write the denominator.

$$Ex. \quad a+b-\frac{a^2+b^2}{a+b} = \frac{(a+b)^2-(a^2+b^2)}{a+b} = \frac{2ab}{a+b}.$$

157. A fraction whose numerator consists of several terms may be split up into an equivalent number of fractions connected with the proper signs.

$$Ex. \quad \frac{a^4+3a^3b+2a^2b^2}{a^3b^2} = \frac{a^4}{a^3b^2} + \frac{3a^3b}{a^3b^2} + \frac{2a^2b^2}{a^3b^2} = \frac{a}{b^2} + \frac{3}{b} + \frac{2}{a}.$$

The truth of the above is evident from the division of polynomials, Art. 82.

EXERCISE XL.

Reduce the following fractions to mixed quantities:—

1. $\dfrac{5x^2+7x-3}{x+2}$.   2. $\dfrac{a^2-ax+x^2}{a+x}$.   3. $\dfrac{3x^3-3y^3+5}{x-y}$.

4. $\dfrac{x^3+y^3}{x-y}$.   5. $\dfrac{2x^4-x^3+2x-3}{x^2-x+1}$.   6. $\dfrac{9a_4^2}{3a+1}$.

7. $\dfrac{3x^3-7x^2+1}{3x^2-4x+3}$.   8. $\dfrac{5x^3-x^2+5}{5x^2+4x-1}$.   9. $\dfrac{x^3-x+1}{x^2+x+1}$.

Reduce to complete fractions

10. $1+\dfrac{x}{1-x}$.   11. $1+x+\dfrac{x^2}{1-x}$.   12. $a-b+\dfrac{b^2}{a+b}$.

13. $1-\dfrac{x-y}{x+y}$.   14. $2a-b-\dfrac{2ab}{a+b}$.   15. $a+b-\dfrac{(a-b)^2}{a+b}$.

16. $a-b+\dfrac{(a+b)^2}{a-b}$.   17. $x+5-\dfrac{2x-15}{x-3}$.   18. $\dfrac{2x^2}{x+y}-(x+y)$.

LOWEST COMMON DENOMINATOR.

19. $x^2 + xy + y^2 + \dfrac{y^3}{x-y}$.

20. $x^2 - xy + y^2 - \dfrac{y^3}{x+y}$.

21. $1 + x + x^2 - \dfrac{x^2 + x^4}{1 - x + x^2}$.

22. $x - a + y + \dfrac{a^2 - ay + y^2}{x+a}$.

23. $x + 1 - \dfrac{x^3 - 4x^2 + 8}{(x-2)^2}$.

24. $1 + 2y + 2y^2 + 2y^3 + \dfrac{2y^4 + 2y^5}{1 - y^2}$.

25. $a + ab + b^2 \left(a + ab + \dfrac{ab^2}{1-b}\right)$.

26. $a - ab + b^2 \left(a - ab + \dfrac{ab^2}{1+b}\right)$.

Separate into fractions with a single term in the numerator.

27. $\dfrac{x^5 + 3ax^4 - 5a^2x^2 + 6a^3x}{3ax^3}$.

28. $\dfrac{a^2xy + ay^2z + ax^2z + xyz^2}{axyz}$.

29. $\dfrac{abc + bcd + cda + dab}{abcd}$.

30. $\dfrac{15x^3 - 40x^2 + 75x - 150}{100x}$.

31. $\dfrac{(a+b)(m-n) - (a-b)(m+n)}{(a^2-b^2)(m^2-n^2)}$.

32. $\dfrac{(a+b)x + (a-2b)y}{x^2 - xy - 2y^2}$.

LOWEST COMMON DENOMINATOR.

158. Since the value of a fraction is unchanged by multiplying both numerator and denominator by the same factor, it follows that two or more fractions may be replaced by equivalent fractions having a common denominator by the following

RULE.—*Find the L. C. M. of the denominators. Multiply both numerator and denominator of each fraction by the quotient obtained by dividing the L. C. M. by its own denominator.*

Ex. 1.—Reduce to equivalent fractions with the lowest common denominator

$$\dfrac{2a}{3bc}, \quad \dfrac{3b}{5ca}, \quad \dfrac{4c}{10ab}.$$

L. C. M. of denominators, $30abc$.

Quotients of $30abc$ by denominators, $10a$, $6b$, $3c$.

Fractions required, $\dfrac{20a^2}{30abc}, \dfrac{18b^2}{30abc}, \dfrac{12c^2}{30abc}$.

ADDITION AND SUBTRACTION OF FRACTIONS.

Ex. 2.—Reduce to equivalent fractions with the lowest common denominator

$$\frac{2}{x+1}, \quad \frac{3}{x-1}, \quad \frac{x-3}{x^2-1}.$$

L. C. M. of denominators, x^2-1.
Quotients of x^2-1 by denominators, $x-1$, $x+1$, 1.

Fractions required, $\dfrac{2x-2}{x^2-1}, \dfrac{3x+3}{x^2-1}, \dfrac{x-3}{x^2-1},$

EXERCISE XLI.

Reduce to equivalent fractions with the lowest common denominator

1. $\dfrac{2x}{3}, \dfrac{4x}{5}, \dfrac{7x}{10}.$ 2. $\dfrac{8x-3}{4x^2}, \dfrac{2-3x}{6x}.$ 3. $\dfrac{a}{bc}, \dfrac{b}{ca}, \dfrac{c}{ab}.$

4. $\dfrac{2}{1-x}, \dfrac{3}{1-x^2}.$ 5. $\dfrac{a}{a-b}, \dfrac{b}{a+b}.$ 6. $\dfrac{x}{xy-y^2}, \dfrac{y}{x^2-xy}.$

7. $\dfrac{a+b}{a-b}, \dfrac{a-b}{a+b}, \dfrac{ab}{a^2-b^2}.$ 8. $\dfrac{5}{2x-4}, \dfrac{7}{3x+6}, \dfrac{8}{5x^2-20}.$

9. $\dfrac{x+4}{x^2-5x+6}, \dfrac{x+2}{x^2-7x+12}.$ 10. $\dfrac{x+2}{1-x}, \dfrac{x+3}{x-x^2}, \dfrac{x+4}{x-x^3}.$

11. $\dfrac{b+c}{(a-b)(a-c)}, \dfrac{c+a}{(b-a)(b-c)}.$ 12. $\dfrac{c(a+b)}{ab(a-c)(b-c)}, \dfrac{a(b+c)}{bc(b-a)(c-a)}.$

ADDITION AND SUBTRACTION OF FRACTIONS.

159. The rule for the addition and subtraction of fractions follows at once from the principle of Division. To divide the sum or difference of quantities we divide the quantities in succession and connect the quotients with the proper signs. Conversely, the sum or the difference of two quotients is equal to the quotient of the sum or the difference of the dividends by the common divisor; therefore

ADDITION AND SUBTRACTION OF FRACTIONS. 95

160. To add two or more fractions we have the following

RULE.—*Reduce the fractions to equivalent ones having the lowest common denominator. Add their numerators, and under their sum as numerator write the common denominator.*

161. To subtract one fraction from another we have the following

RULE.—*Reduce the fractions to equivalent ones with the lowest common denominator. Subtract the numerator of the subtrahend from that of the minuend, and write the difference as numerator over the common denominator.*

Ex. 1.—Simplify $\dfrac{1}{1+x} + \dfrac{1}{1-x} + \dfrac{x-3}{1-x^2}$.

Reducing the fractions to equivalent ones having the lowest common denominator we get

$$\dfrac{1}{1+x} + \dfrac{1}{1-x} + \dfrac{x-3}{1-x^2} = \dfrac{1-x}{1-x^2} + \dfrac{1+x}{1-x^2} + \dfrac{x-3}{1-x^2}$$

$$= \dfrac{1-x+1+x+x-3}{1-x^2} = \dfrac{x-1}{1-x^2} = -\dfrac{1-x}{1-x^2} = -\dfrac{1}{1+x}.$$

Ex. 2.—Simplify $\dfrac{a}{(a+b)b} - \dfrac{b}{(a-b)a}$.

$$\dfrac{a}{(a+b)b} - \dfrac{b}{(a-b)a} = \dfrac{a^2(a-b)}{ab(a^2-b^2)} - \dfrac{b^2(a+b)}{ab(a^2-b^2)}$$

$$= \dfrac{a^2(a-b) - b^2(a+b)}{ab(a^2-b^2)} = \dfrac{a^3 - a^2b - ab^2 - b^3}{ab(a^2-b^2)}.$$

EXERCISE XLII.

Simplify

1. $\dfrac{x}{7} + \dfrac{2x}{7} + \dfrac{3x}{7} + \dfrac{5x}{14}$.

2. $\dfrac{3a}{8} + \dfrac{5a}{8} + \dfrac{7a}{16} + \dfrac{5b}{24}$.

3. $\dfrac{a+x}{3} + \dfrac{2a-x}{3} + \dfrac{x-a}{2}$.

4. $\dfrac{2x+y}{5} + \dfrac{x-2y}{10} + \dfrac{4x-5y}{15}$.

ADDITION AND SUBTRACTION OF FRACTIONS.

5. $\dfrac{3x-8}{7}+\dfrac{2x+4}{14}+\dfrac{6}{7}.$

6. $\dfrac{4x-5}{10}+\dfrac{2x}{5}+\dfrac{4x+2}{25}.$

7. $\dfrac{2x-9}{14}+\dfrac{4x-3}{21}+\dfrac{33-14x}{42}.$

8. $\dfrac{1}{x}+\dfrac{1}{2x}+\dfrac{1}{3x}+\dfrac{1}{6x}.$

9. $\dfrac{2}{5y}+\dfrac{1}{2y}+\dfrac{1}{10y}+\dfrac{y-1}{y}.$

10. $\dfrac{a-b}{ab}+\dfrac{b-c}{bc}+\dfrac{c-a}{ca}.$

11. $\dfrac{1}{x^3}+\dfrac{1}{y^3}+\dfrac{2}{xy}.$

12. $\dfrac{a+2b}{bc}+\dfrac{b+2c}{ca}+\dfrac{c+2a}{ab}.$

13. $\dfrac{a}{a+b}+\dfrac{b}{a-b}+\dfrac{2ab}{a^2-b^2}.$

14. $\dfrac{x-1}{x-2}+\dfrac{x-2}{x-3}+\dfrac{5-2x}{x^2-5x+6}.$

15. $\dfrac{1}{x-5}+\dfrac{2}{x-7}+\dfrac{2}{x^2-12x+35}.$

16. $\dfrac{y}{x^2+xy}+\dfrac{x}{xy+y^2}+\dfrac{2}{x+y}.$

17. $\dfrac{x}{y}+\dfrac{y}{x+y}+\dfrac{x^2}{x^2+xy}.$

18. $\dfrac{x-y}{x^2-xy+y^2}+\dfrac{x+y}{x^2+xy+y^2}.$

19. $\dfrac{b^2-ab}{ba-a^2}+\dfrac{a^2+ab}{ba+b^2}.$

20. $\dfrac{a}{a+b}+\dfrac{ab-b^2}{a^2-b^2}+\dfrac{a^2-b^2}{(a+b)^2}.$

21. $\dfrac{x^2+xy+y^2}{x^2-xy+y^2}+\dfrac{x^2-xy+y^2}{x^2+xy+y^2}.$

22. $\dfrac{a^2+b^2}{a^2-b^2}+2+\dfrac{a^2-b^2}{a^2+b^2}.$

23. $\dfrac{1}{x-y}+\dfrac{x-y}{x^2+xy+y^2}+\dfrac{xy-2x^2}{x^3-y^3}.$

24. $\dfrac{x+1}{x^2+x+1}+\dfrac{x-1}{x^2-x+1}+\dfrac{2}{x^4+x^2+1}.$

25. $\dfrac{a+b}{ax+by}+\dfrac{a-b}{ax-by}+\dfrac{2(a^2x+b^2y)}{a^2x^2+b^2y^2}.$

26. $\dfrac{1}{x^2+3x+2}+\dfrac{2x}{x^2+4x+3}+\dfrac{1}{x^2+5x+6}.$

27. $\dfrac{1-2x}{3(x^2-x+1)}+\dfrac{1+x}{2(x^2+1)}+\dfrac{1}{6(x+1)}.$

28. $\dfrac{x-1}{x+1}+\dfrac{x+1}{x-1}+\dfrac{3x-2}{x+3}+\dfrac{3x-4}{x-2}.$

EXERCISE XLIII.

Simplify

1. $\dfrac{3x}{7} - \dfrac{5x}{14}$.

2. $\dfrac{4x-7}{6} - \dfrac{2x-9}{3}$.

3. $\dfrac{1}{x-1} - \dfrac{1}{x+1}$.

4. $\dfrac{a}{a-b} - \dfrac{b}{a+b}$.

5. $\dfrac{x+y}{x-y} - \dfrac{x-y}{x+y}$.

6. $\dfrac{a-c}{a-b} - \dfrac{b+c}{a+b}$.

7. $\dfrac{a}{x(a-x)} - \dfrac{x}{a(a-x)}$.

8. $\dfrac{m+n}{mn-n^2} - \dfrac{m+n}{m^2-mn}$.

9. $\dfrac{x^2+y^2}{x^2-y^2} - \dfrac{y}{x-y}$.

10. $\dfrac{a+x}{a(a-x)} - \dfrac{ax+x^2}{a^2(a-x)}$.

11. $\dfrac{a+x}{2(a-x)} - \dfrac{a^2+x^2}{2(a^2-x^2)}$.

12. $\dfrac{1+x}{1+x+x^2} - \dfrac{1-x}{1-x+x^2}$.

13. $\dfrac{1}{x^2-xy+y^2} - \dfrac{1}{x^2+xy+y^2}$.

14. $\dfrac{1}{x^2-8x+15} - \dfrac{1}{x^2-5x+6}$.

15. $\dfrac{1}{x^2-19x+84} - \dfrac{1}{x^2-12x+35}$.

16. $\dfrac{31}{12x^2-x-20} - \dfrac{7}{12x^2-25x+12}$.

17. $\dfrac{x^2+8x+15}{x^2+7x+10} - \dfrac{x-1}{x+2}$.

18. $\dfrac{(x-3a)(x-7a)}{x^2-10ax+24a^2} - \dfrac{x-5a}{x-4a}$.

19. $\dfrac{3a+b}{a^2+3ab+2b^2} - \dfrac{a+7b}{a^2+5ab+6b^2}$.

20. $\dfrac{x^2+x-5}{2x^2-11x+12} - \dfrac{x^2+x-1}{2x^2+5x-12}$.

21. $\dfrac{a}{(a-b)(a-c)} - \dfrac{b}{(a-b)(b-c)}$.

22. $\dfrac{c}{x^2-(a+b)x+ab} - \dfrac{(c-a)(x-c-a)}{(x-a)(x-b)(x-c)}$.

23. $\dfrac{4a-3b}{7ab(a-b)-2(a^3-b^3)} - \dfrac{8a-b}{3ab(a+b)-2(a^3+b^3)}$.

24. $\dfrac{3x^2-2x+6}{(2x^2-3x+4)^3} - \dfrac{2x^4-x^3+5x^2-2x+8}{(2x^2-3x+4)^4}$.

25. From $\dfrac{x}{x-a} + \dfrac{x}{x-b} + \dfrac{x}{x-c}$ take $\dfrac{a}{x-a} + \dfrac{b}{x-b} + \dfrac{c}{x-c}$.

162. In simplifying algebraical expressions two formulæ are frequently of service, viz.:—

1. $(a+b)^2 + (a-b)^2 = 2(a^2 + b^2)$.
2. $(a+b)^2 - (a-b)^2 = 4ab$.

They should be remembered in words as well as in symbols, thus:—

1. *The square of the sum of two quantities, plus the square of their difference, is equal to twice the sum of their squares.*

2. *The square of the sum of two quantities, minus the square of their difference, is equal to four times their product.*

The following examples will show that a proper grouping of the fractions to be combined lessens the work required:—

Ex. 1.—Simplify $\dfrac{a+b}{a-b} + \dfrac{a-b}{a+b} - \dfrac{2(a^2-b^2)}{a^2+b^2}$.

$$\dfrac{a+b}{a-b} + \dfrac{a-b}{a+b} = \dfrac{2(a^2+b^2)}{a^2-b^2};$$

$$\dfrac{2(a^2+b^2)}{a^2-b^2} - \dfrac{2(a^2-b^2)}{a^2+b^2} = \dfrac{8a^2b^2}{a^4-b^4}.$$

If we had found the L. C. M. of the three denominators and combined all the fractions at one operation the labor would have been considerably increased.

Ex. 2.—Simplify $\dfrac{x^3}{y(x-y)^2} + \dfrac{y^3}{x(x-y)^2} + \dfrac{2xy}{(x-y)^2} - \dfrac{x}{y} - \dfrac{y}{x} - 2$.

Grouping the first three together, and also the last three, and then combining the results with the proper sign, we get

$$\dfrac{x^3}{y(x-y)^2} + \dfrac{y^3}{x(x-y)^2} + \dfrac{2xy}{(x-y)^2} = \dfrac{x^4+y^4+2x^2y^2}{xy(x-y)^2} = \dfrac{(x^2+y^2)^2}{xy(x-y)^2};$$

$$\dfrac{x}{y} + \dfrac{y}{x} + 2 = \dfrac{x^2+y^2+2xy}{xy} = \dfrac{(x+y)^2}{xy};$$

$$\dfrac{(x^2+y^2)^2}{xy(x-y)^2} - \dfrac{(x+y)^2}{xy} = \dfrac{(x^2+y^2)^2 - (x^2-y^2)^2}{xy(x-y)^2} = \dfrac{4x^2y^2}{xy(x-y)^2} = \dfrac{4xy}{(x-y)^2}.$$

ADDITION AND SUBTRACTION OF FRACTIONS.

Ex. 3.—Simplify $\dfrac{2}{(x-2)(x-5)} + \dfrac{3}{(x-3)(5-x)} + \dfrac{1}{(3-x)(2-x)}$.

Observe that $x-3$ and $3-x$, $x-5$ and $5-x$ are not different factors except in sign; they stand to each other in the same relation as a and $-a$, and the quotient of one by the other is -1. The L. C. M. of the denominators is $(x-2)(x-3)(x-5)$; the quotient of the L. C. M. by the first denominator is $x-3$; by the second, $-(x-2)$; by the third, $x-5$. In the second case one factor in the denominator, $5-x$, is the negative of the corresponding factor in the L. C. M., which gives the negative sign before the quotient; in the third case there are *two* such factors, and consequently the quotient is positive. We get, therefore, the following result:—

$$\dfrac{2(x-3)-3(x-2)+(x-5)}{(x-2)(x-3)(x-5)} = \dfrac{-5}{(x-2)(x-3)(x-5)}.$$

The result may be written with a positive sign in the numerator by reversing the terms of one factor in the denominator, thus:

$$\dfrac{5}{(x-2)(x-3)(5-x)}.$$

In connection with this example read Arts. 55 and 151 (5).

EXERCISE XLIV.

Simplify

1. $\dfrac{1}{x-y} + \dfrac{1}{x+y} - \dfrac{2y}{x^2-y^2}$.

2. $\dfrac{1}{1-x} - \dfrac{1}{1+x} + \dfrac{2x}{1+x^2}$.

3. $\dfrac{x}{1-x} - \dfrac{x^2}{1-x^2} + \dfrac{x}{1+x^2}$.

4. $\dfrac{y}{x-y} - \dfrac{x}{x+y} - \dfrac{2xy}{x^2-y^2}$.

5. $\dfrac{1}{x-y} - \dfrac{1}{x+y} - \dfrac{2y}{x^2+y^2} - \dfrac{4y^3}{x^4+y^4}$.

6. $\dfrac{x}{x-y} - \dfrac{y}{x+y} - \dfrac{x^2-y^2}{x^2+y^2}$.

7. $\dfrac{3}{4(x+1)} - \dfrac{1}{4(x-1)} - \dfrac{x-2}{2(x^2+1)}$.

8. $\dfrac{1}{2(x-1)} - \dfrac{1}{2(x+1)} + \dfrac{1}{(x^2-1)^2}$.

9. $\dfrac{5}{x} - \dfrac{1}{x^2} - \dfrac{5}{x+1} - \dfrac{4}{(x+1)^2}$.

10. $\dfrac{1}{x^3} - \dfrac{1}{x^2} + \dfrac{3}{x} - \dfrac{3}{x+1}$.

ADDITION AND SUBTRACTION OF FRACTIONS.

11. $\dfrac{1}{3(x-1)} - \dfrac{1}{x} + \dfrac{2x+1}{3(x^2+x+1)}$.   12. $\dfrac{1}{3(x+1)} - \dfrac{1}{x} + \dfrac{2x-1}{3(x^2-x+1)}$.

13. $\dfrac{3}{1-2x} - \dfrac{7}{1+2x} - \dfrac{4-20x}{4x^2-1}$.   14. $\dfrac{a}{a-x} + \dfrac{3a}{a+x} - \dfrac{2ax}{a^2-x^2}$.

15. $\dfrac{x(16-x)}{x^2-4} + \dfrac{2x+3}{2-x} - \dfrac{2-3x}{x+2}$.   16. $\dfrac{x-a}{x-b} + \dfrac{x-b}{x-a} - \dfrac{(a-b)^2}{(x-a)(x-b)}$.

17. $\dfrac{x+y}{2x} - \dfrac{y}{x+y} + \dfrac{xy^2-y^3}{2x(y^2-x^2)}$.   18. $\dfrac{x-y}{x} + \dfrac{2y}{x-y} - \dfrac{y^3+xy^2}{x^3-xy^2}$.

19. $\dfrac{x^2-(y-z)^2}{(z+x)^2-y^2} + \dfrac{y^2-(z-x)^2}{(x+y)^2-z^2} + \dfrac{z^2-(x-y)^2}{(y+z)^2-x^2}$.

20. $\dfrac{a+b}{(b-c)(c-a)} + \dfrac{b+c}{(c-a)(a-b)} + \dfrac{c+a}{(a-b)(b-c)}$.

21. $\dfrac{a^2-bc}{(a-b)(a-c)} + \dfrac{b^2+ac}{(b-a)(b+c)} + \dfrac{c^2+ab}{(c-a)(c+b)}$.

22. $\dfrac{2-3x}{6-4x} + \dfrac{3-4x}{6+4x} - \dfrac{1-2x+x^2}{9-12x+4x^2} + \dfrac{2-5x^2}{36-16x^2}$.

23. $\dfrac{1}{x^3} + \dfrac{1}{x^2} - \dfrac{1}{x} - \dfrac{1}{(x^2+1)^2} + \dfrac{x-1}{x^2+1} - \dfrac{3}{x^2(x^2+1)^2}$.

24. $\dfrac{2}{a-b} + \dfrac{2}{b-c} + \dfrac{2}{c-a} + \dfrac{(a-b)^2+(b-c)^2+(c-a)^2}{(a-b)(b-c)(c-a)}$.

25. $\dfrac{5x+13}{(3x-5)(4x-12)} + \dfrac{3x+5}{(12-4x)(5x-13)} - \dfrac{4x+12}{(5-3x)(13-5x)}$.

26. $\dfrac{1}{a-x} - \dfrac{1}{a-y} + \dfrac{x}{(a-x)^2} - \dfrac{y}{(a-y)^2} + \dfrac{a(x-y)(x+y-2a)}{(a-x)^2(a-y)^2}$.

27. $\dfrac{1}{4(x+1)^3} - \dfrac{1}{4(x+1)^2} - \dfrac{1}{16(x+1)} + \dfrac{1}{16(x-1)} + \dfrac{1}{8(x-1)^2}$.

28. $\dfrac{1}{16(x-1)^2} + \dfrac{1}{16(x-1)} + \dfrac{1}{16(x+1)^2} - \dfrac{1}{16(x+1)} + \dfrac{1}{4(x^2+1)^2}$
$\qquad\qquad - \dfrac{1}{4(x^2+1)}$.

MULTIPLICATION OF FRACTIONS.

163. To find the product of $\dfrac{a}{b}$ and $\dfrac{c}{d}$.

Since $\dfrac{a}{b}$ means that a is to be divided by b,

$$\therefore \dfrac{a}{b} \times b = a; \text{ similarly } \dfrac{c}{d} \times d = c.$$

Let $\dfrac{a}{b} = x$, $\dfrac{c}{d} = y$, then xy is the product required.

Now, $\qquad a = bx, \ c = dy,$

$$\therefore ac = bdxy \text{ or } xy = \dfrac{ac}{bd}.$$

To find the product of two or more fractions we have therefore the following

RULE.—*Multiply together all the numerators for the numerator of the product, and all the denominators for the denominator of the product.*

164. If a numerator and a denominator contain the same factor this factor will be found in both numerator and denominator of the product, and must be removed to reduce it to its lowest terms. The process may be shortened by cancelling such factors before multiplying, which will evidently give the same result.

Ex. 1. $\dfrac{3a^2}{5bc} \times \dfrac{10b^2}{27ca} \times \dfrac{9c^2}{16ab} = \dfrac{3 \times 10 \times 9 \times a^2 b^2 c^2}{5 \times 27 \times 16 \times a^2 b^2 c^2} = \dfrac{1}{8}.$

Ex. 2. $\dfrac{x^2+3x+2}{x^2-5x+6} \times \dfrac{x^2-7x+12}{x^2+x} = \dfrac{(x+1)(x+2)}{(x-3)(x-2)} \times \dfrac{(x-4)(x-3)}{x(x+1)}$

$$= \dfrac{(x+2)(x-4)}{x(x-2)} = \dfrac{x^2-2x-8}{x^2-2x}.$$

MULTIPLICATION OF FRACTIONS.

Ex. 3. $\dfrac{a^3+x^3}{a^2-2ax+x^2} \times \dfrac{1}{a+x} \times \dfrac{a-x}{a^2+ax+x^2} \times \dfrac{a^3-x^3}{a^2-ax+x^2}$

$= \dfrac{(a+x)(a^2-ax+x^2)}{(a-x)^2} \times \dfrac{1}{a+x} \times \dfrac{a-x}{a^2+ax+x^2} \times \dfrac{(a-x)(a^2+ax+x^2)}{a^2-ax+x^2} = 1,$

since for every factor in any numerator there is an equal factor in some denominator.

Ex. 4. $\left(1 - \dfrac{2bc}{b^2+c^2}\right)\left(b - c + \dfrac{2bc}{b-c}\right)\left(1 - \dfrac{bc-c^2}{b^2}\right)\dfrac{b}{b^3+c^3}$

$= \dfrac{(b-c)^2}{b^2+c^2} \cdot \dfrac{b^2+c^2}{b-c} \cdot \dfrac{b^2-bc+c^2}{b^2} \cdot \dfrac{b}{b^3+c^3} = \dfrac{b-c}{b(b+c)}.$

EXERCISE XLV.

1. $\dfrac{3a}{4b} \times \dfrac{2a}{3b}.$ 2. $\dfrac{5a}{6b} \times \dfrac{3b}{10a}.$ 3. $\dfrac{6a^2b}{7ab^2} \times \dfrac{35b^2x}{54a^2y}.$

4. $\dfrac{2x}{3y^2} \times 3y.$ 5. $8a^3b^3 \times \dfrac{1}{4a^2b}.$ 6. $\dfrac{7x^2y^2z}{18a^2b^2c} \times \dfrac{9abc^2}{28xyz^2}.$

7. $\dfrac{2a}{5b} \times \dfrac{4b}{3c} \times \dfrac{5c}{6a}.$ 8. $\dfrac{4x^2}{5yz} \times \dfrac{3y^2}{7zx} \times \dfrac{105z^2}{72xy}.$ 9. $\dfrac{4a^2-6ab}{15xy} \times \dfrac{45xy}{8ab-12b^2}.$

10. $\dfrac{a^2x - aby}{abc - b^2d} \times \dfrac{ac^2 - bcd}{adx - bdy}.$ 11. $\dfrac{a^2-b^2}{x^2-xy} \times \dfrac{xy-y^2}{a^2+ab}.$

12. $\dfrac{2a(x^2-y^2)^2}{cx} \times \dfrac{x^3}{(x-y)(x+y)^3}.$ 13. $\dfrac{x^2-11x+30}{x^2-6x+9} \times \dfrac{x^2-3x}{5x-x^2}.$

14. $\dfrac{a}{bx} \times \left(b + \dfrac{bx}{a}\right) \times \left(1 - \dfrac{a}{a+x}\right).$ 15. $\dfrac{a^3-b^3}{a^3+b^3} \times \dfrac{a+b}{a-b} \times \dfrac{a^2-ab+b^2}{a^2+ab+b^2}.$

16. $\left(a^4 - \dfrac{a^2}{x^2}\right) \times \dfrac{a^2x^2 + abx^3}{ax+1} \times \dfrac{ax}{a^2-b^2} \times \left(ax + 1 + \dfrac{a^2x^2}{1-ax}\right).$

17. $\dfrac{x^3-y^3}{x^2-3xy+2y^2} \times \dfrac{xy-2y^2}{x^2+xy} \times \dfrac{x^2-xy}{(x-y)^2} \times \dfrac{x^2+xy-2y^2}{2x^2+5xy+2y^2}.$

18. $\left\{a + \dfrac{ax}{a-x}\right\} \times \left\{a - \dfrac{ax}{a+x}\right\} \times \dfrac{a^2-x^2}{a^2+x^2} \times \dfrac{(a+x)^2+(a-x)^2}{(a^2+x^2)^2-(a^2-x^2)^2}.$

MULTIPLICATION OF FRACTIONS. 103

19. $\dfrac{a^2 - ab + b^2}{a^2 - 2ab + b^2} \times \dfrac{a^3 - b^3}{a^3 + b^3} \times \dfrac{a^2 - b^2}{a^4 + a^2b^2 + b^4}.$

20. $\dfrac{x^6 - y^6}{x^2 + 2xy + y^2} \times \dfrac{x^2 + y^2}{x^2 - y^2} \times \dfrac{x+y}{x^4 + x^2y^2 + y^4}.$

21. $\dfrac{x^2 + x - 2}{x^2 + 2x - 15} \times \dfrac{x^2 - 8x + 15}{x^2 + 6x - 7} \times \dfrac{x^2 + 12x + 35}{10 + 3x - x^2}.$

22. $\dfrac{(x-y)^2 - z^2}{(x+y)^2 - z^2} \times \dfrac{z^2 - (x+y)^2}{(z-y)^2 - x^2} \times \dfrac{x^2 - (y-z)^2}{x^2 - (y+z)^2} \times \dfrac{x+y+z}{x-y+z}.$

23. $\dfrac{m^2 - mn + n^2}{m^3 - 3mn(m-n) - n^3} \times \dfrac{m^2 - n^2}{m^3 - n^3} \times \dfrac{m^4 + n^4 - mn(m^2 + n^2)}{m^4 + n^4 + mn(m^2 + n^2)}.$

24. $\dfrac{a^4 - 3a^2x^2 + x^4}{a^4 + 3a^2x^2 + 4x^4} \times \dfrac{a^2 - ax + 2x^2}{a^2 - ax - x^2} \times \dfrac{a^4 - 4x^4 + ax(2a^2 + ax)}{a^4 - x^4 + ax(2a^2 + ax)}.$

25. $\left\{3 + 4x - \dfrac{11 + 25x + 12x^2}{4 + 3x}\right\} \times \left\{2 - 3x + \dfrac{13x - 6 - 5x^2}{3 - 2x}\right\}.$

26. $\left\{x\left(1 + \dfrac{3y}{x-y}\right)^2 - y\left(2 + \dfrac{3y}{x-y}\right)^2\right\}\left\{x\left(1 - \dfrac{3y}{x+y}\right)^2 + y\left(2 - \dfrac{3y}{x+y}\right)^2\right\}$

165. The product of two expressions consisting of several fractional terms may be found by multiplying each term of the one by each term of the other and connecting the partial products by the proper sign, as in ordinary multiplication, or the two expressions may be combined into single fractions and their product found as in the last exercise.

Ex. 1. $\left(\dfrac{x^2}{2} + \dfrac{x}{3} + \dfrac{1}{2}\right)\left(\dfrac{x}{2} - \dfrac{1}{3}\right) = \dfrac{x^3}{4} + \dfrac{x^2}{6} + \dfrac{x}{4} - \dfrac{x^2}{6} - \dfrac{x}{9} - \dfrac{1}{6}$

$= \dfrac{x^3}{4} + \dfrac{5x}{36} - \dfrac{1}{6};$

or thus $\left(\dfrac{x^2}{2} + \dfrac{x}{3} + \dfrac{1}{2}\right)\left(\dfrac{x}{2} - \dfrac{1}{3}\right) = \dfrac{3x^2 + 2x + 3}{6} \cdot \dfrac{3x - 2}{6}$

$= \dfrac{9x^3 + 5x - 6}{36} = \dfrac{x^3}{4} + \dfrac{5x}{36} - \dfrac{1}{6}.$

MULTIPLICATION OF FRACTIONS.

Ex. 2. $\left(\dfrac{x^2}{2}+1+\dfrac{2}{x^2}\right)\left(\dfrac{x^2}{2}-1+\dfrac{2}{x^2}\right)=\left(\dfrac{x^2}{2}+\dfrac{2}{x^2}\right)^2-1$

$$=\dfrac{x^4}{4}+2+\dfrac{4}{x^4}-1$$

$$=\dfrac{x^4}{4}+1+\dfrac{4}{x^4}.$$

All the various artifices used in the previous chapters to abbreviate the work in multiplication and division, and also the various forms of factoring, may be applied in the same way to fractional expressions.

EXERCISE XLVI.

1. $\left(\dfrac{a}{b}+\dfrac{x}{y}\right)\left(\dfrac{a}{b}-\dfrac{x}{y}\right).$ 2. $\left(x+\dfrac{1}{x}\right)\left(x-\dfrac{1}{x}\right).$

3. $\left(x+\dfrac{1}{x}\right)\left(x+\dfrac{1}{x}\right).$ 4. $\left(x-\dfrac{1}{x}\right)\left(x-\dfrac{1}{x}\right).$

5. $\left(x+\dfrac{2}{x}\right)\left(x+\dfrac{3}{x}\right).$ 6. $\left(x-\dfrac{9}{x}\right)\left(x+\dfrac{10}{x}\right).$

7. $\left(ax+\dfrac{b}{x}\right)\left(bx-\dfrac{a}{x}\right).$ 8. $\left(x^2+1+\dfrac{1}{x^2}\right)\left(x-\dfrac{1}{x}\right).$

9. $\left(x^2-1+\dfrac{1}{x^2}\right)\left(x+\dfrac{1}{x}\right).$ 10. $\left(\dfrac{x^2}{2}-\dfrac{x}{3}+1\right)\left(\dfrac{x}{3}+\dfrac{1}{4}\right).$

11. $\left(\dfrac{x}{3}-2+\dfrac{1}{x}\right)\left(\dfrac{x}{2}-\dfrac{3}{x}\right).$ 12. $\left(x^2+1+\dfrac{1}{x^2}\right)\left(x^2-1+\dfrac{1}{x^2}\right).$

13. $\left(\dfrac{a^2}{b^2}+\dfrac{b^2}{a^2}+2\right)\left(\dfrac{a^2}{b^2}+\dfrac{b^2}{a^2}-2\right).$ 14. $\left(1+\dfrac{a}{b}+\dfrac{a^2}{b^2}\right)\left(1-\dfrac{a}{b}+\dfrac{a^2}{b^2}\right).$

15. $\left(\dfrac{1}{a}+\dfrac{1}{b}+\dfrac{1}{c}\right)\left(\dfrac{1}{a}-\dfrac{1}{b}+\dfrac{1}{c}\right).$ 16. $\left(\dfrac{a}{b}+\dfrac{b}{c}-\dfrac{c}{a}\right)\left(\dfrac{a}{b}-\dfrac{b}{c}+\dfrac{c}{a}\right).$

17. $\left(a^2+2+\dfrac{2}{a^2}\right)\left(a^2-2+\dfrac{2}{a^2}\right).$ 18. $\left(\dfrac{x}{y^2}+\dfrac{1}{x}+\dfrac{1}{y}\right)\left(\dfrac{y}{x^2}+\dfrac{1}{y}-\dfrac{1}{x}\right).$

DIVISION OF FRACTIONS.

166. To divide $\dfrac{a}{b}$ by $\dfrac{c}{d}$.

Let $\dfrac{a}{b} = x$, $\dfrac{c}{d} = y$, then $\dfrac{x}{y}$ is the quotient required.

Now
$$a = bx, \quad c = dy,$$
$$\therefore \frac{a}{c} = \frac{bx}{dy},$$
$$\therefore \frac{a}{c} \times \frac{d}{b} = \frac{bx}{dy} \times \frac{d}{b} = \frac{x}{y}.$$

To divide one fraction by another we have therefore the following

RULE.—*Invert the divisor and proceed as in multiplication.*

The reciprocal of a number is unity divided by the number; thus the reciprocal of 2 is $\tfrac{1}{2}$. The reciprocal of a fraction is the fraction inverted. This follows at once from the definition and the rule for division.

Ex. 1.—Divide $\dfrac{14x^3}{27y^2}$ by $\dfrac{7x}{9y}$.

$$\frac{14x^3}{27y^2} \div \frac{7x}{9y} = \frac{14x^3}{27y^2} \times \frac{9y}{7x} = \frac{2x}{3y}.$$

Ex. 2.
$$\frac{a^2 - b^2}{a^2 + 2ab + b^2} \div \frac{4(a^2 - ab)}{a^2 + ab} = \frac{a^2 - b^2}{a^2 + 2ab + b^2} \times \frac{a^2 + ab}{4(a^2 - ab)}$$
$$= \frac{(a+b)(a-b)}{(a+b)(a+b)} \times \frac{a(a+b)}{4a(a-b)} = \frac{1}{4}.$$

Ex. 3.
$$\left\{\frac{1}{(x+a)(x-b)} + \frac{1}{(x-a)(x+b)}\right\} \div \left\{\frac{1}{(x+a)(x+b)} + \frac{1}{(x-a)(x-b)}\right\}$$
$$= \frac{2(x^2 - ab)}{(x^2 - a^2)(x^2 - b^2)} \div \frac{2(x^2 + ab)}{(x^2 - a^2)(x^2 - b^2)} = \frac{x^2 - ab}{x^2 + ab}.$$

EXERCISE XLVII.

Simplify

1. $\dfrac{35x^2y}{24a^2b} \div \dfrac{5xy^3}{8ab^3}$.

2. $\dfrac{8x^4y}{15ab^3} \div \dfrac{2x^3}{30ab^2}$.

3. $\dfrac{11p^2q^3}{4x^2y^2} \div 33p^3q^3$.

4. $\dfrac{5x}{3x-3} \div \dfrac{10}{x-1}$.

5. $\dfrac{x^2+xy}{xy-y^2} \div \dfrac{xy+y^2}{x^3-xy}$.

6. $\dfrac{ax-x^2}{a^3+x^3} \div \dfrac{(a-x)^2}{x^2-ax+a^2}$.

7. $\dfrac{a^2-4x^2}{a^2+4ax} \div \dfrac{a^3-2ax}{ax+4x^2}$.

8. $\dfrac{2b^2}{a(a^2-b^2)} \div \dfrac{6(ab-b^2)}{a(a+b)^2}$.

9. $\left(\dfrac{a^3}{b^3}+\dfrac{b^3}{a^3}\right) \div \left(\dfrac{a}{b}+\dfrac{b}{a}\right)$.

10. $\left(\dfrac{x^2}{y^2}+\dfrac{1}{x}\right) \div \left(\dfrac{x}{y^2}-\dfrac{1}{y}+\dfrac{1}{x}\right)$.

11. $\left(\dfrac{x^4}{y^4}+1+\dfrac{y^4}{x^4}\right) \div \left(\dfrac{x^2}{y^2}+1+\dfrac{y^2}{x^2}\right)$.

12. $\left(a^4+\dfrac{1}{a^2}\right) \div \left(1+\dfrac{1}{a^4-a^2}\right)$.

13. $\left(\dfrac{1}{a^3}+\dfrac{1}{b^3}+\dfrac{1}{c^3}-\dfrac{3}{abc}\right) \div \left(\dfrac{1}{a}+\dfrac{1}{b}+\dfrac{1}{c}\right)$.

14. $\dfrac{3x^5}{4}-4x^4+\dfrac{77}{8}x^3-\dfrac{43}{4}x^2-\dfrac{33}{4}x+27 \div \dfrac{x^2}{2}-x+3$.

15. $\dfrac{x}{a}-1-\dfrac{b}{a}-\dfrac{b^2}{a^2}+\dfrac{b}{x}+\dfrac{b^2}{x^2} \div x-a$.

16. $\dfrac{x^3}{a^3}-\dfrac{x}{a}+\dfrac{a}{x}-\dfrac{a^3}{x^3} \div \dfrac{x}{a}-\dfrac{a}{x}$.

17. $\dfrac{a^2+b^2-c^2+2ab}{c^2-a^2-b^2+2ab} \div \dfrac{a+b+c}{b+c-a}$.

18. $\left(\dfrac{x^2+y^2}{x^2-y^2}-\dfrac{x^2-y^2}{x^2+y^2}\right) \div \left(\dfrac{x+y}{x-y}-\dfrac{x-y}{x+y}\right)$.

19. $\dfrac{(a+b)^2-(c+d)^2}{(a+c)^2-(b+d)^2} \div \dfrac{(a-c)^2-(d-b)^2}{(a-b)^2-(d-c)^2}$.

20. $\left(\dfrac{x^4-b^4}{x^2-2bx+b^2} \div \dfrac{x^2+bx}{x-b}\right) \div \left(\dfrac{x^3+b^3}{x^2-b^2x^2} \div \dfrac{x^2-bx+b^2}{x^4-2bx^3+b^2x^2}\right)$.

COMPLEX FRACTIONS.

21. $\dfrac{3(x^2+xy+y^2)}{2x-3y} \div \left(\dfrac{2x+3y}{x^2-y^2} \div \dfrac{4x^2-9y^2}{x^3-y^3}\right).$

22. $\left(\dfrac{a+bx}{a-bx}+\dfrac{b+ax}{b-ax}\right) \div \left(\dfrac{a+bx}{a-bx}-\dfrac{b+ax}{b-ax}\right).$

23. $\left(\dfrac{3abc}{a+b}+\dfrac{a^2b^2}{(a+b)^3}\right) \div \left(3c+\dfrac{b}{a}-\dfrac{(2a+b)b^2}{a(a+b)^2}\right).$

COMPLEX FRACTIONS.

167. The division of one fraction by another is frequently represented by writing the former above the latter in the form of a complex fraction. Such fractions can always be simplified by the preceding rules.

Ex. 1. $\dfrac{a-\dfrac{b^2}{a}}{1+\dfrac{a}{b}} = \dfrac{a^2-b^2}{a} \div \left(\dfrac{a+b}{b}\right) = \dfrac{a^2-b^2}{a} \times \dfrac{b}{a+b} = \dfrac{(a-b)b}{a}.$

Ex. 2. $1 - \dfrac{1}{1-\dfrac{1}{1-x-\dfrac{1}{1+x}}} = 1 - \dfrac{1}{1+\dfrac{1+x}{x^2}} = 1 - \dfrac{x^2}{1+x+x^2}$

$= \dfrac{1+x}{1+x+x^2}.$

168. It is frequently the simplest way to multiply both numerator and denominator by such a quantity as will cause some of the smaller fractions to disappear. Thus, in Ex. 2,

$\dfrac{1}{1-x-\dfrac{1}{1+x}} = \dfrac{1+x}{1-x^2-1}$ by multiplying numerator and denominator by $1+x$.

$\dfrac{1}{1+\dfrac{1+x}{x^2}} = \dfrac{x^2}{1+x+x^2}$ by multiplying by x^2.

COMPLEX FRACTIONS.

EXERCISE XLVIII.

Simplify

1. $\dfrac{3 - \frac{1}{2}x}{5}$.

2. $\dfrac{2x}{1 - \frac{2}{3}(x-2)}$.

3. $\dfrac{2\frac{2}{3}x - 3\frac{1}{2}y}{\frac{5}{6}(x-y)}$.

4. $\dfrac{6x - \frac{3}{4}(3+5x)}{1\frac{1}{2} - \frac{1}{4}(x-6)}$.

5. $\dfrac{2\frac{1}{3} - \frac{1}{2}(x-2)}{\frac{1}{3}(x+1) - 4\frac{1}{2}}$.

6. $\dfrac{\frac{1}{4}(3x+1) - \dfrac{x}{6}}{\frac{5}{4}(x+1) + \dfrac{x-1}{2}}$.

7. $\dfrac{1 + \dfrac{(a-b)^2}{4ab}}{1 - \dfrac{b^2 + a^2}{2ab}}$.

8. $\dfrac{\dfrac{1+x}{1-x} + \dfrac{1-x}{1+x}}{\dfrac{1+x}{1-x} - \dfrac{1-x}{1+x}}$.

9. $\dfrac{\dfrac{a+2b}{a+b} + \dfrac{a}{b}}{\dfrac{a+2b}{b} - \dfrac{a}{a+b}}$.

10. $\dfrac{\dfrac{a^2+b^2}{2a^2} - \dfrac{2b^2}{a^2+b^2}}{\dfrac{a^2+b^2}{2b^2} - \dfrac{2a^2}{a^2+b^2}}$.

11. $\dfrac{\dfrac{1}{ab} + \dfrac{1}{bc} + \dfrac{1}{ca}}{\dfrac{a^2 - (b+c)^2}{abc}}$.

12. $\dfrac{\left(\dfrac{a}{x} - \dfrac{x}{a}\right)\left(\dfrac{a}{x} + \dfrac{x}{a}\right)}{\left(\dfrac{1}{x} - \dfrac{1}{a}\right)\left(\dfrac{1}{x} + \dfrac{1}{a}\right)}$.

13. $\dfrac{\dfrac{1}{x-y} - \dfrac{x}{x^2-y^2}}{\dfrac{x}{xy+y^2} - \dfrac{y}{x^2+xy}}$.

14. $\dfrac{\dfrac{a+b}{c+d} + \dfrac{a-b}{c-d}}{\dfrac{a+b}{c-d} + \dfrac{a-b}{c+d}}$.

15. $\dfrac{1}{x - 1 + \dfrac{1}{1 + \dfrac{x}{4-x}}}$.

16. $\dfrac{x}{x-a} - \dfrac{x}{x+a} - \dfrac{\dfrac{x+a}{x-a} - \dfrac{x-a}{x+a}}{\dfrac{x+a}{x-a} + \dfrac{x-a}{x+a}}$.

17. $\dfrac{\dfrac{1}{a} + \dfrac{1}{b+c}}{\dfrac{1}{a} - \dfrac{1}{b+c}}\left\{1 + \dfrac{b^2 + c^2 - a^2}{2bc}\right\}$.

18. $\dfrac{3abc}{bc + ca - ab} - \dfrac{\dfrac{a-1}{a} + \dfrac{b-1}{b} + \dfrac{c+1}{c}}{\dfrac{1}{a} + \dfrac{1}{b} - \dfrac{1}{c}}$.

19. $\dfrac{a + \dfrac{b}{1 + \dfrac{a}{b}}}{a - \dfrac{b}{1 - \dfrac{a}{b}}} \times \dfrac{b - \dfrac{a}{1 - \dfrac{a}{b}}}{b + \dfrac{a}{1 + \dfrac{a}{b}}}$.

COMPLEX FRACTIONS.

20. $\dfrac{\dfrac{1}{a-x}-\dfrac{1}{a-y}+\dfrac{x}{(a-x)^2}-\dfrac{y}{(a-y)^2}}{\dfrac{1}{(a-y)(a-x)^2}-\dfrac{1}{(a-y)^2(a-x)}}.$

21. $\dfrac{\dfrac{(1-2m)^2+(1+2m)^2}{(1-4m^2)-(1-2m)^2}}{\dfrac{(1+2m)^2-(1-4m^2)}{(1-2m)^2-(1+2m)^2}}.$

22. $\dfrac{\dfrac{1+x}{1-x}+\dfrac{4x}{1+x^2}+\dfrac{8x}{1+x^4}-\dfrac{1-x}{1+x}}{\dfrac{1+x^2}{1-x^2}+\dfrac{4x^2}{1+x^4}-\dfrac{1-x^2}{1+x^2}}.$

23. $\dfrac{\dfrac{(x-2y)^3}{y(x+y)^3}+\dfrac{(2x-y)^3}{x(x+y)^3}}{\dfrac{(x+2y)^3}{y(x-y)^3}-\dfrac{(2x+y)^3}{x(x-y)^3}}.$

EXERCISE XLIX.

MISCELLANEOUS EXAMPLES IN FRACTIONS.

1. Divide $\dfrac{x^3}{a^3}-\dfrac{y^3}{b^3}-\dfrac{3xy}{ab}\left(\dfrac{x}{a}-\dfrac{y}{b}\right)$ by $bx-ay$.

2. Divide $\left(\dfrac{a^3-b^3}{a-b}\right)^2+\left(\dfrac{a^3+b^3}{a+b}\right)^2-2b^4$ by $(a+b)^3+(a-b)^3$.

3. Find the value of $\dfrac{1}{x-a}+\dfrac{1}{x-b}$ when $x=\dfrac{2ab}{a+b}$.

Simplify

4. $\left\{\dfrac{a-b}{a+b}+\dfrac{a+b}{a-b}-\dfrac{2(a^2+b^2)}{a^2-b^2}\right\}\dfrac{x+y}{a+b}.$

5. $\dfrac{(ax^2-ay^2+2bxy)^2+(bx^2-by^2-2axy)^2}{(ax+by)^2+(bx-ay)^2}.$

6. $\left(\dfrac{x^2}{y^2}-1\right)\left(1-\dfrac{x}{x+y}\right)+\left(\dfrac{x^3}{y^3}-1\right)\left(\dfrac{1}{\dfrac{x^2}{y^2}+\dfrac{x}{y}+1}\right).$

7. $\dfrac{\dfrac{1}{c}\left(\dfrac{a}{b}-1\right)+\dfrac{1}{a}\left(\dfrac{b}{c}-1\right)+\dfrac{1}{b}\left(\dfrac{c}{a}-1\right)}{\dfrac{1}{b}\left(1-\dfrac{b}{a}\right)\left(\dfrac{b}{c}-1\right)+\dfrac{1}{c}\left(1-\dfrac{c}{b}\right)\left(\dfrac{c}{a}-1\right)+\dfrac{1}{a}\left(1-\dfrac{a}{c}\right)\left(\dfrac{a}{b}-1\right)}.$

COMPLEX FRACTIONS.

8. $\dfrac{\{a^2(x^2+y^2)+b^2(x^2-y^2)\}^2 - \{a^2(x^2-y^2)+b^2(x^2+y^2)\}^2}{\{a(x+y)+b(x-y)\}^2 - \{a(x-y)+b(x+y)\}^2}.$

9. $\dfrac{\{(a+b)^2x^2+(a-b)^2y^2\}^2 - \{(a-b)^2x^2+(a+b)^2y^2\}^2}{\{(x+y)^2a^2+(x-y)^2b^2\}^2 - \{(x-y)^2a^2+(x+y)^2b^2\}^2}.$

10. $\dfrac{\{(a-b)(x+y)+(a+b)(x-y)\}^2 + \{(a+b)(x+y)-(a-b)(x-y)\}^2}{(ax-ay+bx-by)^2+2xy(a+b)^2-2ab(x-y)^2-4abxy}.$

11. $\dfrac{\left(\dfrac{m}{n}+\dfrac{n}{m}+1\right)\left(\dfrac{m}{n}+\dfrac{n}{m}-1\right)+1}{\left\{\dfrac{(m+n)^2+(m-n)^2}{(m+n)^2-(m-n)^2}\right\}\left\{\dfrac{m^6+n^6}{m^2+n^2}+3m^2n^2\right\}}.$

12. $\left(\dfrac{a^2-ab}{a^3-b^3}\right)\left(\dfrac{a^2+ab+b^2}{a+b}\right) + \left(\dfrac{2a^3}{a^3+b^3}-1\right)\left(1-\dfrac{2ab}{a^2+ab+b^2}\right).$

13. $\dfrac{x^3+x}{x^3+1} \times \dfrac{(x^2+x)^2}{x^4+x^2+1} \times \dfrac{x^6-1}{(x^3-x)^2} \times \dfrac{x^3-1-3x(x-1)}{x^3-x^2+x-1}.$

14. $\left(\dfrac{x-2a}{y+a}\right)^3 + \dfrac{(x-a)(y+b)}{(y-a)(x-b)}$ when $x=\dfrac{a+b}{2}, y=\dfrac{a-b}{2}.$

15. $\dfrac{x+2a}{2b-x} + \dfrac{x-2a}{2b+x} - \dfrac{4ab}{4b^2-x^2}$ when $x=\dfrac{ab}{a+b}.$

16. $\dfrac{(ay-bx)^3-(ax-by)^3}{(a+b)(y-x)} + \dfrac{(ay-bx)^3+(ax-by)^3}{(a-b)(y+x)}.$

17. $\dfrac{1+a}{1-a} \cdot \dfrac{1+b}{1-b} \cdot \dfrac{1+c}{1-c}$ when $a=\dfrac{x-y}{x+y}, b=\dfrac{y-z}{y+z}, c=\dfrac{z-x}{z+x}.$

18. $\left\{\dfrac{p^2x^3}{qs}+\dfrac{pqx^2}{r^2s}-\dfrac{prx^3}{s^2}-\dfrac{q^2x}{rs^2}+\dfrac{p^2x}{qr}-\dfrac{p}{s}\right\} \div \left(\dfrac{px}{r}-\dfrac{q}{s}\right).$

19. $\dfrac{3a(x^2+ax+a^2)}{x^3-a^3} - \dfrac{2x^3+3ax^2-5a^3}{x^3+a^3} - \dfrac{3a}{x-a} + \dfrac{3ax-3a^2}{x^2-ax+a^2}.$

20. $\dfrac{(a+b)(1-ab)}{(1-ab)^2-(a+b)^2} - \dfrac{a(1-b^2)+b(1-a^2)}{(1-a^2)(1-b^2)-4ab}.$

21. $\dfrac{(c-d)a^2+6(bc-bd)a+9(b^2c-b^2d)}{(bc-bd+c^2-cd)a+3(b^2c+bc^2-b^2d-bcd)}.$

COMPLEX FRACTIONS.

22. $\dfrac{x^3 + y^3 + z^3 - 3xyz}{a^3 + b^3 + c^3 - 3abc}$ when $x = b+c$, $y = c+a$, $z = a+b$.

23. $\dfrac{(2b-c-a)^3 - (2c-a-b)^3}{(c-a)^3 - (a-b)^3}$.

24. If $y+z+u = ax$, $z+u+x = by$, $u+x+y = cz$, $x+y+z = du$,

    prove $\dfrac{1}{a+1} + \dfrac{1}{b+1} + \dfrac{1}{c+1} + \dfrac{1}{d+1} = 1$.

25. If $x^2 - yz = a^2$, $y^2 - zx = b^2$, $z^2 - xy = c^2$, find the value of

    $\dfrac{a^2 x + b^2 y + c^2 z}{x+y+z}$ in terms of a, b, c.

26. If $\dfrac{a-b}{1+ab} + \dfrac{c-d}{1+cd} = 0$ prove $\dfrac{b-c}{1+bc} + \dfrac{d-a}{1+ad} = 0$.

27. Find the value of

    $\dfrac{(ab+1)(x^2+1)}{(xy+1)(a^2+1)} - \dfrac{x+1}{y+1}$ when $x = \dfrac{1+a}{1-a}$, $y = \dfrac{1+b}{1-b}$.

28. Find the value of

    $\dfrac{(x+a)(x+mb) - (x-ma)(x-b)}{(mx+a)(x+b) - (x-a)(mx-b)}$ when $x = \dfrac{2ab}{a+b}$.

29. If $s = a+b+c+d$ prove

    $\dfrac{s-a}{a} + \dfrac{s-b}{b} + \dfrac{s-c}{c} + \dfrac{s-d}{d} = s\left(\dfrac{1}{a} + \dfrac{1}{b} + \dfrac{1}{c} + \dfrac{1}{d}\right) - 4$.

30. If $2s = a+b+c$ prove

    $\dfrac{1}{s-a} + \dfrac{1}{s-b} + \dfrac{1}{s-c} - \dfrac{1}{s} = \dfrac{abc}{s(s-a)(s-b)(s-c)}$.

31. If $x = \dfrac{a+b-c}{a+b+c}$ prove $\dfrac{a+bx^2}{b+ax^2} = \dfrac{(a-b+c)^2 + 4ab}{(b-a+c)^2 + 4ab}$.

32. If $x+y+z = 0$ prove $\dfrac{x(y^3 - z^3)}{y-z} + \dfrac{y(z^3 - x^3)}{z-x} + \dfrac{z(x^3 - y^3)}{x-y} = 0$.

33. If $t = \dfrac{2}{2-w}$, $w = \dfrac{2}{2-z}$, $z = \dfrac{2}{2-y}$, $y = \dfrac{2}{2-x}$, prove that $t = x$.

112 THEOREMS IN FRACTIONS.

34. If $b = \dfrac{2ac}{a+c}$ prove $\dfrac{1}{a-b} + \dfrac{1}{b-c} + \dfrac{4}{c-a} = \dfrac{1}{c} - \dfrac{1}{a}$.

35. If $x = y - \dfrac{1}{y}$, $y = z + \dfrac{1}{z}$, prove

$$\dfrac{z}{x} + \dfrac{1}{y+1} + \dfrac{y(1-z)}{z(y^2-1)} = \left(\dfrac{a}{a+b} + \dfrac{b}{a-b}\right) \div \left(\dfrac{a}{a-b} - \dfrac{b}{a+b}\right).$$

THEOREMS IN FRACTIONS.

169. The following theorems are of great practical value:—

I. Let $\dfrac{a}{b}$ and $\dfrac{c}{d}$ be any two fractions equal to each other, then

$$\dfrac{a+b}{a-b} = \dfrac{c+d}{c-d}.$$

In other words, *if two fractions are equal, the sum of the numerator and denominator of the one fraction, divided by their difference, will be equal to the sum of the numerator and denominator of the other fraction divided by their difference.*

Proof.—Let $\dfrac{a}{b} = \dfrac{c}{d}$. Adding 1 to each side,

$$\dfrac{a}{b} + 1 = \dfrac{c}{d} + 1 \text{ or } \dfrac{a+b}{b} = \dfrac{c+d}{d}. \qquad (1)$$

Again, $\because \dfrac{a}{b} = \dfrac{c}{d}$, subtracting 1 from each side,

$$\dfrac{a}{b} - 1 = \dfrac{c}{d} - 1 \text{ or } \dfrac{a-b}{b} = \dfrac{c-d}{d}, \qquad (2)$$

$\therefore$ from (1) and (2), $\dfrac{\frac{a+b}{b}}{\frac{a-b}{b}} = \dfrac{\frac{c+d}{d}}{\frac{c-d}{d}}.$

Simplifying, $\dfrac{a+b}{a-b} = \dfrac{c+d}{c-d}.$

THEOREMS IN FRACTIONS.

II. If $\quad \dfrac{a}{b} = \dfrac{c}{d} = \dfrac{e}{f}$, then $\dfrac{a}{b} = \dfrac{a+c+e}{b+d+f}$.

For, let $\quad \dfrac{a}{b} = x, \therefore \dfrac{c}{d} = x$ and $\dfrac{e}{f} = x;$

$\therefore a = bx, \; c = dx, \; e = fx,$

and $\quad a + c + e = bx + dx + fx$

$= x(b + d + f),$

$\therefore \dfrac{a+c+e}{b+d+f} = x = \dfrac{a}{b} = \dfrac{c}{d} = \dfrac{e}{f}.$

Hence we see that, *if any number of fractions be equal, the algebraic sum of their numerators divided by the algebraic sum of their denominators will be equal to each fraction.*

170. This, however, is only a particular case of a more general theorem, which may be stated as follows:—

If $\quad \dfrac{a}{b} = \dfrac{c}{d} = \dfrac{e}{f} = \ldots$

then each fraction $\quad = \dfrac{ma \pm nc \pm pe \ldots}{mb \pm nd \pm pf \ldots}$

when $m, n, p \ldots$ are any multipliers whatever.

Proof.—As before, let $\dfrac{a}{b} = \dfrac{c}{d} = \dfrac{e}{f} \ldots = x,$

then $\quad a = bx, \; c = dx, \; e = fx \ldots$

$\therefore ma = mbx, \; nc = ndx, \; pe = pfx \ldots$

$\therefore ma \pm nc \pm pe \ldots = mbx \pm ndx \pm pfx \ldots$

$= x(mb \pm nd \pm pf \ldots).$

Dividing, $\quad \dfrac{ma \pm nc \pm pe \ldots}{mb \pm nd \pm pf \ldots} = x = \dfrac{a}{b} = \dfrac{c}{d} = \dfrac{e}{f} \ldots$

NOTE—Theorem I. might have been proved in the same way as Theorem II., by letting $\dfrac{a}{b} = \dfrac{c}{d} = x$, and then substituting for a, bx, and for c, dx, in $\dfrac{a+b}{a-b}$ and $\dfrac{c+d}{c-d}$.

THEOREMS IN FRACTIONS.

171. Another proof is sometimes given for Theorem II.:—

Since $\quad a = b\left(\dfrac{a}{b}\right),\ c = d\left(\dfrac{c}{d}\right),\ e = f\left(\dfrac{e}{f}\right),$

$$\therefore \frac{a+c+e}{b+d+f} = \frac{b\left(\dfrac{a}{b}\right) + d\left(\dfrac{c}{d}\right) + f\left(\dfrac{e}{f}\right)}{b+d+f}.$$

But $\quad\dfrac{a}{b} = \dfrac{c}{d} = \dfrac{e}{f},$

$$\therefore b\left(\frac{a}{b}\right) + d\left(\frac{c}{d}\right) + f\left(\frac{e}{f}\right) = b\left(\frac{a}{b}\right) + d\left(\frac{a}{b}\right) + f\left(\frac{a}{b}\right) = (b+d+f)\frac{a}{b},$$

$$\therefore \frac{a+c+e}{b+d+f} = \frac{(b+d+f)\dfrac{a}{b}}{b+d+f} = \frac{a}{b} = \frac{c}{d} = \frac{e}{f}.$$

Ex. 1.—If $\dfrac{1+x}{1-x} = \dfrac{b}{a}\left(\dfrac{1+x+x^2}{1-x+x^2}\right)$, prove $x^3 = \dfrac{b-a}{b+a}$.

If $\quad\dfrac{1+x}{1-x} = \dfrac{b}{a}\left(\dfrac{1+x+x^2}{1-x+x^2}\right),$

then $\quad\dfrac{(1+x)(1-x+x^2)}{(1-x)(1+x+x^2)} = \dfrac{b}{a}$

or $\quad\dfrac{1+x^3}{1-x^3} = \dfrac{b}{a}.$

Adding numerator and denominator of each fraction, and dividing by their difference:

$$\frac{1+x^3+1-x^3}{1+x^3-(1-x^3)} = \frac{b+a}{b-a}$$

or $\quad\dfrac{2}{2x^3} = \dfrac{b+a}{b-a};$

that is, $\quad\dfrac{1}{x^3} = \dfrac{b+a}{b-a};$

and inverting each fraction: $\quad x^3 = \dfrac{b-a}{b+a}.$

THEOREMS IN FRACTIONS.

Ex. 2.—If $\dfrac{a}{b} = \dfrac{c}{d}$ prove $\dfrac{a^2 - ab + b^2}{ab - 4b^2} = \dfrac{c^2 - cd + d^2}{cd - 4d^2}.$

$$\dfrac{a^2 - ab + b^2}{ab - 4b^2} = \dfrac{\dfrac{a^2}{b^2} - \dfrac{ab}{b^2} + 1}{\dfrac{ab}{b^2} - 4} = \dfrac{\dfrac{a^2}{b^2} - \dfrac{a}{b} + 1}{\dfrac{a}{b} - 4}.$$

But since $\dfrac{a}{b} = \dfrac{c}{d}$,

$$\dfrac{\dfrac{a^2}{b^2} - \dfrac{a}{b} + 1}{\dfrac{a}{b} - 4} = \dfrac{\left(\dfrac{c}{d}\right)^2 - \dfrac{c}{d} + 1}{\dfrac{c}{d} - 4} = \dfrac{\dfrac{c^2 - cd + d^2}{d^2}}{\dfrac{cd - 4d^2}{d^2}} = \dfrac{c^2 - cd + d^2}{cd - 4d^2}.$$

Another proof.—Let $\dfrac{a}{b} = \dfrac{c}{d} = x$, $\therefore a = bx$ and $c = dx$,

$$\therefore \dfrac{a^2 - ab + b^2}{ab - 4b^2} = \dfrac{b^2 x^2 - b^2 x + b^2}{b^2 x - 4b^2} = \dfrac{x^2 - x + 1}{x - 4};$$

also, $\dfrac{c^2 - cd + d^2}{cd - 4d^2} = \dfrac{d^2 x^2 - d^2 x + d^2}{d^2 x - 4d^2} = \dfrac{x^2 - x + 1}{x - 4};$

$\therefore \dfrac{a^2 - ab + b^2}{ab - 4b^2} = \dfrac{c^2 - cd + d^2}{cd - 4d^2}$, since each fraction $= \dfrac{x^2 - x + 1}{x - 4}$.

Ex. 3. If $\dfrac{a-b}{ay + bx} = \dfrac{b-c}{bz + cx} = \dfrac{c-a}{cy + az} = \dfrac{a+b+c}{ax + by + cz}$

then each fraction $= \dfrac{1}{x+y+z}$, when $a+b+c$ is not $= 0$.

For each fraction equals the sum of the numerators of all the fractions divided by the sum of their denominators;

$\therefore$ each fraction $= \dfrac{a - b + b - c + c - a + a + b + c}{ay + bx + bz + cx + cy + az + ax + by + cz}$

$$= \dfrac{a+b+c}{(a+b+c)(x+y+z)} = \dfrac{1}{x+y+z}$$

when $a+b+c$ is not $= 0$.

THEOREMS IN FRACTIONS.

If $a+b+c=0$ the sum of the numerators divided by the sum of the denominators $=\dfrac{0}{0}$, the value of which we do not know Hence the necessity of the condition that $a+b+c$ is not $=0$.

Ex. 4. If $\dfrac{a+b}{a-b}=\dfrac{b+c}{2(b-c)}=\dfrac{c+a}{3(c-a)}$ then $8a+9b+5c=0$.

$(a-b)+(b-c)+(c-a)=0$, $\therefore$ $m(a-b)+m(b-c)+m(c-a)=0$.

Hence, if the quantities $(a-b)$, $(b-c)$, $(c-a)$, in the denominators of these fractions can be made to have the same coefficient, the sum of the new denominators will be $=0$. Multiplying both numerator and denominator of each fraction in turn by that quantity which will make the common coefficient of $(a-b)$, $(b-c)$, $(c-a)$ the L. C. M. of the given coefficients, we obtain

$$\frac{a+b}{a-b}=\frac{6(a+b)}{6(a-b)}=\frac{3(b+c)}{6(b-c)}=\frac{2(c+a)}{6(c-a)}=k;$$

$$\therefore\ 6(a+b)=6k(a-b),$$
$$3(b+c)=6k(b-c),$$
$$2(c+a)=6k(c-a).$$

Adding $\qquad (8a+9b+5c)=6k(a-b+b-c+c-a)$
$$=6k(0)=0;$$
$$\therefore\ 8a+9b+5c=0.$$

Ex. 5.—If $\dfrac{x}{a+b-c}=\dfrac{y}{b+c-a}=\dfrac{z}{c+a-b}$, then

$$(a-b)x+(b-c)y+(c-a)z=0.$$

Since we require $(a-b)x$, let us multiply both numerator and denominator of the first fraction by $(a-b)$. For an analogous reason multiply both numerator and denominator of second fraction by $(b-c)$, and numerator and denominator of third fraction by $(c-a)$.

THEOREMS IN FRACTIONS.

$$\therefore \frac{x}{a+b-c} = \frac{(a-b)x}{a^2-b^2-c(a-b)} = \frac{(b-c)y}{b^2-c^2-a(b-c)} = \frac{(c-a)z}{c^2-a^2-b(c-a)} = k;$$

$$\therefore (a-b)x = k\{a^2-b^2-c.(a-b)\},$$
$$(b-c)y = k\{b^2-c^2-a.(b-c)\},$$
$$(c-a)z = k\{c^2-a^2-b.(c-a)\}.$$

But $a^2-b^2-c(a-b)+b^2-c^2-a(b-c)+c^2-a^2-b(c-a)$
$= a^2-b^2-ac+bc+b^2-c^2-ab+ac+c^2-a^2-bc+ab = 0;$

$$\therefore (a-b)x+(b-c)y+(c-a)z = k(0) = 0.$$

EXERCISE L.

1. If $\dfrac{x}{a} = \dfrac{y}{b} = \dfrac{z}{c}$, prove $\dfrac{x+y}{a+b} = \dfrac{x+z}{a+c} = \dfrac{y+z}{b+c}$.

2. If $\dfrac{x}{a} = \dfrac{y}{b} = \dfrac{z}{c}$, prove $\dfrac{x}{a} = \dfrac{mx+ny+pz}{ma+nb+pc}$.

3. If $\dfrac{x}{5} = \dfrac{y}{8}$, find the value of $\dfrac{x+5}{y+8}$.

4. If $\dfrac{a}{b} = \dfrac{c}{d}$, prove $\dfrac{ax+b}{cx+d}$ has always the same value, whatever be the value of x.

5. If $\dfrac{a}{b} = \dfrac{c}{d}$, prove $\dfrac{a^2+b^2}{c^2+d^2} = \dfrac{(a-b)^2}{(c-d)^2}$.

6. If $\dfrac{a}{b} = \dfrac{c}{d}$, prove $\dfrac{a^2+c^2}{b^2+d^2} = \dfrac{(ab+dc)^2}{(b^2+d^2)^2}$.

7. If $\dfrac{a}{b} = \dfrac{c}{d}$, prove $\dfrac{a^2+3ab+b^2}{c^2+3cd+d^2} = \dfrac{2ab+3b^2}{2cd+3d^2}$.

8. If $\dfrac{x}{a} = \dfrac{y}{b} = \dfrac{z}{c}$, prove

$$\frac{x^2+y^2+z^2}{a^2+b^2+c^2} = \frac{(x+y+z)^2}{(a+b+c)^2} = \frac{xy+yz+zx}{ab+ac+bc} = \frac{x^2}{a^2}.$$

THEOREMS IN FRACTIONS.

9. If $\dfrac{x}{a} = \dfrac{y}{b} = \dfrac{z}{c}$, prove $\dfrac{ax+by+cz}{a^2+b^2+c^2} = \dfrac{x^2+y^2+z^2}{ax+by+cz}$.

10. If $\dfrac{x}{a} = \dfrac{y}{b} = \dfrac{z}{c}$, prove $\dfrac{ax^2+by^2+cz^2}{a^3+b^3+c^3} = \dfrac{x^3+y^3+z^3}{a^2x+b^2y+c^2z}$.

11. If $\dfrac{x}{a} = \dfrac{y}{b} = \dfrac{z}{c}$, prove $\dfrac{x^3+y^3+z^3-3xyz}{a^3+b^3+c^3-3abc} = \dfrac{x^3}{a^3}$.

12. If $\dfrac{ad-bc}{a-b-c+d} = \dfrac{ac-bd}{a-b+c-d}$, then shall $a+b = c+d$, and each ratio $= \dfrac{a+b+c+d}{4}$.

13. If $\dfrac{x-y}{a} = \dfrac{y-z}{b} = \dfrac{z-x}{c}$, then shall $a+b+c = 0$.

14. If $\dfrac{a+b}{x+y} = \dfrac{b+c}{y+z} = \dfrac{c+a}{z+x}$,

show that each fraction is equal to $\dfrac{a+b+c}{x+y+z}$, and that $\dfrac{a}{x} = \dfrac{b}{y} = \dfrac{c}{z}$.

15. If $\dfrac{cy+bz}{b-c} = \dfrac{az+cx}{c-a} = \dfrac{bx+ay}{a-b}$, then

$$(a+b+c)(x+y+z) = ax+by+cz.$$

16. If $\dfrac{a}{b+c} = \dfrac{b}{c+a} = \dfrac{c}{a+b}$, prove $a = b = c$.

17. If $\dfrac{b^2+c^2-a^2}{bc} = \dfrac{c^2+a^2-b^2}{ca} = \dfrac{a^2+b^2-c^2}{ab}$, prove each fraction $= 1$.

18. If $\dfrac{a_1+a_2x}{a_2+a_3y} = \dfrac{a_2+a_3x}{a_3+a_1y} = \dfrac{a_3+a_1x}{a_1+a_2y}$,

each of these fractions $= \dfrac{1+x}{1+y}$, supposing $a_1+a_2+a_3$ not $= 0$.

19. If $\dfrac{ay-bx}{c} = \dfrac{cx-az}{b} = \dfrac{bz-cy}{a}$, then $\dfrac{x}{a} = \dfrac{y}{b} = \dfrac{z}{c}$.

20. If $\dfrac{x+y}{3a-b} = \dfrac{y+z}{3b-c} = \dfrac{z+x}{3c-a}$, prove $\dfrac{x+y+z}{ax+by+cz} = \dfrac{a+b+c}{a^2+b^2+c^2}$.

THEOREMS IN FRACTIONS. 119

21. If $\dfrac{x}{b+c-a} = \dfrac{y}{c+a-b} = \dfrac{z}{a+b-c}$, prove

$$(a+b+c)(yz+zx+xy) = (x+y+z)(ax+by+cz).$$

22. If $\dfrac{a^2}{x^2-yz} = \dfrac{b^2}{y^2-zx} = \dfrac{c^2}{z^2-xy}$, prove

$$a^2x + b^2y + c^2z = (a^2+b^2+c^2)(x+y+z).$$

23. If $\dfrac{bz-cy}{a} = \dfrac{cx-az}{b} = \dfrac{ay-bx}{c}$, prove $ax+by+cz=0$.

24. If $\dfrac{a+b}{3(a-b)} = \dfrac{b+c}{4(b-c)} = \dfrac{c+a}{5(c-a)}$, prove $32a+35b+27c=0$.

25. If $\dfrac{x}{a(y-z)} = \dfrac{y}{b(z-x)} = \dfrac{z}{c(x-y)}$, prove $\dfrac{x}{a} + \dfrac{y}{b} + \dfrac{z}{c} = 0$.

26. If $\dfrac{x}{a(y+z)} = \dfrac{y}{b(z+x)} = \dfrac{z}{c(x+y)}$, prove

$$\dfrac{x}{a}(y-z) + \dfrac{y}{b}(z-x) + \dfrac{z}{c}(x-y) = 0.$$

27. If $\dfrac{x^2-yz}{a^2} = \dfrac{y^2-zx}{b^2} = \dfrac{z^2-xy}{c^2} = 1$, show that

$$x+y+z = \dfrac{a^2x + b^2y + c^2z}{a^2+b^2+c^2}.$$

28. If $\dfrac{m}{x} = \dfrac{n}{y} = \dfrac{r}{z}$ and $\dfrac{x^2}{a^2} + \dfrac{y^2}{b^2} + \dfrac{z^2}{c^2} = 1$, prove

$$\dfrac{m^2}{a^2} + \dfrac{n^2}{b^2} + \dfrac{r^2}{c^2} = \dfrac{m^2+n^2+r^2}{x^2+y^2+z^2}.$$

29. If $\dfrac{hx}{a^2} = \dfrac{ky}{b^2} = \dfrac{lz}{c^2}$ and $\dfrac{x^2}{a^2} + \dfrac{y^2}{b^2} + \dfrac{z^2}{c^2} = 1$, prove

$$\left(\dfrac{x}{h} + \dfrac{y}{k} + \dfrac{z}{l}\right)^2 = \dfrac{a^2}{h^2} + \dfrac{b^2}{k^2} + \dfrac{c^2}{l^2}.$$

30. If $\dfrac{x^2+2x+1}{x^2-2x+3} = \dfrac{y^2+2y+1}{y^2-2y+3}$, each fraction $= \dfrac{xy-1}{xy-3}$.

CHAPTER IX.

FRACTIONAL SIMPLE EQUATIONS.

172. In Chapter VI. we gave examples of simple equations of *one* unknown, without fractional terms. We now proceed to give examples of equations involving such terms.

173. A simple equation involving fractional terms can be solved by multiplying both sides of the equation by the Lowest Common Denominator (L. C. D.) of the different fractional terms, and then proceeding according to the rule laid down for solving equations without fractions. The object of multiplying through by the L. C. D. is to *clear of fractions*.

If a fraction is preceded by a *minus* sign the sign of every term of the numerator must be changed when the fraction is multiplied by the L. C. D.

Thus, $-\dfrac{a-b+c}{d} \times de = e(-a+b-c)$.

174. The different methods usually adopted for solving such equations can best be made clear by examples.

Ex. 1.—Solve $\quad \dfrac{5x}{2} - \dfrac{5x}{4} = \dfrac{9}{4} - \dfrac{3-x}{2}$.

Multiply both sides by 4, the L. C. D.,

Then $\qquad 10x - 5x = 9 - (6 - 2x)$

or $\qquad 10x - 5x = 9 - 6 + 2x;$

$\therefore 3x = 3$ or $x = 1$.

FRACTIONAL SIMPLE EQUATIONS.

Ex. 2.—Solve $\dfrac{x+4}{3} - \dfrac{x-4}{5} = 2 + \dfrac{3x-1}{15}$.

Multiply by 15,

Then $\quad 5x + 20 - 3x + 12 = 30 + 3x - 1$.

Collecting and transposing, $x = 3$.

NOTE.—The student can always test the correctness of his result by substituting in the given equation the value found for x, when, if correct, the two sides will become identical.

EXERCISE LI.

Solve

1. $x + \dfrac{x-3}{3} = \dfrac{17}{3}$.

2. $2x - \dfrac{5x-4}{6} = 7 - \dfrac{1-2x}{5}$.

3. $\dfrac{5x+3}{8} - \dfrac{3-4x}{3} = \dfrac{31}{2} - \dfrac{9-5x}{6} - \dfrac{x}{2}$.

4. $\dfrac{10x+3}{3} - \dfrac{6x-7}{2} = 10(x-1)$.

5. $\dfrac{5x-7}{2} - \dfrac{2x+7}{3} = 3x - 14$.

6. $\dfrac{7x+5}{6} - \dfrac{5x-6}{4} = \dfrac{8-5x}{12}$.

7. $\dfrac{3x+5}{7} - \dfrac{2x+7}{3} + 10 = \dfrac{3x}{5}$.

8. $\dfrac{3x-4}{7} + \dfrac{5x+3}{3} = 43 - 5x$.

9. $\dfrac{1}{2}(27-2x) = \dfrac{9}{2} - \dfrac{1}{10}(7x-54)$.

10. $\dfrac{2x+7}{7} - \dfrac{9x-8}{11} = \dfrac{x-11}{2}$.

11. $\dfrac{8x-15}{3} - \dfrac{11x-1}{7} = \dfrac{7x+2}{13}$.

12. $2x - \dfrac{1}{3} = \dfrac{2}{5}(3-2x) + \dfrac{1}{2}x$.

13. $\dfrac{1}{12}(2x-3) - \dfrac{1}{5}(3x-2) = \dfrac{1}{8}(4x-3)$.

14. $\dfrac{15x}{4} - \dfrac{10}{3} + \dfrac{x}{6} - \dfrac{x}{2} = 1\tfrac{1}{2} + x$.

15. $\dfrac{8}{6} - \dfrac{7}{3x} - \dfrac{3}{4} + \dfrac{5}{2x} = 1$.

16. $7x - \dfrac{3}{8}x = \dfrac{7}{10}x + \dfrac{x}{15} - 6\tfrac{1}{3}$.

17. $\dfrac{3-x}{4} - \dfrac{3-1}{x} + \dfrac{6+x}{3} = \dfrac{x}{12}$.

18. $\dfrac{x+1}{2} - \dfrac{3}{x} = \dfrac{x}{3} - \dfrac{5-x}{6}$.

19. $\dfrac{1}{2}(11-x) - \dfrac{1}{3}(13+2x) = \dfrac{1}{4}(15-3x)$.

20. $\dfrac{3}{8}(x) + \dfrac{5}{6}(11-x) = 1 - \dfrac{5}{24}(x-2)$.

FRACTIONAL SIMPLE EQUATIONS.

175. If the denominators contain both simple and compound expressions it is frequently best to first combine the simple expressions and then clear of fractions.

Ex. 1.—Solve $\quad -\dfrac{9x+20}{36} = \dfrac{4(x-3)}{5(x-4)} + \dfrac{x}{4}.$

Transposing $\dfrac{x}{4}$, and combining, we get

$$\dfrac{9x+20-9x}{36} = \dfrac{20}{36} = \dfrac{4(x-3)}{5(x-4)};$$

$$\therefore \dfrac{5}{9} = \dfrac{4}{5}\left(\dfrac{x-3}{x-4}\right).$$

Clearing of fractions, $25x - 100 = 36x - 108$,

from which we obtain $\quad x = \dfrac{8}{11}.$

Ex. 2.—Solve $\quad \dfrac{9x+15}{14} + \dfrac{8x-7}{6x+2} = \dfrac{36x+15}{56} + \dfrac{41}{56}.$

Transposing, $\dfrac{9x+15}{14} - \dfrac{36x+15}{56} - \dfrac{41}{56} = -\dfrac{8x-7}{6x+2}.$

Combining fractions on the left side,

$$\dfrac{4}{56} = \dfrac{1}{14} = -\dfrac{8x-7}{6x+2}.$$

Clearing of fractions, $\quad 6x+2 = -112x+98;$

$$\therefore 118x = 96 \text{ or } x = \dfrac{96}{118} = \dfrac{48}{59}.$$

EXERCISE LII.

Solve

1. $\dfrac{8x-5}{14} + \dfrac{7x-3}{6x+2} = \dfrac{4x+6}{7}.$

2. $\dfrac{9(2x-3)}{14} + \dfrac{11x-1}{3x+1} = \dfrac{9x+11}{7}.$

3. $\dfrac{10x+17}{18} - \dfrac{12x+2}{13x-16} = \dfrac{5x-4}{9}.$

4. $\dfrac{6x+13}{15} - \dfrac{3x+5}{5x-25} = \dfrac{2x}{5}.$

5. $\dfrac{6x+7}{15} - \dfrac{2x-2}{7x-6} = \dfrac{2x+1}{5}.$

6. $\dfrac{6x+1}{15} - \dfrac{2x-4}{7x-16} = \dfrac{2x-1}{5}.$

7. $\dfrac{7x-6}{35} - \dfrac{x-5}{6x-101} = \dfrac{x}{5}.$

8. $\dfrac{4x+3}{9} = \dfrac{8x+19}{18} - \dfrac{7x-29}{5x-12}.$

9. $\dfrac{4x+5}{10} - \dfrac{9x-6}{7x+4} = \dfrac{2x-3}{5}.$

10. $\dfrac{x}{3} - \dfrac{x^2-5x}{3x-7} = \dfrac{2}{3}.$

176. Complex fractions generally should be simplified before proceeding to find the lowest common multiple of denominators

Ex. 1.—Solve $\quad \dfrac{3 - \dfrac{4x}{9}}{4} = \dfrac{1}{4} - \dfrac{\dfrac{7x}{9} - 3}{10}.$

Simplifying first and third fractions,

$$\dfrac{27-4x}{36} = \dfrac{1}{4} - \dfrac{7x-27}{90}.$$

Multiplying by L. C. D.,

$$135 - 20x = 45 - 14x + 54.$$

Transposing and collecting like terms,

$$6x = 36, \therefore x = 6.$$

Ex. 2.—Solve $\quad \dfrac{25 - \dfrac{1}{3}x}{x+1} + \dfrac{16x + 4\frac{1}{5}}{3x+2} = \dfrac{23}{x+1} + 5.$

Simplifying complex fractions,

$$\dfrac{75-x}{3(x+1)} + \dfrac{80x+21}{5(3x+2)} = \dfrac{23}{x+1} + 5.$$

Transposing, $\quad \dfrac{75-x}{3(x+1)} - \dfrac{23}{x+1} + \dfrac{80x+21}{5(3x+2)} = 5;$

$$\therefore \dfrac{75-x-69}{3(x+1)} + \dfrac{80x+21}{5(3x+2)} = 5.$$

Multiplying both sides by $15(x+1)(3x+2)$,

$$5(3x+2)(6-x) + 3(x+1)(80x+21) = 75(x+1)(3x+2).$$

Multiplying out and collecting terms,

$$8x = 27, \therefore x = 3\tfrac{3}{8}.$$

Solve   EXERCISE LIII.

1. $\dfrac{5x}{9} = \dfrac{4\tfrac{1}{2}x + 11}{16} - \dfrac{7x+5}{12}.$

2. $10\tfrac{1}{2}\left(x - \dfrac{2}{3}\right) + \dfrac{1}{2}(3\tfrac{3}{4}x + 7) = 21\tfrac{1}{4}.$

3. $\dfrac{2x+5}{21} - \dfrac{x-15}{3\tfrac{1}{2}} = \dfrac{x(5 - 2\tfrac{2}{3})}{14}.$

4. $\dfrac{x}{6\tfrac{1}{4}} + \dfrac{56 - 2x}{5\tfrac{1}{2}x + 5\tfrac{1}{4}} = x - \dfrac{7x-2}{8\tfrac{1}{3}}.$

5. $\dfrac{2\tfrac{3}{4}x - 1}{8\tfrac{3}{4}} - \dfrac{1\tfrac{1}{6}x - 13\tfrac{1}{2}}{7\tfrac{1}{2}} - \dfrac{1\tfrac{2}{3}x - 2}{10\tfrac{1}{2}} = \dfrac{x + 46}{x - 46}.$

6. $\dfrac{3x - 2}{4\tfrac{1}{2}x + \tfrac{1}{2}} - \dfrac{2 - 3x}{1 + 2x} = \dfrac{8\tfrac{2}{3}x - 5}{4x + 2}.$

7. $\dfrac{x + 1\tfrac{1}{2}}{3} - \dfrac{10 - x}{3\tfrac{2}{3}} = \dfrac{4 - \tfrac{2}{3}x}{11} - \dfrac{1}{11}.$

8. $\dfrac{x - \tfrac{1}{3}(x - 1)}{3} + \dfrac{31}{36} = \dfrac{3 - \tfrac{1}{4}(x - 2)}{5}.$

9. $\dfrac{3x - \tfrac{2}{3}(1 + x)}{4} + \dfrac{1 - \tfrac{1}{5}x}{5\tfrac{1}{2}} = \dfrac{2\tfrac{2}{3} + \dfrac{x - 1}{25}}{2\tfrac{1}{3}}.$

FRACTIONAL SIMPLE EQUATIONS.

10. $\dfrac{1}{2}x - \dfrac{\dfrac{2x-3}{3} - \dfrac{3x-1}{4}}{\dfrac{x-1}{2}} = \dfrac{3}{2} \cdot \dfrac{x^2 - \dfrac{x}{3} + 2}{3x-2}$.

11. $\dfrac{2-3x}{1.5} + \dfrac{5x}{1.25} - \dfrac{2x-3}{9} = \dfrac{x-2}{1.8} + 2\tfrac{7}{5}$.

12. $.5x + \dfrac{.45x + .75}{.6} = \dfrac{1.2}{.2} - \dfrac{.3x - .6}{.9}$.

13. $.5 - \dfrac{3.5x}{x-2} - \dfrac{24-3x}{8} = .375x$.

14. $.15x + \dfrac{.135x - .225}{.6} = \dfrac{.36}{.2} - \dfrac{.09x - .18}{.9}$.

15. $\dfrac{4x-17}{9} - \dfrac{3\tfrac{2}{3} - 22x}{33} = x - \dfrac{6}{x}\left(1 - \dfrac{x^2}{54}\right)$.

177. Other artifices are often employed to lessen the labor of solving equations. In a certain class of examples the *actual* division of each numerator by its denominator assists greatly in reducing the equation to a simple form. In others a judicious combination of fractions according to their denominators is of great value in facilitating the work.

Ex. 1.—Solve $\dfrac{8x+25}{2x+5} + \dfrac{16x+93}{2x+11} = \dfrac{18x+86}{2x+9} + \dfrac{6x+26}{2x+7}$.

Divide each numerator by its denominator,

$\therefore\ 4 + \dfrac{5}{2x+5} + 8 + \dfrac{5}{2x+11} = 9 + \dfrac{5}{2x+9} + 3 + \dfrac{5}{2x+7}$;

$\therefore\ \dfrac{5}{2x+5} + \dfrac{5}{2x+11} = \dfrac{5}{2x+9} + \dfrac{5}{2x+7}$

Dividing by 5, $\quad \dfrac{1}{2x+5} + \dfrac{1}{2x+11} = \dfrac{1}{2x+9} + \dfrac{1}{2x+7}$.

FRACTIONAL SIMPLE EQUATIONS.

Combining in pairs,
$$\frac{4x+16}{(2x+5)(2x+11)} = \frac{4x+16}{(2x+9)(2x+7)}$$

or
$$\frac{4x+16}{4x^2+32x+55} = \frac{4x+16}{4x^2+32x+63}.$$

In these fractions the numerators are equal, the fractions equal, but the denominators *unequal*. How are we to reconcile the apparent inconsistency? Only by putting the numerator of each fraction $= 0$; the fractions will then be equal for all values of the denominators except zero. $\therefore 4x+16=0$ or $x=-4$.

This result may be obtained otherwise. Bring both fractions to the same side of the equation, then
$$\frac{4x+16}{4x^2+32x+35} - \frac{4x+16}{4x^2+32x+63} = 0$$

or
$$(4x+16)\left\{\frac{1}{4x^2+32x+35} - \frac{1}{4x^2+32x+63}\right\} = 0.$$

But the second factor is not $=0$, $\therefore 4x+16=0$ and $x=-4$.

Ex. 2.—Solve $\quad \dfrac{3x+1}{2x+3} - \dfrac{x+2}{6x+9} + \dfrac{1}{3x} = \dfrac{4}{3}.$

Transposing, $\quad \dfrac{3x+1}{2x+3} - \dfrac{x+2}{3(2x+3)} = \dfrac{4}{3} - \dfrac{1}{3x}.$

Combining in pairs, $\quad \dfrac{9x+3-x-2}{3(2x+3)} = \dfrac{4x-1}{3x}.$

Simplifying, $\quad \dfrac{8x+1}{3(2x+3)} = \dfrac{4x-1}{3x}.$

Clearing of fractions, $\quad 8x^2+x = 8x^2+10x-3.$

$\therefore 9x = 3$ and $x = \dfrac{1}{3}.$

FRACTIONAL SIMPLE EQUATIONS. 127

Ex. 3.—Solve $\dfrac{1}{x+1}+\dfrac{1}{x+2}-\dfrac{3(x+5)}{(x+1)(x+2)}=6-\dfrac{6x+17}{x+2}.$

Transposing, $\dfrac{1}{x+1}+\dfrac{1}{x+2}+\dfrac{6x+17}{x+2}=6+\dfrac{3(x+5)}{(x+1)(x+2)}.$

Combining, $\dfrac{1}{x+1}+\dfrac{6x+18}{x+2}=6+\dfrac{3(x+5)}{(x+1)(x+2)}.$

Transposing, $\dfrac{1}{x+1}+\dfrac{6x+18}{x+2}-6=\dfrac{3(x+5)}{(x+1)(x+2)}.$

Combining, $\dfrac{1}{x+1}+\dfrac{6x+18-6x-12}{x+2}=\dfrac{3(x+5)}{(x+1)(x+2)}.$

Simplifying, $\dfrac{1}{x+1}+\dfrac{6}{x+2}=\dfrac{3(x+5)}{(x+1)(x+2)}.$

Combining, $\dfrac{7x+8}{(x+1)(x+2)}=\dfrac{3x+15}{(x+1)(x+2)}.$

$\therefore\ 7x+8=3x+15,$ or $4x=7$ and $x=\dfrac{7}{4}.$

EXERCISE LIV.

Solve

1. $\dfrac{5x+3}{x-1}+\dfrac{2x-3}{2x-2}=9.$

2. $\dfrac{6x+8}{2x+1}-\dfrac{2x+38}{x+12}=1.$

3. $\dfrac{5x^2+x-3}{5x-4}=\dfrac{7x^2-3x-9}{7x-10}.$

4. $\dfrac{x}{x-2}+\dfrac{x-9}{x-7}=\dfrac{x+1}{x-1}+\dfrac{x-8}{x-6}.$

5. $\dfrac{x^3+ax^2-bx+c}{x^3-ax^2+bx+c}=\dfrac{x^2+ax-b}{x^2-ax+b}.$

6. $\dfrac{1+x+x^2}{1-x+x^2}=\dfrac{62}{63}\cdot\dfrac{1+x}{1-x}.$

7. $\dfrac{x+4a+b}{x+a+b}+\dfrac{4x+a+2b}{x+a-b}=5.$

8. $\dfrac{3x-1}{2x-1}-\dfrac{4x-2}{3x-2}=\dfrac{1}{6}.$

FRACTIONAL SIMPLE EQUATIONS.

9. $\dfrac{2}{2x-3}+\dfrac{1}{x-2}=\dfrac{6}{3x+2}.$

10. $\dfrac{3+x}{3-x}-\dfrac{2+x}{2-x}-\dfrac{1+x}{1-x}=1.$

11. $\dfrac{x^2-x+1}{x-1}+\dfrac{x^2+x+1}{x+1}=2x.$

12. $\dfrac{4}{x-8}+\dfrac{3}{2x-16}-\dfrac{29}{24}=\dfrac{2}{3x-24}.$

13. $\dfrac{4}{x+2}+\dfrac{7}{x+3}=\dfrac{37}{x^2+5x+6}.$

14. $\dfrac{x-4}{x-5}-\dfrac{x-5}{x-6}=\dfrac{x-7}{x-8}-\dfrac{x-8}{x-9}.$

15. $\dfrac{3}{x-1}-\dfrac{x+1}{x-1}=\dfrac{x^2}{1-x^2}.$

16. $\dfrac{x-8}{x-10}-\dfrac{x-5}{x-7}=\dfrac{x-7}{x-9}-\dfrac{x-4}{x-6}.$

17. $\dfrac{1}{x+1}+\dfrac{4}{2x-1}+\dfrac{9}{3x-1}=\dfrac{36}{6x-1}.$

18. $\dfrac{4}{x-6}-\dfrac{x-2}{x-3}=\dfrac{x+4}{x-5}-\dfrac{2(x-1)}{x-4}.$

19. $\dfrac{4x-17}{x-4}+\dfrac{10x-13}{2x-3}=\dfrac{8x-30}{2x-7}+\dfrac{5x-4}{x-1}.$

178. In some equations known numbers are represented by letters. These are called *literal* equations. The same artifices and methods are employed in solving these equations as when figures are used.

NOTE.—Usually the *first* letters of the alphabet are employed to represent *known* numbers, the *last* letters *unknown* numbers. But this rule does not always hold good, as any letter may be used in either way.

Ex. 1.—Solve $\quad ax+bc=bx+ac.$

Transposing, $\quad ax-bx=ac-bc.$

Factoring, $\quad x(a-b)=c(a-b).$

Dividing by $(a-b)$, $\quad x=\dfrac{c(a-b)}{a-b}=c.$

FRACTIONAL SIMPLE EQUATIONS. 129

Ex. 2.—Solve $(a+x+b)(a+b-x) = (a+x)(b-x) - ab$.

Multiplying, $\quad (a+b)^2 - x^2 = ab + x(b-a) - x^2 - ab$.

$\therefore (a+b)^2 = x(b-a)$ and $x = \dfrac{(a+b)^2}{b-a}$.

Ex. 3.—Solve $\quad \dfrac{3ax-2b}{3b} - \dfrac{ax-a}{2b} = \dfrac{ax}{b} - \dfrac{2}{3}$.

Transposing, $\quad \dfrac{3ax-2b}{3b} - \dfrac{ax-a}{2b} - \dfrac{ax}{b} = -\dfrac{2}{3}$.

Combining, $\quad \dfrac{6ax - 4b - 3ax + 3a - 6ax}{6b} = -\dfrac{2}{3}$

or $\quad \dfrac{-3ax - 4b + 3a}{6b} = -\dfrac{2}{3}$.

Clearing of fractions, $\quad -3ax - 4b + 3a = -4b$,

$\therefore ax = a$ and $x = 1$.

EXERCISE LV.

Solve

1. $a(x-b) = b(a-x) - (a+b)x$.   2. $(a+x)(b+x) = (c+x)(d+x)$.

3. $\dfrac{6x-a}{4x-b} = \dfrac{3x+b}{2x+a}$.   4. $\dfrac{a(a^2+x^2)}{a+x} = ax + b^2$.

5. $\dfrac{x-a}{b+c} + \dfrac{x-b}{c+a} + \dfrac{x-c}{a+b} = 3$.   6. $\dfrac{1}{ab-ax} + \dfrac{1}{bc-bx} = \dfrac{1}{ac-ax}$.

7. $\dfrac{a}{x+a} - \dfrac{c}{x-c} = \dfrac{a-c}{x+a-c}$.   8. $x+a+b+c = \dfrac{x^2+a^2+b^2+c^2}{a+b-c+x}$.

9. $\dfrac{ax^2+bx+c}{ax+b} = \dfrac{px^2+qx+r}{px+q}$.   10. $\dfrac{a+b}{x-c} = \dfrac{a}{x-a} + \dfrac{b}{x-b}$.

11. $\dfrac{(x-a)^3}{(x+b)^3} = \dfrac{x-2a-b}{x+a+2b}$.   12. $\dfrac{x+a}{b+c} + \dfrac{x+b}{c+a} + \dfrac{x+c}{a+b} = -3$.

13. $\dfrac{2x+a}{b} - \dfrac{x-b}{a} = \dfrac{3ax+(a-b)^2}{ab}$. 14. $\dfrac{m(x+a)}{x+b} + \dfrac{n(x+b)}{x+a} = m+n$.

15. $\dfrac{a+c}{(a-b)(x-a)} - \dfrac{b+c}{(a-b)(x-b)} = \dfrac{x+c}{(x-a)(x-b)} + x - a$.

16. $(a+x)(b+x) - a(b+x) = \dfrac{a^2c}{b} + x^2$.

17. $\dfrac{x-a}{b} + \dfrac{x-b}{c} + \dfrac{x-c}{a} = \dfrac{x-a-b-c}{abc}$.

18. $(x-a)^3 + (x-b)^3 + (x-c)^3 = 3(x-a)(x-b)(x-c)$.

19. $\dfrac{1}{(x-a)(x-b)} - \dfrac{2}{(x-a)(x-c)} + \dfrac{1}{(x-b)(x-c)}$
$= \dfrac{1}{(x+a)(x+b)} - \dfrac{2}{(x+a)(x+c)} + \dfrac{1}{(x+b)(x+c)}$.

EXERCISE LVI.

MISCELLANEOUS EXAMPLES.

Solve

1. $(3x-1)^2 + (4x-2)^2 = (5x-3)^2$. 2. $(x+2a)(x-a)^2 = (x+2b)(x-b)^2$.

3. $\left(\dfrac{x+1}{2x+1}\right)^2 = \dfrac{x+a}{4x+a}$. 4. $\left(\dfrac{x-a}{x-b}\right)^2 = \dfrac{x-2a}{x-2b}$.

5. $\left(\dfrac{x^2-11x+19}{x^2+x-11}\right)^2 + \dfrac{3(x-2)}{x+2} = 0$. 6. $\dfrac{x+6}{x-1} + \dfrac{x-6}{x+1} = \dfrac{2(x-6)^2}{x^2-1}$.

7. $\dfrac{1}{x-1} + \dfrac{2}{x-2} = \dfrac{3}{x-3}$. 8. $\dfrac{2x^3-3x+1}{x^2-2x+2} = \dfrac{2x-3}{x-2}$.

9. $1 - x - \dfrac{1+x}{1-x} = \dfrac{3+2x^2}{2-2x}$. 10. $\dfrac{a+b}{x+c} = \dfrac{a}{x+a} + \dfrac{b}{x+b}$.

11. $(x+1)(x+2)(x+3) = (x-3)(x+4)(x+5)$.

12. $(x+1)(x+2)(x+3) = (x-1)(x-2)(x-3) + 3(4x-1)(x+1)$.

13. $(x+b)(x+c) + (x+c)(x+a) = (2x+a)(x+b)$.

14. $(x-a)(x-2a) = (x-3a)(x-4a)$.

FRACTIONAL SIMPLE EQUATIONS. 131

15. $\dfrac{3}{(3x+1)(2x-3)} = \dfrac{5}{(5x-4)(2x+5)}$.

16. $\dfrac{1}{(3x-1)(x-2)} = \dfrac{2}{(2x-1)(x-4)}$.

17. $(x-9)(x-7)(x-5)(x-1) = (x-2)(x-4)(x-6)(x-10)$.

18. $(x-2a)^3 + (x-2b)^3 = 2(x-a-b)^3$.

19. $\dfrac{3abc}{a+b} + \dfrac{a^2b^2}{(a+b)^3} + \dfrac{(2a+b)b^2x}{a(a+b)^2} = 3cx + \dfrac{bx}{a}$.

20. $\dfrac{2x-3}{x-4} + \dfrac{3x-2}{x-8} = \dfrac{5x^2-29x-4}{x^2-12x+32}$.

21. $(x+a)(x+b) + (x+c)(x+a) = (x+b)(x+d) + (x+d)(x+c)$.

22. $\dfrac{m}{x-a} + \dfrac{n}{x-b} + \dfrac{p}{x-c} = \dfrac{m}{x-c} + \dfrac{n}{x-a} + \dfrac{p}{x-b}$.

23. $\dfrac{x-2a}{b+c-a} + \dfrac{x-2b}{c+a-b} + \dfrac{x-2c}{a+b-c} = 3$.

24. $\dfrac{x-2a}{b+c-a} + \dfrac{x-2b}{c+a-b} + \dfrac{x-2c}{a+b-c} = \dfrac{3x}{a+b+c}$.

25. $\dfrac{a-x}{a^2-bc} + \dfrac{b-x}{b^2-ac} + \dfrac{c-x}{c^2-ab} = \dfrac{3}{a+b+c}$.

26. $\dfrac{a-b}{x-c} + \dfrac{b-c}{x-a} + \dfrac{c-a}{x-b} = \dfrac{a+b}{x-a} + \dfrac{b+c}{x-b} - \dfrac{a+2b+c}{x-c}$.

27. $\dfrac{ab-bc}{x+c} + \dfrac{bc-ca}{x+a} + \dfrac{ac-ab}{x+b} = \dfrac{a^2-b^2}{x+a} + \dfrac{b^2-c^2}{x+b} + \dfrac{c^2-a^2}{x+c}$.

28. $\dfrac{a^2-bc}{x+a} + \dfrac{b^2-ca}{x+b} + \dfrac{c^2-ab}{x+c} = \dfrac{\dfrac{a^3-abc}{x+a} + \dfrac{b^3-abc}{x+b} + \dfrac{c^3-abc}{x+c}}{a+b+c}$.

29. $\dfrac{a}{x-md} + \dfrac{b}{x-mc} + \dfrac{c}{x+mb} + \dfrac{d}{x+ma} = 0$ if $a+b+c+d=0$.

PROBLEMS PRODUCING FRACTIONAL EQUATIONS.

179. The solution of problems resulting in fractional equations is accomplished by employing the same methods and principles as were used in solving problems not producing fractional equations. It will be well, however, to give specimen solutions of different types of problems frequently occurring in practice.

Ex. 1.—A can do a piece of work in 6 days, and B can do the same work in 8 days. How long will it take A and B working together to do it?

Let $x =$ number of days it takes A and B to do it.

Then $\dfrac{1}{x} =$ amount A and B can together do in a day.

But A can do it in 6 days, $\therefore$ A can do $\dfrac{1}{6}$ in 1 day,

and B can do it in 8 days, $\therefore$ B can do $\dfrac{1}{8}$ in 1 day;

therefore A and B can together do $\dfrac{1}{6} + \dfrac{1}{8}$ in one day.

But A and B can together do $\dfrac{1}{x}$ in one day, $\therefore \dfrac{1}{6} + \dfrac{1}{8} = \dfrac{1}{x}$.

Simplifying, $\dfrac{7}{24} = \dfrac{1}{x}$, $\therefore x = \dfrac{24}{7} = 3\frac{3}{7}$ days.

Ex. 2.—A tank can be filled by two pipes in 24 minutes and 30 minutes respectively, and emptied by a third in 20 minutes. In what time will it be filled if all three are working together?

Let $x =$ number of minutes it will be filled by all three together.

Then $\dfrac{1}{x} =$ amount filled in 1 minute by all three.

PROBLEMS PRODUCING FRACTIONAL EQUATIONS.   133

But $\dfrac{1}{24}$ = amount filled by first tap in 1 minute,

and $\dfrac{1}{30}$ = amount filled by second tap in 1 minute,

also $\dfrac{1}{20}$ = amount emptied by third tap in one minute;

∴ $\dfrac{1}{24}+\dfrac{1}{30}-\dfrac{1}{20}$ = amount filled by three taps in 1 minute.

But $\dfrac{1}{x}$ = amount filled in 1 minute by 3 taps,

∴ $\dfrac{1}{x}=\dfrac{1}{24}+\dfrac{1}{30}-\dfrac{1}{20}$.

Simplifying, $\dfrac{1}{x}=\dfrac{3}{120}=\dfrac{1}{40}$, ∴ $x=40$ minutes.

Ex. 3.—A person walks to the top of a mountain at the rate of $2\frac{1}{3}$ miles an hour, and down the same way at the rate of $3\frac{1}{2}$ miles an hour, and is out 5 hours. How far is it to the top of the mountain?

Let x = distance to the top of the mountain.

Then $\dfrac{x}{2\frac{1}{3}}$ = time to walk up

and $\dfrac{x}{3\frac{1}{2}}$ = time to walk down,

∴ $\dfrac{x}{2\frac{1}{3}}+\dfrac{x}{3\frac{1}{2}}$ = whole time.

But whole time of walking up and down is 5 hours,

∴ $\dfrac{x}{2\frac{1}{3}}+\dfrac{x}{3\frac{1}{2}}=5$ or $\dfrac{3x}{7}+\dfrac{2x}{7}=5$, ∴ $\dfrac{5x}{7}=5$, ∴ $x=5$ miles.

Ex. 4.—Find the time between 3 and 4 o'clock when the hands of a clock are at right angles.

134 PROBLEMS PRODUCING FRACTIONAL EQUATIONS.

The minute-hand of a clock moves twelve times faster than the hour-hand. When the hands are at right angles one hand is 15 minute-spaces ahead of the other. At 3 o'clock the minute-hand is at 12, and therefore the hands are at right angles at that time. There is, however, another solution; for the minute-hand may get 15 minute-spaces ahead of the hour-hand. To find the time when that occurs: The minute-hand at 3 o'clock is 15 minute-spaces behind the hour-hand, and it has to gain this space and get 15 minute-spaces ahead of the hour-hand, so that it has to gain altogether 30 minute-spaces.

Let $x =$ number of units of space moved by hour-hand from 3 o'clock to time required.

Then $12x =$ number of units of space moved by minute-hand;

$\therefore 12x - x = 11x =$ spaces gained by minute-hand.

But spaces gained is 30 minute-spaces,

$\therefore 11x = 30$ minute-spaces,

$\therefore x = 2\frac{8}{11}$ minute-spaces.

But time is shown by minute-hand, which has passed over $12x$ spaces,

$\therefore 12x = 12 \times 2\frac{8}{11} = \frac{360}{11} = 32\frac{8}{11}$ minutes.

Therefore the time is $32\frac{8}{11}$ minutes past 3 o'clock.

EXERCISE LVII.

1. Find a number whose third part exceeds its fourth part by 14.

2. The half, fourth and fifth of a certain number are together equal to 76. Find the number.

3. Divide 60 into two such parts that a seventh of one part may be equal to an eighth of the other.

4. Divide 45 into two such parts that the first part divided by 2 shall be equal to the second part multiplied by 2.

5. In a mixture of wine and water the wine was 25 gallons more than half the mixture, and the water 5 gallons less than one-third of the mixture. How many gallons were there of each ?

PROBLEMS PRODUCING FRACTIONAL EQUATIONS.

6. Divide 46 into two such parts that if one part be divided by 7 and the other by 3 the sum of the quotients shall be 10.

7. A can do a piece of work in 5 days and B can do it in 4 days. How long will it take A and B together to do it?

8. A can do a piece of work in 5 days, B in 6 days, and C in $7\frac{1}{2}$ days. In what time will they do it, all working together?

9. A can do a piece of work in $2\frac{1}{2}$ days, B in $3\frac{1}{3}$ days, and C in $3\frac{3}{4}$ days. In what time will they do it, all working together?

10. Two men who can separately do a piece of work in 15 days and 16 days can, with the help of another, do it in 6 days. How long would it take the third man to do it alone?

11. A does $\frac{5}{8}$ of a piece of work in 10 days, when B comes to help him, and they finish the work in 3 days more. How long would it have taken B alone to do the whole work?

12. A and B together can reap a field in 12 hours, A and C together in 16 hours, and A by himself in 20 hours. In what time can B and C together reap it? In what time can A, B and C together reap it?

13. A and B together can do a piece of work in 12 days, A and C in 15 days, B and C in 20 days. In what time can they do it, all working together?

14. A tank can be filled in 15 minutes by two pipes, A and B, running together. After A has been running by itself for 5 minutes B is also turned on, and the tank is filled in 13 minutes more. In what time may it be filled by each pipe separately?

15. A cistern could be filled by two pipes in 6 hours and 8 hours respectively, and could be emptied by a third in 12 hours. In what time would the cistern be filled if the pipes were all running together?

16. A tank can be filled by three pipes in 1 hour and 20 minutes, 3 hours and 20 minutes and 5 hours respectively. In what time will the tank be filled when all three pipes are running together?

136   PROBLEMS PRODUCING FRACTIONAL EQUATIONS.

17. A fish was caught whose tail weighed 9 pounds, his head weighed as much as his tail and half his body, and his body weighed as much as his head and tail together. Find the weight of the fish.

18. A hare is 50 leaps before a greyhound, and takes 4 leaps to the greyhound's 3 leaps; but 2 of the greyhound's = 3 of the hare's. How many leaps must the greyhound take to catch the hare?

19. Find the time between 2 and 3 o'clock when the hour and minute hands of a watch are, 1st, coincident; 2nd, in exactly opposite directions; 3rd, at right angles to each other.

20. Find the respective times between 7 and 8 o'clock when the hour and minute hands of a watch are, 1st, exactly opposite to each other; 2nd, at right angles to each other; 3rd, coincident.

21. It is between 2 and 3 o'clock, but a person looking at his watch and mistaking the hour-hand for the minute-hand fancies that the time of day is 55 minutes earlier than it really is. What is the true time?

22. A horse was sold at a loss for $200, but if it had been sold for $250 the gain would have been $\frac{3}{4}$ of the loss when sold for $200. Find the value of the horse.

23. A merchant adds yearly to his capital $\frac{1}{3}$ of it, but takes from it, at the end of each year, $5000 for expenses. At the end of the third year, after deducting the last $5000, he has twice his original capital. How much had he at first?

24. A trader maintained himself for three years at an expense of $250 a year, and each year increased that part of his stock which was not so expended by $\frac{1}{3}$ of it. At the end of the third year his original stock was doubled. What was his original stock?

25. A cask contains 12 gallons of wine and 18 gallons of water; another contains 9 gallons of wine and 3 gallons of water. How many gallons must be drawn from each cask to produce a mixture containing 7 gallons of wine and 7 gallons of water?

PROBLEMS PRODUCING FRACTIONAL EQUATIONS. 137

26. A man rowed down the river a distance of 11 miles in $1\frac{1}{2}$ hours with the stream, and on his return rowed back again in $3\frac{3}{4}$ hours. Find the rate of the stream per hour.

27. A boatman moves 5 miles in $\frac{3}{4}$ of an hour, rowing with the tide; and in returning it takes him $1\frac{1}{2}$ hours, rowing against a tide one-half as strong. What is the velocity of the stronger tide?

28. A boatman rowing with the tide moves n miles in t hours. Returning he uses t_1 hours to accomplish the same distance, rowing against a tide m times as strong as the first. What is the velocity of the stronger tide?

29. A train which travels 32 miles an hour is $\frac{3}{4}$ of an hour in advance of a second train which travels 42 miles an hour. In how long a time will the last overtake the first?

30. A train travelling b miles per hour is m hours in advance of a second train which travels a miles per hour. In how long a time will the last overtake the first? Discuss the result when $a > b;\ a = b;\ a < b$.

31. An express train which travels 42 miles per hour starts 50 minutes after a freight train, which it overtakes in 2 hours 5 minutes. What is the velocity of the freight train?

32. If A, who is travelling, makes $\frac{1}{2}$ of a mile more per hour he will employ only $\frac{4}{5}$ of the time, but if he makes $\frac{1}{2}$ of a mile less per hour he will be on the route $2\frac{1}{2}$ hours more. Find the length of the route and the speed.

33. At 12 o'clock the hands of a watch are together. At what hour will they be opposite to each other?

34. A and B accomplish a piece of work in m days; A and C can do it in n days, and B and C in p days. How many days will it take each to do the work alone, and how many if they work together?

35. If a men or b boys can dig m acres in n days, required the number of boys whose assistance will be required to enable $(a-p)$ men to dig $(m+p)$ acres in $(n-p)$ days.

10

138 PROBLEMS PRODUCING FRACTIONAL EQUATIONS.

36. If A can do a piece of work in $2m$ days, and B and A in n days, and A and C in $m + \dfrac{n}{2}$ days, find the number of days in which A, B and C together would do the work.

37. Two friends at a distance of 78 miles agree to meet in an intermediate locality, and set out at the same moment, one from A, travelling $5\tfrac{1}{4}$ miles per hour; the other from B, travelling $7\tfrac{3}{4}$ miles per hour. When and where do they meet?

38. A person, after paying a poor rate and also an income tax of 7d. in the £, has £486 remaining. The poor rate amounts to £22 10s. more than the income tax. Find the original income and the number of pence in the £ in the poor rate.

39. What must be the value of n in order that $\dfrac{2a + n}{3n + 69a}$ may be equal to $\tfrac{1}{33}$ when a is $\tfrac{1}{3}$?

40. A person, after paying an income tax of 6d. in the £, gave away $\tfrac{1}{13}$ of his remaining income, and had £540 left. What was his original income?

41. I bought a certain number of eggs at 2 a penny and the same number at 3 a penny. I sold them at 5 for twopence and lost a penny. How many eggs did I buy?

42. The sum of £330 is laid out in two investments, by one of which 15% is gained and by the other 8% is lost, and the amount of the returns is £345. Find each investment.

43. Find the weight of a mass of copper and tin, 40 pounds more copper than tin, to which if a quantity of copper $\tfrac{5}{8}$ heavier than the tin be added there will be 11 pounds of copper for every 3 pounds of tin.

44. The first digit of a certain number exceeds the second digit by 4, and when the number is divided by the sum of the digits the quotient is 7. Find it.

45. One-half of a population can read; of the remainder 42% can read and write; of the remainder again 16% can read, write and cipher; while 243,600 can neither read, write nor cipher. What is the population?

PROBLEMS PRODUCING FRACTIONAL EQUATIONS. 139

46. Divide £607 1s. 8d. into two sums such that the simple interest of the greater sum for 2 years at $3\tfrac{1}{2}\%$ shall exceed that of the less for $2\tfrac{1}{2}$ years at $3\tfrac{1}{4}\%$ by £18 16s.

47. A person possessed of £5222 invested a part in 5% stock at 105, and the remainder in 3% stock at 96. How much did he invest in each kind of stock if his whole income amounts to £191 16s. 8d.?

48. The hour is between 2 and 3 o'clock, and the minute-hand is in advance of the hour-hand by $14\tfrac{1}{2}$ minute-spaces of the dial. What o'clock is it?

49. 112 pounds of bronze contains by weight 70% of copper and 30% of tin. With how much copper must it be melted in order that it may contain 84% of copper?

50. Find the time between h and $h+1$ o'clock when the minute hand is m minute-divisions before the hour-hand.

CHAPTER X.

SIMULTANEOUS EQUATIONS OF THE FIRST DEGREE.

180. If *one* equation contain *two* or more unknown quantities an *indefinite number* of values may be found that will satisfy the equation.

Thus, when $x + y = 6$ any value may be given to y, and a corresponding value will be found for x.

If $x = 2$ then $y = 4$,
" $x = 3$ " $y = 3$,
" $x = 1$ " $y = 5$, and so on.

Any connected pair of these values substituted for x and y will satisfy the equation, and are called its roots.

181. But if, in connection with the preceding examples, another equation be given, expressing a *different* relation between x and y, then only *one* pair of values for x and y can be found which will satisfy *both* the equations.

Thus, if not only $x + y = 6$
but $x - y = 2$,

then the only values of x and y that will satisfy both equations are $x = 4$ and $y = 2$.

Such equations are called **simultaneous** because they are both true at *the same time, i.e.,* they are satisfied by the same values of x and y.

182. If there are *two* unknown numbers to be found then *two independent* equations must be given. Equations are said to be *independent* when they express *different* relations between the unknown quantities.

Thus $x+y=4$
and $2x+2y=8$

are not independent, since they express the same relation between x and y.

183. If *three* unknowns are given *three* independent equations will be required. Generally, if n unknown numbers are to be found it is necessary to have n independent equations to obtain a definite solution.

184. Different methods are adopted in solving simple simultaneous equations, the more common of which we now proceed to illustrate:—

FIRST METHOD,

Ex. 1.—Solve $\qquad x+y=8,$ $\qquad$ (1)

$\qquad\qquad\qquad 2x-3y=2.$ $\qquad$ (2)

Multiply (1) by 3, then

$\qquad\qquad\qquad 3x+3y=24,$ $\qquad$ (3)

but $\qquad\qquad 2x-3y=2.$ $\qquad$ (2)

Adding, $\qquad\qquad 5x=26, \therefore x=5\frac{1}{5}.$

To find y, multiply (1) by 2 and subtract from the result (2).

Then $\qquad\qquad 5y=14, \therefore y=2\frac{4}{5}.$

The object of this method is to make in turn the coefficients of x and y the same in both equations. Thus (1) was multiplied by 3 to make the coefficients of y the same in both equations; also (1) was multiplied by 2 to make the coefficients of x the same. Then, adding in the first instance and subtracting in the second, one of the unknowns disappears, or is *eliminated*, and the remaining one can be found. This process is called *elimination by addition or subtraction*.

185. Having found the value of either x or y the remaining value can be found by substitution. Thus, in the preceding example,

$$x + y = 8, \qquad (1)$$
$$2x - 3y = 2, \qquad (2)$$

we found $x = \dfrac{26}{5}$. Then substituting this value of x in (1) we get

$$\dfrac{26}{5} + y = 8,$$
$$\therefore y = 8 - \dfrac{26}{5} = \dfrac{14}{5}.$$

SECOND METHOD.

Ex. 2.—Solve
$$2x + 3y = 7, \qquad (1)$$
$$3x - y = 5. \qquad (2)$$

From (1) $2x = 7 - 3y$ or $x = \dfrac{7 - 3y}{2}$.

Substitute this value of x in (2), then

$$\dfrac{3(7 - 3y)}{2} - y = 5.$$

Simplifying, $\qquad 21 - 9y - 2y = 10$
or $\qquad\qquad\qquad -11y = -11$
or $\qquad\qquad\qquad y = 1.$

Since $y = 1$, $\qquad \therefore x = \dfrac{7 + 3y}{2}$

$$= \dfrac{7 - 3}{2} = 2.$$

This method is called *elimination by substitution.*

THIRD METHOD.

186. *Ex. 3.*—Solve
$$5x + 4y = 58, \qquad (1)$$
$$3x + 7y = 67. \qquad (2)$$

SIMULTANEOUS EQUATIONS OF THE FIRST DEGREE. 143

From (1) $\quad 5x = 58 - 4y, \therefore x = \dfrac{58 - 4y}{5},$

From (2) $\quad 3x = 67 - 7y, \therefore x = \dfrac{67 - 7y}{3},$

since the two values of x must be the same;

$$\therefore \dfrac{58 - 4y}{5} = \dfrac{67 - 7y}{3}.$$

Simplifying, $\quad 174 - 12y = 335 - 35y,$
or $\quad\quad\quad\quad 35y - 12y = 335 - 174,$
or $\quad\quad\quad\quad 23y = 161$
and $\quad\quad\quad\quad y = 7.$

If $y = 7$ then $\quad x = \dfrac{58 - 4y}{5}$

$$x = \dfrac{58 - 4 \times 7}{5} = 6.$$

This method is called *elimination by comparison.*

187. The following examples should be carefully noted:—

Ex. 4.—Solve $\quad\quad \dfrac{2}{x} + \dfrac{3}{y} = 12,$ $\quad\quad\quad$ (1)

$$\dfrac{5}{x} + \dfrac{4}{y} = 15. \quad\quad\quad (2)$$

To find x, multiply (1) by 4 and (2) by 3.

Then $\quad\quad\quad \dfrac{8}{x} + \dfrac{12}{y} = 48 \quad\quad\quad$ (3)

and $\quad\quad\quad \dfrac{15}{x} + \dfrac{12}{y} = 45. \quad\quad\quad$ (4)

Subtracting (4) from (3), $\quad -\dfrac{7}{x} = 3;$

$$\therefore 3x = -7$$

and $\quad\quad\quad x = -\dfrac{7}{3}.$

Similarly y can be found.

Ex. 5.—Solve
$$2x + 3y = 15xy, \quad (1)$$
$$3x + 4y = 21xy. \quad (2)$$

Divide both (1) and (2) by xy,

Then
$$\frac{2}{y} + \frac{3}{x} = 15, \quad (3)$$
$$\frac{3}{y} + \frac{4}{x} = 21. \quad (4)$$

From (3) and (4) the values of x and y can be obtained as in Ex. 4.

Ex. 6.—Solve
$$x(y + 7) = y(x + 1), \quad (1)$$
$$2x + 20 = 3y + 1. \quad (2)$$

Clearing (1) of brackets, $xy + 7x = xy + y;$ (3)

$$\therefore 7x = y.$$

Substituting this value of y in (2) x can be readily found.

EXERCISE LVIII.

Solve

1. $4x + 3y = 31,$
   $3x + 2y = 22.$

2. $3x - 2y = 7,$
   $8x + 2y = 48.$

3. $7x + 3y = 17.$
   $5x + 3y = 13.$

4. $5x + 7y = 43,$
   $11x + 9y = 69.$

5. $3y - 2x = 11,$
   $13x - 5y = 1.$

6. $8x - 9y = 1,$
   $6x - 3y = 4x.$

7. $\frac{1}{2}x + \frac{1}{3}y = 6,$
   $\frac{1}{3}x + \frac{1}{2}y = 6\frac{1}{2}.$

8. $10x - \frac{y}{5} = 69,$
   $10y - \frac{x}{7} = 49.$

SIMULTANEOUS EQUATIONS OF THE FIRST DEGREE. 145

9. $\dfrac{x}{3} + \dfrac{y}{4} = 2,$
$3x + 4y = 25.$

10. $3x + 7y = 79,$
$2y - \dfrac{x}{2} = 9.$

11. $2x - \dfrac{5y}{7} = 3x - 1,$
$3y + \dfrac{2x}{5} = y + 2.$

12. $1\tfrac{1}{2}x - 1\tfrac{1}{4}y = 5,$
$\dfrac{2}{3}x = \dfrac{5}{6}y.$

13. $4\tfrac{1}{3}x - 3\tfrac{1}{2}y = 46,$
$4\tfrac{1}{2}x + 3\tfrac{1}{3}y = 215.$

14. $2\tfrac{1}{3}x + 3\tfrac{1}{4}y = 74,$
$4\tfrac{1}{2}x - 5\tfrac{1}{6}y = 1.$

15. $\dfrac{1}{2}x + \dfrac{1}{3}y = 13,$
$\dfrac{1}{5}x + \dfrac{1}{8}y = 5.$

16. $\dfrac{1}{9}x + \dfrac{1}{8}y = 43,$
$\dfrac{1}{8}x + \dfrac{1}{9}y = 42.$

17. $\dfrac{2x + 3y}{6} + \dfrac{x}{3} = 8,$
$\dfrac{7y - 3x}{2} - y = 11.$

18. $\dfrac{2x - y}{4} + 1 = \dfrac{7 + x}{5},$
$\dfrac{3 - 4x}{6} + 3 = \dfrac{5y - 7}{2}.$

19. $\dfrac{2(15x + 13y)}{7}$
$= 7(x - 2y + 1)$
$= 12\{y - 3 + \dfrac{2}{3}(y + 7\tfrac{1}{2})\}.$

20. $x - \dfrac{y - 2}{7} = 5,$
$4y - \dfrac{x + 10}{3} = 3.$

21. $2x - \dfrac{y + 3}{4} = 7 + \dfrac{3y - 2x}{5},$
$4y - \dfrac{8 - x}{3} = 24\tfrac{1}{2} - \dfrac{2y + 1}{2}.$

22. $5x - \dfrac{5y + 2}{4} = 32,$
$3y + \dfrac{x - 2}{3} = 9.$

23. $ax + y = b,$
$x + by = a.$

24. $x + y = a,$
$ax + by = b^2.$

25. $ax = by$,
$x + y = c$.

26. $\dfrac{x}{a} + \dfrac{y}{b} = 1$,
$\dfrac{x}{b} - \dfrac{y}{a} = 1$.

27. $\dfrac{x}{a} - \dfrac{y}{b} = m$,
$\dfrac{ax}{c} - \dfrac{by}{a} = 0$.

28. $\dfrac{x}{a} + \dfrac{y}{b} = 1 - \dfrac{x}{c}$,
$\dfrac{y}{a} + \dfrac{x}{b} = 1 + \dfrac{y}{c}$.

29. $ax + by = c^2$,
$\dfrac{a}{b+y} - \dfrac{b}{a+x} = 0$.

30. $ax + by = c$,
$(a-b)x + (a+b)y = 2c$.

31. $(a+b)x - (a-b)y = c$,
$(a-b)x + (a+b)y = c$.

32. $\dfrac{x+y}{x-y} = \dfrac{5}{3}$,
$x + 5y = 36$.

33. $\dfrac{4x-2}{3} + 4y - 3 = 19$,
$\dfrac{8y-6}{13} + 9x - 7 = 40$.

34. $\dfrac{3x - 7y}{3} = \dfrac{2x+y-1}{5}$,
$8 - \dfrac{x-y}{5} = 6$.

35. $\dfrac{x+5}{y-3} = 3$,
$\dfrac{5x-4}{11} + \dfrac{4y-8}{4} = \dfrac{2x-5}{9} + 17$.

36. $(x+7)(y-3) + 6 = (y+7)(x+3) - 64$,
$4x - 11y = 0$.

37. $\dfrac{5}{x} + \dfrac{2}{y} = \dfrac{20}{y} - \dfrac{15}{x} = \dfrac{13}{16}$.

38. $\dfrac{42}{x} - \dfrac{24}{y} = 17$,
$\dfrac{24}{x} + \dfrac{42}{y} = 19$.

39. $\dfrac{16}{x} + \dfrac{y}{12} = 45$,
$\dfrac{14}{x} - \dfrac{y}{20} = 32$.

40. $\dfrac{3}{x} - \dfrac{2}{y} = \dfrac{1}{4}$,
$\dfrac{7}{x} + \dfrac{4}{y} = \dfrac{5}{3}$.

SIMULTANEOUS EQUATIONS OF THE FIRST DEGREE.

41. $\dfrac{1}{7x} + \dfrac{y}{9} = 11,$

$\dfrac{1}{9x} + \dfrac{y}{2} = 16.$

42. $3x - \dfrac{1}{3y} + \dfrac{1}{9} = 6,$

$\dfrac{5}{y} - \dfrac{x}{3} = 1.$

43. $\dfrac{x-m}{y-m} = \dfrac{m-n}{m+n},$

$\dfrac{x}{y} = \dfrac{m^3 - n^3}{m^3 + n^3}.$

44. $\dfrac{x+y+1}{x-y+1} = \dfrac{p+1}{p-1},$

$\dfrac{x+y+1}{x-y-1} = \dfrac{1+m}{1-m}.$

45. $\dfrac{x+1}{y+1} = \dfrac{m+n+p}{m-n+p},$

$\dfrac{x-1}{y-1} = \dfrac{m+n-p}{m-n-p}.$

46. $\dfrac{b}{x-a+c} = \dfrac{c}{y-a+b},$

$\dfrac{x+c}{a+b} = \dfrac{y+b}{a+c}.$

47. $\dfrac{x+y-1}{x-y+1} = m,$

$\dfrac{y-x+1}{x-y+1} = nm.$

48. $\dfrac{bx+ay}{a+b+2} = x + \dfrac{b-a-1}{b},$

$bx - a = ay - b.$

49. $x + y = mxy,$

$y - x = nxy.$

50. $my + x = pxy,$

$ny + x = qxy.$

51. $\dfrac{a}{b+y} = \dfrac{b}{a-x},$

$\dfrac{c}{d-x} = \dfrac{d}{c+y}.$

52. $\dfrac{x+y}{x-y} = \dfrac{a}{b-c},$

$\dfrac{x+c}{y+b} = \dfrac{a+b}{a+c}.$

53. $(m+n)x - (m-n)y = 4mn,$

$\dfrac{x}{m+n} + \dfrac{y}{m-n} = 2.$

54. $\dfrac{x}{q^2-1} - \dfrac{y}{p^2-1} = p^2 - q^2,$

$\dfrac{x}{p^2+1} + \dfrac{y}{q^2+1} = p^2 + q^2 - 2.$

148 SIMULTANEOUS EQUATIONS OF THREE UNKNOWNS.

SIMPLE SIMULTANEOUS EQUATIONS OF THREE UNKNOWNS.

188. If there are *three* unknowns their values may be found if *three independent* equations be given; for between one pair of the equations *one* of the unknowns can be eliminated, and between a different pair the *same* unknown can be eliminated, so that there will be two resulting equations from which to determine the remaining two unknowns.

Ex. 1.—Solve
$$2x - 3y + 4z = 4, \quad (1)$$
$$3x + 5y - 7z = 12, \quad (2)$$
$$5x - y - 8z = 5. \quad (3)$$

To eliminate z between (1) and (2):
Multiply (1) by 7 and (2) by 4, then
$$14x - 21y + 28z = 28, \quad (4)$$
$$12x + 20y - 28z = 48. \quad (5)$$
Adding together (4) and (5),
$$26x - y = 76. \quad (6)$$
Multiply (1) by 2 and add (3), then
$$9x - 7y = 13. \quad (7)$$
But $\qquad 26x - y = 76. \quad (6)$

∴ multiplying (6) by 7 and subtracting (7),
$$173x = 519, \therefore x = 3.$$
But $\qquad 26x - y = 76, \therefore 78 - y = 76, \therefore y = 2.$
Substituting values of x and y in (1), $z = 1$.

189. It is not always necessary to go through this process. Various methods of shortening the work will present themselves in special cases. Thus,

SIMULTANEOUS EQUATIONS OF THREE UNKNOWNS.

Solve
$$x+y=1, \quad (1)$$
$$y+z=9, \quad (2)$$
$$x+z=5. \quad (3)$$

Add together $2x+2y+2z=15$,

$$\therefore x+y+z=\frac{15}{2}, \quad (4)$$

But $x+y=1, \quad (1)$

Subtracting (1) from (4), $z=\frac{13}{2}$.

Similarly, by subtracting (2) and (3) in turn from (4) we get

$$x=\frac{3}{2} \text{ and } y=\frac{5}{2}.$$

In this example the student might add (1) and (2) together, and from the result subtract (3), in which case y would be found. Similarly, by adding (1) and (3) and subtracting (2) x would be found.

EXERCISE LIX.

Solve

1. $5x+3y-6z=4,$
   $3x-y+2z=8,$
   $x-2y+2z=2.$

2. $4x-5y+2z=6,$
   $2x+3y-z=20,$
   $7x-4y+3z=35.$

3. $x+y+z=6,$
   $5x+4y+3z=22,$
   $15x+10y+6z=53.$

4. $4x-3y+z=9,$
   $9x+y-5z=16,$
   $x-4y+3z=2.$

5. $y-x+z=-5,$
   $z-y-x=-25,$
   $x+y+z=35.$

6. $15y=24z-10x+41,$
   $15x=12y-16z+10,$
   $18x=7z+14y-13.$

7. $2x-3y=3,$
   $3y-4z=7,$
   $4z-5x=2.$

8. $7x-3y=30,$
   $9y-5z=34,$
   $x+y+z=33.$

9. $ax+by+cz=a,$
   $ax-by-cz=b,$
   $ax+cy+bz=c.$

10. $bz+cy=a,$
    $az+cx=b,$
    $ay+bx=c.$

150 SIMULTANEOUS EQUATIONS OF THREE UNKNOWNS.

11. $x + \dfrac{y}{2} + \dfrac{z}{3} = 6,$

    $y + \dfrac{z}{2} + \dfrac{x}{3} = -1,$

    $z + \dfrac{x}{2} + \dfrac{y}{3} = 17.$

12. $\dfrac{1}{2}z + \dfrac{2}{3}x + \dfrac{3}{5}y = 17,$

    $\dfrac{1}{3}z + \dfrac{3}{4}x + \dfrac{2}{3}y = 19,$

    $\dfrac{1}{4}z + \dfrac{5}{6}x + \dfrac{4}{5}y = 18.$

13. $\dfrac{3}{x} - \dfrac{4}{5y} + \dfrac{1}{z} = \dfrac{38}{5},$

    $\dfrac{1}{3x} + \dfrac{1}{2y} + \dfrac{2}{z} = \dfrac{61}{6},$

    $\dfrac{4}{5x} - \dfrac{1}{2y} + \dfrac{4}{z} = \dfrac{161}{10}.$

14. $\dfrac{1}{x} + \dfrac{2}{y} = 5,$

    $\dfrac{3}{y} - \dfrac{4}{z} = -6,$

    $\dfrac{3}{z} - \dfrac{4}{x} = 5.$

15. $\dfrac{1}{x} + \dfrac{1}{y} - \dfrac{1}{z} = a,$

    $\dfrac{1}{x} - \dfrac{1}{y} + \dfrac{1}{z} = b,$

    $\dfrac{1}{y} + \dfrac{1}{z} - \dfrac{1}{x} = c.$

16. $\dfrac{2}{x} + \dfrac{1}{y} - \dfrac{3}{z} = 0,$

    $\dfrac{3}{z} - \dfrac{2}{y} - 2 = 0,$

    $\dfrac{1}{x} + \dfrac{1}{z} - \dfrac{4}{3} = 0.$

17. $\dfrac{x+1}{y+1} = 2,$

    $\dfrac{y+2}{z+1} = 4,$

    $\dfrac{z+3}{x+3} = \dfrac{1}{2}.$

18. $\dfrac{3x+y}{z+1} = 2,$

    $\dfrac{3y+z}{x+1} = 2,$

    $\dfrac{3z+x}{y+1} = 2.$

19. $\dfrac{4}{x} - \dfrac{3}{y} = 1,$

    $\dfrac{2}{x} + \dfrac{3}{z} = 4,$

    $\dfrac{3}{y} - \dfrac{1}{z} = 0.$

20. $\dfrac{xy}{x+y} = \dfrac{1}{5},$

    $\dfrac{yz}{y+z} = \dfrac{1}{6},$

    $\dfrac{zx}{z+x} = \dfrac{1}{7}.$

SIMULTANEOUS EQUATIONS OF THREE UNKNOWNS. 151

21. $\dfrac{xy}{4y-3x}=20,$

   $\dfrac{xz}{2x-3z}=15,$

   $\dfrac{yz}{4y-5z}=12.$

22. $\dfrac{9}{x}-\dfrac{2}{y}=4,$

   $\dfrac{10}{z}-\dfrac{6}{x}=8,$

   $\dfrac{21}{y}+\dfrac{45}{2z}=12.$

23. $9xy=20(x+y),$
    $10xz=24(x+z),$
    $11yz=30(y+z).$

24. $xy=3(x+y),$
    $xz=8(x+z),$
    $7yz=24(y+z).$

25. $\dfrac{5xy}{6}=x+y,$

   $\dfrac{3xz}{10}=x+z,$

   $\dfrac{2yz}{15}=y+z.$

26. $\dfrac{x}{a}+\dfrac{y}{b}+\dfrac{z}{c}=1,$

   $\dfrac{x}{a}+\dfrac{y}{c}+\dfrac{z}{b}=1,$

   $\dfrac{x}{b}+\dfrac{y}{a}+\dfrac{z}{c}=1.$

27. $\dfrac{a}{x}+\dfrac{b}{y}+\dfrac{c}{z}=3,$

   $\dfrac{a}{x}-\dfrac{b}{y}+\dfrac{c}{z}=1,$

   $\dfrac{2a}{x}-\dfrac{b}{y}-\dfrac{c}{z}=0.$

28. $\dfrac{xy}{x+y}=a,$

   $\dfrac{xz}{x+z}=b,$

   $\dfrac{yz}{y+z}=c.$

29. $y+z-x=a,$
    $z+x-y=b,$
    $x+y-z=c.$

30. $2np=(p+m)x-(p-m)y,$
    $2mp=(m+n)y-(m-n)z,$
    $2mn=(n+p)z-(n-p)x.$

31. $(x+2)(2y+1)=(2x+9)y,$
    $(x-2)(3z+1)=(x+3)(3z-1),$
    $(y+1)(z+2)=(y+3)(z+1).$

32. $6x(y+z)=4y(z+x)=3z(x+y),$

   $\dfrac{1}{x}+\dfrac{1}{y}+\dfrac{1}{z}=9.$

33. $x+y+z=a+b-c,$
    $bx-cy+az=ay+bz-cx=ab-(a+b)c.$

PROBLEMS PRODUCING SIMPLE SIMULTANEOUS EQUATIONS.

190. It is often convenient and sometimes necessary to use more than one unknown quantity in solving problems. If we use two unknowns, x and y, we must have two *independent* equations resulting from the statement of the problem. If there are three unknowns employed then the problem must admit of three *independent* equations; and generally the problem must furnish as many independent equations as there are unknowns. If there be more equations than unknown numbers some of the equations are superfluous or contradictory, in other words, too much has been given; if there be less equations than unknown numbers then the problem is *indeterminate*, that is, more than one solution can be obtained.

Ex. 1.—The sum of two numbers divided by 2 gives as a quotient 24, and the difference between them divided by 2 gives as a quotient 17. What are the numbers?

Let $x =$ one number
and $y =$ the other number.

Then $$\frac{x+y}{2} = 24, \qquad (1)$$

also $$\frac{x-y}{2} = 17; \qquad (2)$$

$$\therefore x+y = 48 \qquad (3)$$

and $$x-y = 34. \qquad (4)$$

From (3) and (4) we find $x = 41$ and $y = 7$.

PROBLEMS PRODUCING SIMULTANEOUS EQUATIONS. 153

Ex. 2.—A certain fraction equals $\frac{1}{2}$ when 7 is added to its denominator, and equals 2 when 13 is added to its numerator. Find the fraction.

Let $x =$ numerator of fraction
and $y =$ denominator of fraction.

Therefore $\qquad \dfrac{x}{y+7} = \dfrac{1}{2},$ \hfill (1)

also $\qquad \dfrac{x+13}{y} = 2.$ \hfill (2)

Simplifying (1) and (2),

$$2x - y = 7 \qquad (3)$$
and $\qquad x - 2y = -13.$ \hfill (4)

Solving (3) and (4) by usual methods we get $x = 9$, $y = 11$.

Ex. 3.—The sum of the two digits of a number is 8, and if 36 be added to the number the digits will be interchanged. What is the number?

Let $x =$ right-hand digit of the number
and $y =$ left-hand " " "

Then, since y represents the digits in tens' place, $10y + x =$ the number.

If the digits are interchanged $10x + y =$ the new number.

$\qquad \therefore 36 + 10y + x = 10x + y,$ \hfill (1)
also $\qquad x + y = 8.$ \hfill (2)

From (1), $\qquad 9x - 9y = 36$
or $\qquad x - y = 4,$ \hfill (3)
but $\qquad x + y = 8.$ \hfill (2)

Therefore from (3) and (2) we obtain $x = 6$, $y = 2$. Hence the number is 26.

PROBLEMS PRODUCING SIMULTANEOUS EQUATIONS.

Ex. 4.—A crew which can pull at the rate of 12 miles an hour down the stream finds that it takes twice as long to come up the river as to go down. At what rate does the stream flow?

Let x = rate of rowing in still water
and y = " stream.

Then, since rate down stream is rate of rowing in still water + rate of stream, we have for equation

$$x + y = 12. \qquad (1)$$

But rate up stream is rate of rowing in still water − rate of stream; it is also 6 miles an hour.

$$\therefore\ x - y = 6. \qquad (2)$$

From (1) and (2) we find $x = 9$ and $y = 3$. Hence stream flows at the rate of 3 miles an hour.

Ex. 5.—*A* and *B* together earn \$40 in 6 days; *A* and *C* together earn \$54 in 9 days; *B* and *C* together earn \$80 in 15 days. What does each earn a day?

Let x = amount earned by *A* in 1 day,
" y = " " *B* "
" z = " " *C* "

Then from conditions of the problem,

$$6x + 6y = 40, \qquad (1)$$
$$9x + 9z = 54, \qquad (2)$$
$$15y + 15z = 80. \qquad (3)$$

Dividing (1) by 6, (2) by 9, and (3) by (15),

$$x + y = \frac{20}{3}, \qquad (4)$$
$$x + z = 6, \qquad (5)$$
$$y + z = \frac{16}{3}. \qquad (6)$$

PROBLEMS PRODUCING SIMULTANEOUS EQUATIONS. 155

Adding (4), (5) and (6) together,

$$2x + 2y + 2z = 18 \quad (7)$$
or $$x + y + z = 9. \quad (8)$$
But $$x + y = \frac{20}{3}, \quad (4)$$

$$\therefore z = \frac{7}{3}.$$

Similarly, by subtracting (5) and (6) in turn from (8) we get $y = 3$ and $x = \frac{11}{3}$. Hence A earns $3\frac{2}{3}$, B $3, and C $2\frac{1}{3}$.

Ex. 6.—A sum of money at simple interest amounted in 6 years to $26,000, and in 10 years to $30,000. Find the sum and the rate of interest.

Let x = sum of money
and y = rate of interest per dollar for 1 year.

Then $6y$ = interest on $1 for 6 years
and $10y =$ "   "   10   "
Also, $6xy =$ " x dollars for 6 years,
and $10xy =$ "   "   "   10   "
$\therefore x + 6xy = 26000$ (1), amount in 6 years,
and $x + 10xy = 30000$ (2),   "   10   " .

Subtracting (1) from (2),

$$4xy = 4000,$$
$$\therefore xy = 1000.$$
Hence $x + 6xy = x + 6000 = 26000,$
$$\therefore x = 20000.$$

Again, because $xy = 1000$
and $x = 20000,$
$$y = \frac{1000}{20000} = \frac{1}{20} = .05.$$

Hence sum of money is $20,000, and rate per cent. is 5.

EXERCISE LX.

1. Two numbers are such that three times the first plus five times the second equals 44, but three times the second plus six times the first equals 60. What are the numbers?

2. The sum of two numbers is 210, and their sum is to the first as 7:4. Find the numbers.

3. Find two numbers whose sum is 54, and whose sum and difference are in the ratio of 9:5.

4. A fraction is such that if 3 be added to each of its terms it equals $\frac{2}{3}$, and if 3 be subtracted from each of its terms it equals $\frac{1}{2}$. What is the fraction?

5. A number consists of two digits whose difference is 3. If the order of the digits be changed the number obtained will be $\frac{4}{7}$ of the first number. What is the first number?

6. A number consists of two digits whose sum is 8. If the order of the digits be changed the new number will be $\frac{3}{4}$ of the sum of the digits plus twice the first number. What is the first number?

7. The sum of two digits of which a number is composed is 9. If 3 be subtracted from each of the digits the result is $\frac{1}{2}$ of the first number diminished by 6. What is the first number?

8. The sum of two numbers is 26. The third of the first and three-fourths of the second are equal. What are the numbers?

9. If $28 are paid to 6 artificers and 2 laborers for a day's work, and if for another day's work the same sum is paid to 5 artificers and 4 laborers, what is a day's pay of an artificer and of a laborer?

10. Find two numbers whose sum and difference are as 5:1, and their sum and product as 5:8.

11. Two numbers are such that the smaller divided by the greater gives for a quotient $\frac{4}{11}$, and the greater divided by the smaller gives 2 for a quotient and 5 for a remainder. Find the number.

PROBLEMS PRODUCING SIMULTANEOUS EQUATIONS. 157

12. Find two numbers such that the first added to four times the second equals 29, and the second added to six times the first equals 36?

13. A servant receives $9 to buy 4 kilogrammes of butter and 7 of soap, and 15 cents ought to be returned; but she makes a mistake and buys 7 kilogrammes of butter and 4 of soap, and ought to receive 30 cents more money. What was the price of the butter and of the soap?

14. A man has two vessels, and for the two a single cover worth 90 cents. If he puts the cover upon the first vessel it will be worth $1\frac{1}{2}$ times as much as the other. If he puts the cover upon the second vessel it will be worth $1\frac{1}{5}$ times as much as the first. What is the value of each vessel?

15. Two persons each owe $1200. The first said to the second, "If you give me $\frac{3}{4}$ of what you have I shall have enough to pay my debt." The second replied, "If you give me $\frac{8}{9}$ of what your purse contains I can pay my debt." How much do they each have?

16. Two friends, A and B, in reviewing their libraries, said that $\frac{1}{3}$ of A's library plus $\frac{1}{4}$ of B's would be 780 vols.; but that $\frac{1}{6}$ of A's library added to $\frac{1}{3}$ of B's would be 20 vols. less. How many volumes did each have?

17. Two women buy velvet and silk. One buys $3\frac{1}{2}$ yards of velvet and $12\frac{3}{4}$ yards of silk; the other took $4\frac{1}{2}$ yards of velvet and 5 yards of silk. They each pay $63.80. How much per yard did each cost?

18. Three women go to market to sell eggs. If the first gives $\frac{1}{4}$ of hers and the third $\frac{1}{13}$ of hers to the second they will all have equal numbers. They all together carry 360. How many has each?

19. A number is composed of three figures whose sum is 17. The figure of the hundreds is double that of the units. When 396 is subtracted the order of the figures is reversed. What is the number?

20. Find three numbers such that the first plus ½ of the second, the second plus ⅓ of the third, and the third plus ¼ of the first, are each 1000.

21. A grocer bought tea at 60 cents a pound and coffee at 40 cents a pound to the amount altogether of $120. He sold the tea at 75 cents a pound and the coffee at 48 cents, and gained altogether $28. How many pounds of each did he buy?

22. A's money with ½ of B's would be ⅓ as much again as before; and if 2s. be taken from A's present sum and added to B's the latter amount will be ⅓ of the former. What had they each at first?

23. A cistern has three pipes, A, B and C. By A and B together it can be filled in 36 minutes, and emptied by C in 45 minutes; whereas if A and C were opened together it would be emptied in 1½ hours. In what time would it be filled by A, or by B, or by all opened together?

24. Two vessels, A and B, contain each a mixture of water and wine, A in the ratio of $2:3$, B in that of $3:7$. What quantity must be taken from each to form a mixture which shall consist of 5 gallons of water and 11 of wine?

25. A pound of tea and 3 pounds of sugar cost together 6s.; but if sugar were to rise 50 per cent. and tea 10 per cent. they would cost 7s. Find the prices.

26. A and B can reap a field of wheat in m days, B and C in n days, and A can do p times as much as C in the same time. In what time would the three reap it together?

27. If A's money were increased by ½ of B's it would amount to £54; and if B's present sum were trebled it would exceed three times the difference of their original sum by £6. What had each at first?

28. A vintner would mix wine at 10s. a gallon with another sort at 6s. a gallon to make 100 gallons to be sold at 7s. a gallon. How much of each sort must he take?

29. Find that number of two figures to which, if the number

PROBLEMS PRODUCING SIMULTANEOUS EQUATIONS. 159

formed by changing the places of the digits be added, the sum is 121, and if the same two numbers be subtracted the remainder is 9.

30. A man and his wife could drink a barrel of beer in 15 days. After drinking together 6 days the woman alone drank the remainder in 30 days. In what time would either alone drink it?

31. A farmer mixes barley at 2s. 4d. a bushel with rye at 3s. a bushel and wheat at 4s. a bushel, so that the whole is 100 bushels and worth 3s. 4d. a bushel. Had he put double as much rye and 10 bushels more wheat the whole would have been worth exactly the same per bushel. How much of each kind was there?

32. What fraction is that which, when its numerator is increased by 7, becomes equal to $\frac{2}{3}$, and when its denominator is increased by 10, equal to $\frac{1}{2}$?

33. A, B and C together possess £60; A, B and D together £72; A, C and D together £90; and B, C and D together £102. Required what each possesses.

34. Income and assessed taxes together amount to £30; but if income tax be increased 20 per cent. and assessed taxes diminished 25 per cent. the taxes will together amount to £32 2s. 6d. Required each.

35. A and B, with C pulling against them, would raise a weight in 5 hours; A and C, with B pulling against them, in 7; and B and C, with A pulling against them, in 8. Required the time which it would occupy each alone to raise it.

36. A person rows down a stream, which runs at the rate of 4 miles an hour, for a certain distance in 1 hour and 40 minutes. In returning it takes him 4 hours and 15 minutes to arrive at a point 3 miles short of his starting place. Find the distance he pulled down the stream and the rate of his pulling.

37. A person rows down a stream a distance of 20 miles and back again in 10 hours. He finds he can row 2 miles against the stream in the same time he can row 3 miles with it. Find

the time of his rowing down and of his rowing up the stream, also the rate of the stream.

38. A grocer mixed tea that cost him 42 cents a pound with tea that cost him 54 cents a pound. He had 30 pounds of the mixture, and by selling it at the rate of 60 cents a pound he gained as much as 10 pounds of the cheaper tea cost him. How many pounds of each did he put into the mixture?

39. A grocer mixes tea that cost him 90 cents a pound with tea that cost him 28 cents a pound. The cost of the mixture is $61.20. He sells the mixture at 50 cents a pound and gains $3.80. How many pounds of each did he put into the mixture?

40. A farmer has 28 bushels of barley worth 84 cents a bushel. With his barley he wishes to mix rye worth $1.08 a bushel and wheat worth $1.44 a bushel, so that the mixture may be 100 bushels and be worth $1.20 a bushel. How many bushels of rye and of wheat must he take?

41. A cistern has three pipes, A, B and C A and B will fill it in 1 hour and 10 minutes, A and C in 1 hour and 24 minutes, and B and C in 2 hours and 20 minutes. How long will it take each to fill it?

42. A piece of work can be completed by A, B and C together in 10 days, by A and B together in 12 days, by B and C if B work 15 days and C 30 days. How long will it take each alone to do the work?

43. A cistern has three pipes, A, B and C. A and B will fill it in a minutes, A and C in b minutes, and B and C in c minutes. How long will it take each alone to fill it?

44. A rectangular room having been measured it was observed that if it were 5 feet broader and 4 feet longer it would contain 116 square feet more; but if it were 4 feet broader and 5 feet longer it would contain 113 square feet more. Required its length and breadth.

45. If the sides of a rectangular field were each increased by 2 yards the area would be increased by 220 square yards; if the

length were increased and the breadth were diminished each by 5 yards the area would be diminished by 185 square yards. What is its area?

46. If a given rectangular floor had been 3 feet longer and 2 feet broader it would have contained 64 square feet more; but if it had been 2 feet longer and 3 feet broader it would have contained 68 square feet more. Find the length and breadth of the floor.

47. A cask, B, contains 12 gallons of wine and 4 gallons of water; another cask, C, contains 8 gallons of wine and 12 gallons of water. How many gallons must be drawn from each cask so as to produce by their mixture 7 gallons of wine and 7 gallons of water?

48. A cask, A, contains 12 gallons of wine and 18 gallons of water, and another cask, B, contains 9 gallons of wine and 3 gallons of water. How many gallons must be drawn from each cask so as to produce by their mixture 7 gallons of wine and 7 gallons of water?

CHAPTER XI.

SQUARE AND CUBE ROOT.

SQUARE ROOT.

191. The **Square Root** of an algebraical expression is one of the two *equal* factors of which the expression is composed.

Thus the square root of $16a^2$ is either $+4a$ or $-4a$; for $(+4a) \times (+4a) = 16a^2$, and $(-4a) \times (-4a) = 16a^2$.

From this example we see that the square root of an algebraical expression may be either *positive* or *negative*.

In this chapter the roots will be taken with *positive* signs.

192. First let us proceed to find the square root of a monomial.

Since $ab \times ab = a^2b^2$, the square root of a^2b^2 can be found by multiplying together the square roots of the different factors of a^2b^2.

Again, $a^2b^2c^2 = abc \times abc$, $\therefore$ the square root of $a^2b^2c^2$ is abc or the product of the square roots of the different factors, a^2, b^2, c^2.

Generally, then, to find the square root of a monomial, extract the square root of each factor and multiply together the quantities so obtained; the result will be the square root of the given quantity.

Ex. 1.—Extract square root of $81a^6b^8$.

$$\sqrt{81} = 9, \quad \sqrt{a^6} = a^3, \quad \sqrt{b^8} = b^4,$$

$$\therefore \sqrt{81a^6b^8} = 9a^3b^4.$$

Ex. 2.—Find $\sqrt{64a^4b^{10}c^2}$.

$$\sqrt{64} = 8, \quad \sqrt{a^4} = a^2, \quad \sqrt{b^{10}} = b^5, \quad \sqrt{c^2} = c,$$

$$\therefore \sqrt{64a^4b^{10}c^2} = 8a^2b^5c.$$

SQUARE ROOT.

193. To find the square root of a fraction.

$$\frac{a}{b} \times \frac{a}{b} = \frac{a^2}{b^2}, \quad \therefore \sqrt{\frac{a^2}{b^2}} = \frac{a}{b}.$$

Hence, to find the square root of a fraction:—
Find the square root of the numerator for a new numerator, and square root of the denominator for a new denominator; the new fraction thus obtained will be the result required.

Ex. 1. $\sqrt{\dfrac{4a^2}{81b^2}} = \dfrac{2a}{9b}.$

Ex. 2. $\sqrt{\dfrac{9a^2b^2}{16c^2d^2}} = \dfrac{3ab}{4cd}.$

Ex. 3. $\sqrt{\dfrac{25(a-b)^2}{36(a+b)^2}} = \dfrac{5(a-b)}{6(a+b)}.$

194. We now proceed to explain the method of extracting the square root of a multinomial.

The following mode of arranging the square of any expression should be carefully noticed:—

$$(a+b)^2 = a^2 + (2a+b)b, \qquad (1)$$
$$(a+b+c)^2 = a^2 + (2a+b)b + (2a+2b+c)c, \qquad (2)$$
$$(a+b+c+d)^2 = a^2 + (2a+b)b + (2a+2b+c)c + (2a+2b+2c+d)d, \quad (3)$$

and so on for the square of any number of terms.

From (1) it is seen that $a+b$ is the square root of $a^2+(2a+b)b$ or $a^2+2ab+b^2$.

To find the *first* term, a, of the root, it is necessary to extract the square root of a^2, the *first* term of $a^2+(2a+b)b$.

To find the *second* term, b, we subtract the square of a from $a^2+2ab+b^2$, and into the first term of the remainder, $2ab+b^2$ or $(2a+b)b$, divide $2a$ or double the *first* term of the root; the quotient will be b or *second* term required. We now add b to $2a$, and multiply the sum by b; this product subtracted from the remainder $2ab+b^2$ will leave no remainder, hence the root has been found.

195. The process of finding a square root consisting of two terms may now be stated in a general form:—

Arrange the terms of the given expression in the order of magnitude of indices of one of the letters involved; then take the square root of the first term, and set down the result as the first term of the root. Subtract its square from the given expression and bring down the remainder. Double the first term of the root and set down the result as the first term of a trial-divisor; divide the first term of the remainder by the first term of this divisor, and add the result to the first term of the root and also to the first term of the divisor. Multiply the complete divisor by the second term of the root, and subtract the result from the first remainder; then if there be no remainder the root has been found.

Ex. 1.—Extract the square root of $4a^2 + 12ab + 9b^2$.

$$\begin{array}{r} 4a^2 + 12ab + 9b^2 \ (2a + 3b \\ 4a^2 \\ \hline 4a + 3b)\ \ +12ab + 9b^2 \\ +12ab + 9b^2 \\ \hline \end{array}$$

EXPLANATION.

The square root of $4a^2$ is $2a$. Squaring $2a$ and subtracting the result from $4a^2 + 12ab + 9b^2$ the first remainder is $12ab + 9b^2$. Doubling $2a$ for a trial-divisor we find that $4a$ will divide $12ab$ and give the quotient $3b$, $\therefore$ second term of root is $3b$. Adding $3b$ to $4a$ and multiplying the sum by $3b$ we get $12ab + 9b^2$. As there is no remainder $2a + 3b$ is the required square root.

Ex. 2.—Find the square root of $a^2b^2 + 162ab + 6561$.

$$\begin{array}{r} a^2b^2 + 162ab + 6561\ (ab + 81 \\ a^2b^2 \\ \hline 2ab + 81)\ \ +162ab + 6561 \\ +162ab + 6561 \\ \hline \end{array}$$

Remainder = 0, $\therefore$ square root = $ab + 81$.

SQUARE ROOT.

196. Again, from (2) it is seen that $a+b+c$ is the square root of $a^2+(2a+b)b+(2a+2b+c)c$ or $a^2+b^2+c^2+2ab+2bc+2ca$.

To find the root $a+b+c$ from $a^2+(2a+b)b+(2a+2b+c)c$, we find first $(a+b)$ as in preceding case, and then treat $(a+b)$ as one term, and proceed as before. For after finding $a+b$ there will be a remainder, $(2a+2b+c)c$ or $2ac+2bc+c^2$. Doubling $a+b$ and dividing the first term of the product into $2ac$ we obtain the quotient, c, the third term required. Adding c to $2(a+b)$ for a complete divisor, and multiplying the sum by c, we find there is no remainder, and $\therefore$ $a+b+c$ is the root required.

Similarly the square root $a+b+c+d$ of

$$a^2+(2a+b)b+(2a+2b+c)c+(2a+2b+2c+d)d$$

can be found.

197. The method of the extraction of the square root of an expression of more than three terms can best be made clear to the beginner by a few examples.

Ex. 1.—Extract the square root of $a^4+4a^3+2a^2-4a+1$.

$$\begin{array}{l} a^4+4a^3+2a^2-4a+1\ \left(a^2+2a-1\ \textit{Ans.}\right.\\ \underline{a^4} \\ 2a^2+2a\,)\ \ +4a^3+2a^2-4a+1 \\ \ \ \ \ \ \ \ \ \ \ \ \ \underline{+4a^3+4a^2} \\ 2a^2+4a-1\,)\ -2a^2-4a+1 \\ \ \ \ \ \ \ \ \ \ \ \ \ \ \ \ \ \underline{-2a^2-4a+1} \end{array}$$

Ex. 2.—Extract the square root of

$$9x^6-12x^3y^2+16x^2y^4-24x^4y^2+4y^6+16xy^5.$$

Re-arrange as follows:—

$$\begin{array}{l} 9x^6-24x^4y^2-12x^3y^3+16x^2y^4+16xy^5+4y^6 \\ \underline{9x^6} \\ 6x^3-4xy^2)\ -24x^4y^2-12x^3y^3+16x^2y^4\ \ \left(3x^3-4xy^2-2y^3\ \textit{Ans.}\right. \\ \ \ \ \ \ \ \ \ \ \ \ \ \ \underline{-24x^4y^2+16x^2y^4} \\ 6x^3-8xy^2-2y^3)\ -12x^3y^3+16xy^5\ +4y^6 \\ \underline{-12x^3y^3+16xy^5\ +4y^6} \end{array}$$

EXERCISE LXI.

Extract the square root of

1. $4a^6b^2$, $16a^{12}b^4c^8$, $25a^2b^4c^2$, $16x^4z^2$, $169a^{10}b^8c^{12}$.

2. $\dfrac{25a^4b^4}{121x^8y^{10}}$, $\dfrac{256x^{12}}{289y^4}$, $\dfrac{625a^2}{324\,b^2}$, $\dfrac{49a^2b^4c^6}{64c^6d^4}$.

3. $x^2+12x+36$, $x^2-8x+16$, $4a^2x^2+4abx+b^2$.
4. $x^4+2x^3+3x^2+2x+1$, $4x^4-4x^3-3x^2+2x+1$.
5. $x^4-10ax^3+33a^2x^2-40a^3x+16a^4$.
6. $49x^4+56x^3y+30x^2y^2+8xy^3+y^4$.
7. $x^4-2x^3y+3x^2y^2-2xy^3+y^4$.
8. $4a^6-12a^5x+5a^4x^2+6a^3x^3+a^2x^4$.
9. $16x^4-16abx^2+16b^2x^2+4a^2b^2-8ab^3+4b^4$.
10. $4x^4+9-30x-20x^3+37x^2$.
11. $x^6+25x^2+10x^4-4x^5-20x^3+16-24x$.
12. $4a^8+16c^8+16a^6c^2-32a^2c^6$.
13. $4-12a-11a^4+5a^2-4a^5+4a^6+14a^3$.
14. $x^6+8x^4y^2-4x^5y-4xy^5+8x^2y^4-10x^3y^3+y^6$
15. $25x^6-31x^4y^2+34x^3y^3-30x^5y+y^6-8xy^5+10x^2y^4$.
16. $4y^4-12y^3z+25y^2z^2-24yz^3+16z^4$.
17. $a^2+4ab+4b^2+9c^2+6ac+12bc$.
18. $a^6+2a^5b+3a^4b^2+4a^3b^3+3a^2b^4+2ab^5+b^6$.
19. $9-24x+58x^2-116x^3+129x^4-140x^5+100x^6$.
20. $9a^2-12ab+24ac-16bc+4b^2+16c^2$.
21. $25x^4y^2-30x^3y^3+29x^2y^4-12xy^5+4y^6$.
22. $4y^4x^2-12y^3x^3+17y^2x^4-12yx^5+4x^6$.
23. $25x^2-20xy+4y^2+9z^2-12yz+30xz$.
24. $4x^2(x^2-y)+y^3(y-2)+y^2(4x^2+1)$.
25. $m^8-4m^7+10m^6-20m^5-44m^3+35m^4+46m^2-40m+25$.
26. $4a^2b^2+(a^2+b^2)^2+4ab(a^2+b^2)$.

SQUARE ROOT.

198. When fractional terms occur in an expression its square root can be obtained in the same manner as when the terms are not fractional, but the beginner will require to exercise more care.

Ex.—Find the square root of $\dfrac{a^4}{9} - \dfrac{a^3x}{2} + \dfrac{43}{48}a^2x^2 - \dfrac{3}{4}ax^3 + \dfrac{x^4}{4}$.

$$\dfrac{a^4}{9} - \dfrac{a^3x}{2} + \dfrac{43}{48}a^2x^2 - \dfrac{3}{4}ax^3 + \dfrac{x^4}{4} \left(\dfrac{a^2}{3} - \dfrac{3}{4}ax + \dfrac{x^2}{2} \right.$$

$$\dfrac{a^4}{9}$$

$$\dfrac{2a^2}{3} - \dfrac{3}{4}ax \bigg) \; -\dfrac{a^3x}{2} + \dfrac{43}{48}a^2x^2$$

$$-\dfrac{a^3x}{2} + \dfrac{9}{16}a^2x^2$$

$$\dfrac{2a^2}{3} - \dfrac{3}{2}ax + \dfrac{x^2}{2} \bigg) \; + \dfrac{1}{3}a^2x^2 - \dfrac{3}{4}ax^3 + \dfrac{x^4}{4}$$

$$+ \dfrac{1}{3}a^2x^2 - \dfrac{3}{4}ax^3 + \dfrac{x^4}{4}$$

EXERCISE LXII.

Find the square root of

1. $x^4 - x^3y - \dfrac{7}{4}x^2y^2 + xy^3 + y^4$.

2. $x^4 - 4x^3y + 6x^2y^2 - 6xy^3 + 5y^4 - \dfrac{2y^5}{x} + \dfrac{y^6}{x^2}$.

3. $1 + \dfrac{4}{x} + \dfrac{10}{x^2} + \dfrac{20}{x^3} + \dfrac{25}{x^4} + \dfrac{24}{x^5} + \dfrac{16}{x^6}$.

4. $x^4 + x^3 - \dfrac{5x^2}{12} - \dfrac{x}{3} + \dfrac{1}{9}$.

5. $\dfrac{a^2}{b^2} - \dfrac{2a}{b} + 3 - \dfrac{2b}{a} + \dfrac{b^2}{a^2}$.

6. $4a^2 - 12ab + ab^2 + 9b^3 - \dfrac{3b^3}{2} + \dfrac{b^4}{16}.$

7. $x^4 + 8x^2 + 24 + \dfrac{16}{x^2} + \dfrac{32}{x^3}.$

8. $\dfrac{1}{x^2} + \dfrac{4}{y^2} + \dfrac{9}{z^2} - \dfrac{4}{xy} + \dfrac{6}{xz} - \dfrac{12}{yz}.$

9. $\dfrac{4x^2}{z^2} + \dfrac{z^2}{x^2} + \dfrac{9y^2}{z^2} + 4 - \dfrac{6y}{x} - \dfrac{12xy}{z^2}.$

10. $\dfrac{a^2}{9} + \dfrac{b^2}{16} + \dfrac{c^2}{25} + \dfrac{d^2}{4} - \dfrac{ab}{6} + \dfrac{2ac}{15} - \dfrac{ad}{3} - \dfrac{bc}{10} + \dfrac{bd}{4} - \dfrac{cd}{5}.$

199. The following are examples of a more difficult character. No fixed rule can be given for facilitating the extraction of the roots; much must be left to the teacher and the ingenuity of the pupil. Sometimes it will be necessary to remove the brackets and then proceed in the usual manner; sometimes the root can be obtained by inspection and sometimes by factoring.

Ex. 1.—Extract the square root of
$$16a^3(a+b+c) + 4abc(b+c) + 4a^2(b^2+c^2) + 16a^2bc + b^2c^2.$$

Remove brackets and arrange according to powers of a.

$$16a^4 + 16a^3(b+c) + 4a^2(b^2+c^2+4bc) + 4abc(b+c) + b^2c^2$$

$8a^2 + 2a(b+c) \,)\, \overline{16a^3(b+c) + 4a^2(b^2+c^2+4bc)} \quad \left(4a^2 + 2a(b+c) + bc\right.$

$16a^3(b+c) + 4a^2(b^2+c^2+2bc)$

$8a^2 + 4a(b+c) + bc \,)\, \overline{+ 4a^2(2bc) + 4abc(b+c) + b^2c^2}$

$+ 4a^2(2bc) + 4abc(b+c) + b^2c^2$

Ex. 2.—Extract the square root of
$$a^4 + b^4 + c^4 + d^4 + 2a^2(b^2+d^2) + 2b^2(c^2+d^2) + 2c^2(a^2+d^2).$$

Re-arranging we get the equivalent expression,
$$a^4 + 2a^2(b^2+c^2+d^2) + b^4 + 2b^2(c^2+d^2) + c^4 + 2c^2d^2 + d^4,$$
which is evidently the square of $a^2 + b^2 + c^2 + d^2$.

SQUARE ROOT.

Ex. 3.—Extract the square root of
$$(x^2-yz)^3+(y^2-zx)^3+(z^2-xy)^3-3(x^2-yz)(y^2-zx)(z^2-xy).$$

This expression may be arranged as follows:—

$(x^2-yz)^3-(x^2-yz)(y^2-zx)(z^2-xy)+(y^2-zx)^3-(x^2-yz)(y^2-zx)(z^2-xy)$
$\qquad\qquad\qquad\qquad +(z^2-xy)^3-(x^2-yz)(y^2-zx)(z^2-xy)$

$= (x^2-yz)\{(x^2-yz)^2-(y^2-zx)(z^2-xy)\} + \text{anal.} + \text{anal.}$
$= (x^2-yz)\{x^4-2x^2yz+y^2z^2-y^2z^2+xy^3+xz^3-x^2yz\} + \text{anal.} + \text{anal}$
$= (x^2-yz)\{x^4+xy^3+xz^3-3x^2yz\} + \text{anal.} + \text{anal.}$
$= x(x^2-yz)(x^3+y^3+z^3-3xyz) + \text{anal.} + \text{anal.}$
$= (x^3+y^3+z^3-3xyz)\{x(x^2-yz)+y(y^2-zx)+z(z^2-xy)\}$
$= (x^3+y^3+z^3-3xyz)(x^3+y^3+z^3-3xyz)$
$= (x^3+y^3+z^3-3xyz)^2.$

∴ Root required is $x^3+y^3+z^3-3xyz.$

EXERCISE LXIII.

Extract the square root of

1. $3(3a^2-2ab+b^2)(a^2+3b^2)+b^2(a+4b)^2.$
2. $a^2(a-5b)(a-b)+b^2(3a-b)^2-3a^2b^2.$
3. $(a-b)^4+2(a^4+b^4)-2(a^2+b^2)(a-b)^2.$
4. $ax(ax+1)(ax+2)(ax+3)+1.$ Hence show that the product of any four consecutive numbers plus one is a perfect square.
5. $a^2(a^2+b^2-c^2)+2(a+b)(b+c)ac+2a^2(ab+ac+bc)+b^2c^2.$
6. $(a^2+b^2+c^2)^3+2(ab+bc+ca)^3-3(a^2+b^2+c^2)(ab+ac+bc)^2.$
7. $4\{(a^2-b^2)cd+ab(c^2-d^2)\}^2+\{(a^2-b^2)(c^2-d^2)-4abcd\}^2.$
8. $a^4+b^4+c^4+d^4-2a^2(b^2+d^2)-2b^2(c^2-d^2)+2c^2(a^2-d^2).$
9. $(ab+ac+bc)^2-4abc(a+c).$
10. $\left(x+\dfrac{1}{x}\right)^2-4\left(x-\dfrac{1}{x}\right).$
11. $x^2(x^2+y^2+z^2)+y^2z^2+2x(y+z)(yz-x^2).$

CUBE ROOT.

200. The **Cube Root** of an algebraical expression is one of the three *equal* factors of which the expression is composed.

Thus the cube root of a^6 is a^2, for $a^2 \times a^2 \times a^2 = a^6$. The cube root of $-a^6$ is $-a^2$, for $(-a^2) \times (-a^2) \times (-a^2) = -a^6$. Hence we see that the cube root of a *positive* quantity is *positive*, and the cube root of a *negative* quantity is *negative*.

201. We find the cube root of a monomial by extracting the cube root of its different factors and then multiplying the different quantities so obtained together. The product is the cube root required.

202. To find the cube root of a fraction, extract the cube root of the numerator, and divide the result by the cube root of the denominator. The resulting fraction is the root required.

Ex. 1.—Find the cube root of $27a^3b^6c^9$.

$$\sqrt[3]{27a^3b^6c^9} = 3ab^2c^3.$$

Ex. 2.—Find the cube root of $\dfrac{8a^3b^6c^{12}}{27c^3d^{12}e^{15}}$.

$$\sqrt[3]{\dfrac{8a^3b^6c^{12}}{27c^3d^{12}e^{15}}} = \dfrac{2ab^2c^4}{3cd^4e^5}.$$

203. To find the cube root of a polynomial is a more tedious process. Let it be required to find the cube root of $a^3 + 3a^2b + 3ab^2 + b^3$. Now, we know that $(a+b)^3 = a^3 + 3a^2b + 3ab^2 + b^3$ or its equivalent, $a^3 + (3a^2 + 3ab + b^2)b$. (1) Therefore $a+b$ is the cube root of (1). To find a, or first term of cube root, we extract the cube root of first term, a^3. Subtracting a^3 from the given expression there remains $3a^2b + 3ab^2 + b^3$ or $b(3a^2 + 3ab + b^2)$. To find second term, b, of the cube root, we must square a, multiply the result by 3, and divide the product into the first term of the remainder, $3a^2b$. To obtain the complete divisor, $3a^2+3ab+b^2$, multiply the product of first and second terms of the root by 3, and add the result to three times the square of the first term of the root; then add the square of the second term of the root to the previous sum. Multiply the complete divisor now obtained by the second term of the root, and subtract the product from the first remainder. If the second remainder is zero the expression we have obtained is the cube root.

Ex. 1.—Find the cube root of $8x^3+36x^2y+54xy^2+27y^3$.

$$\begin{array}{r|l} 8x^3+36x^2y+54xy^2+27y^3 & (2x+3y \text{ } Ans. \\ \underline{8x^3} & \\ 3 \times (2x)^2 = 12x^2 \quad & +36x^2y+54xy^2+27y^3 \\ 3 \times 2x \times 3y = 18xy \quad & +36x^2y+54xy^2+27y^3 \\ (3y)^2 = 9y^2 & \\ \hline 12x^2+18xy+9y^2 & \end{array}$$

Here $2x$ is the first term, $12x^2$ the trial-divisor, and $12x^2+18xy+9y^2$ the complete divisor.

204. To find the cube root of a quantity whose root consists of more than two terms.

Since $\{(a+b)+c\}^3 = (a+b)^3 + 3(a+b)^2c + 3(a+b)c^2 + c^3$
$= (a+b)^3 + \{3(a+b)^2 + 3(a+b)c + c^2\}c$,

we see that the third term, c, can be found by the rule employed in finding the second term, that is, we first find $a+b$ and then treat $a+b$ as one term to find c. The second trial-divisor is now $3(a+b)^2$, and the complete divisor $3(a+b)^2+3(a+b)c+c^2$. The divisor, $3(a+b)^2+3(a+b)c+c^2$, is multiplied by the third term, c, and the product subtracted from the second remainder, $3(a+b)^2c+3(a+b)c^2+c^3$. There is now no remainder, hence the cube root has been found.

205. Similarly, to find a cube root consisting of four terms, $a+b+c+d$, we first find $a+b+c$, then treat $(a+b+c)$ as one term to find the fourth term, d. The student must bear in mind that the method employed to find a cube root of three, four, etc., terms is exactly the same as that employed in finding the first two terms.

Ex. 2.—Find the cube root of $x^6 - 3x^5 + 5x^3 - 3x - 1$.

$$
\begin{array}{r|l}
 & x^6 - 3x^5 + 5x^3 - 3x - 1\ (x^2 - x - 1 \\
 & \underline{x^6} \\
3x^4 - 3x^3 + x^2)\ & -3x^5 + 5x^3 - 3x - 1 \\
 & \underline{-3x^5 + 3x^4 - x^3} \\
3(x^2-x)^2 - 3(x^2-x)(-1) + (-1)^2\ & -3x^4 + 6x^3 - 3x - 1 \\
= 3x^4 - 6x^3 + 3x + 1\ & -3x^4 + 6x^3 - 3x - 1
\end{array}
$$

In this example, to find *second* trial-divisor, we treated $x^2 - x$ as one term, and then proceeded to find third term in precisely the same way as we found the second term.

EXERCISE LXIV.

Find the cube root of

1. $x^3 + 6x^2y + 12xy^2 + 8y^3$.
2. $x^3 + 12x^2 + 48x + 64$.
3. $a^3 - 9a^2 + 27a - 27$.
4. $x^6 - 3ax^5 + 5a^3x^3 - 3a^5x - a^6$.
5. $x^6 + 3x^5 + 6x^4 + 7x^3 + 6x^2 + 3x + 1$.
6. $x^6 - 6x^5 + 15x^4 - 20x^3 + 15x^2 - 6x + 1$.
7. $8a^6 - 36a^4b^2 + 54a^2b^4 - 27b^6$.
8. $a^3 - b^3 + c^3 - 3a^2b + 3a^2c + 3b^2a + 3b^2c + 3c^2a - 3c^2b - 6abc$.
9. $8x^6 - 36x^5 + 66x^4 - 63x^3 + 33x^2 - 9x + 1$.
10. $1 - 9x + 39x^2 - 99x^3 + 156x^4 - 144x^5 + 64x^6$.
11. $64x^6 + 192x^5 + 144x^4 - 32x^3 - 36x^2 + 12x - 1$.

12. $a^6 + 9a^5b - 135a^3b^3 + 729ab^5 - 729b^6$.

13. $c^6 - 12bc^5 + 60b^2c^4 - 160b^3c^3 + 240b^4c^2 - 192b^5c + 64b^6$.

14. $x^6 - \dfrac{8}{x^3} + 12 - 6x^3$.

15. $\dfrac{a^3}{8} - \dfrac{8}{27a^6} + \dfrac{2}{3a^3} - \dfrac{1}{2}$.

16. $x^3 + \dfrac{1}{x^3} + 3\left(x + \dfrac{1}{x}\right)$.

17. $\dfrac{a^3}{b^3} - \dfrac{b^3}{a^3} - 3\left(\dfrac{a^2}{b^2} + \dfrac{b^2}{a^2}\right) + 5$.

206. The square root of a^4 is a^2, and the square root of a^2 is a; therefore the fourth root, a, of the given expression, a^4, can be found by extracting the square root of the expression and then the square root of the result.

207. To find the sixth root of an expression we notice that a, the sixth root of a^6, is the cube root of a^2, which is the square root of a^6. Therefore we can find the sixth root by taking the square root of the expression and then the cube root of the result, or *vice versa*.

CHAPTER XII.

THEORY OF INDICES.

208. In Art. 10 a meaning has been attached to the expression a^n when n is a *positive integer*. We have now to *assign* meanings to such expressions as a^0, $a^{\frac{p}{q}}$, a^{-m}, for it is evident that the definition in Art. 10 is not applicable to quantities with such indices as 0, $\frac{p}{q}$, $-m$. Before, however, we proceed to interpret such symbols it is necessary to prove what are known as "Index Laws."

209. If m and n are positive integers,

(1) $a^m \times a^n = a^{m+n}$;

(2) $(a^m)^n = a^{mn}$;

(3) $(ab)^n = a^n b^n$;

(4) $\left(\dfrac{a}{b}\right)^n = \dfrac{a^n}{b^n}$.

(1), (2), (3) and (4) are known as "Index Laws."

210. To prove the Index Laws:—

I. $a^m \times a^n = (a \times a \times a \ldots$ to m factors$) \times (a \times a \ldots$ to n factors$)$
$= a \times a \times a \ldots$ to $(m+n)$ times factors
$= a^{m+n}$ (by definition).

II. $(a^m)^n = a^m \times a^m \times a^m \ldots$ to n factors
$= a^{m+m+m\ldots \text{ to } n \text{ terms}}$ (From I.)
$= a^{mn}$.

III. $(ab)^n = ab \times ab \times ab \ldots$ to n factors
$= (a \times a \times a \ldots$ to n factors$) \times (b \times b \ldots$ to n factors$)$
$= a^n \times b^n = a^n b^n$.

THEORY OF INDICES.

IV. $\left(\dfrac{a}{b}\right)^n = \dfrac{a}{b} \times \dfrac{a}{b} \times \dfrac{a}{b} \dots$ to n factors

$= \dfrac{a \times a \times a \dots \text{to } n \text{ factors}}{b \times b \times b \dots \text{to } n \text{ factors}}$

$= \dfrac{a^n}{b^n}$.

211. We proceed now to assign meanings to a^0, $a^{\frac{p}{q}}$ and a^{-m}. Assuming that the laws which are proved to be true when the indices are *positive integers* to hold good when the indices are *zero, negative* or *fractional*, we arrive at the following results:—

(a) $a^0 = 1$.

From (1), $a^m \times a^0 = a^{m+0} = a^m$;

$\therefore a^m \times a^0 = a^m$.

Dividing by a^m, $a^0 = \dfrac{a^m}{a^m} = 1$.

(b) $a^{-m} = \dfrac{1}{a^m}$.

From (1), $a^m \times a^{-m} = a^{m-m} = a^0$.

But, from (a), $a^0 = 1$,

$\therefore a^m \times a^{-m} = 1$.

Dividing by a^m, $a^{-m} = \dfrac{1}{a^m}$.

(c) $a^{\frac{p}{q}} = \sqrt[q]{a^p}$.

It has been proved that $(a^m)^n = a^{mn}$ when m and n are positive integers; and as we assume the same law to hold good for fractional indices,

$\therefore (a^{\frac{p}{q}})^q = a^{\frac{p}{q} \times q} = a^p$.

$\therefore (a^{\frac{p}{q}})^q = a^p$.

Extracting the qth root of each side,

$\therefore a^{\frac{p}{q}} = \sqrt[q]{a^p}$.

THEORY OF INDICES.

Hence the equivalent of any expression with *zero* for an index is unity; with a *negative* index, unity divided by the same expression with a corresponding *positive* index; and with a *fractional* index, the expression raised to a power indicated by the *numerator* of the fraction, and then the root extracted indicated by the *denominator*. For instance,

$$2^0 = 1, \quad 3^{-2} = \frac{1}{3^2} = \frac{1}{9}, \quad a^{\frac{1}{3}} = \sqrt[3]{a}, \quad a^{\frac{2}{3}} = \sqrt[3]{a^2}, \quad 2^{\frac{5}{2}} = \sqrt{2^5} = \sqrt{32}.$$

212. Again, since $\quad a^m \times a^n = a^{m+n}$,

$$\therefore a^{m-n} \times a^n = a^{m-n+n} = a^m;$$

Dividing by a^n, $\quad a^{m-n} = \dfrac{a^m}{a^n}$, or conversely,

$$\frac{a^m}{a^n} = a^{m-n}.$$

213. Prove that $a^{\frac{1}{n}} \times b^{\frac{1}{n}} = (ab)^{\frac{1}{n}}$.

$$(a^{\frac{1}{n}} \times b^{\frac{1}{n}})^n = (a^{\frac{1}{n}})^n \times (b^{\frac{1}{n}})^n \qquad \text{Art. 209 (3)}$$
$$= a \times b = ab;$$

$$\therefore (a^{\frac{1}{n}} \times b^{\frac{1}{n}})^n = ab.$$

Extracting the nth root of each side,

$$a^{\frac{1}{n}} \times b^{\frac{1}{n}} = (ab)^{\frac{1}{n}}.$$

This result is of great use in surds. For example, Prove $\sqrt{2} \times \sqrt{3} = \sqrt{6}$.

$$\sqrt{2} = 2^{\frac{1}{2}}, \quad \sqrt{3} = 3^{\frac{1}{2}};$$

$$\therefore \sqrt{2} \times \sqrt{3} = 2^{\frac{1}{2}} \times 3^{\frac{1}{2}} = 6^{\frac{1}{2}} = \sqrt{6}.$$

Again, prove $\sqrt[3]{3} \times \sqrt[3]{4} = \sqrt[3]{12}$.

$$\sqrt[3]{3} = 3^{\frac{1}{3}}, \quad \sqrt[3]{4} = 4^{\frac{1}{3}};$$

$$\therefore \sqrt[3]{3} \times \sqrt[3]{4} = 3^{\frac{1}{3}} \times 4^{\frac{1}{3}} = 12^{\frac{1}{3}} = \sqrt[3]{12}.$$

THEORY OF INDICES. 177

Also, from the above result, we can find the value of such a product as $\sqrt[3]{3} \times \sqrt[3]{2}$.

For
$$\sqrt[3]{3} = 3^{\frac{1}{3}} = 3^{\frac{2}{6}} = (3^2)^{\frac{1}{6}} = 9^{\frac{1}{6}}.$$

and
$$\sqrt[3]{2} = 2^{\frac{1}{3}} = 2^{\frac{2}{6}} = (2^3)^{\frac{1}{6}} = (8)^{\frac{1}{6}};$$

$$\therefore \sqrt[3]{3} \times \sqrt[3]{2} = 9^{\frac{1}{6}} \times 8^{\frac{1}{6}} = (9 \times 8)^{\frac{1}{6}}$$

$$= (72)^{\frac{1}{6}} = \sqrt[6]{72}.$$

Ex. 1.—Find the value of $x^{a+b-c} \times x^{b+c-a} \times x^{c+a-b}$.

$x^{a+b-c} \times x^{b+c-a} \times x^{c+a-b} = x^{a+b-c+b+c-a+c+a-b} = x^{a+b+c}$.

Ex. 2.—Find the value of $x^{a-b} \times x^{b-c} \times x^{c-a}$.

$x^{a-b} \times x^{b-c} \times x^{c-a} = x^{a-b+b-c+c-a} = x^0 = 1$, from (a).

Ex. 3.—Simplify $ab^{\frac{1}{2}}c \times a^{-\frac{1}{2}}bc^{\frac{1}{3}}$.

$$ab^{\frac{1}{2}}c \times a^{-\frac{1}{2}}bc^{\frac{1}{3}} = a^{1-\frac{1}{2}}b^{\frac{1}{2}+1}c^{1+\frac{1}{3}}$$

$$= a^{\frac{1}{2}}b^{\frac{3}{2}}c^{\frac{4}{3}}.$$

Ex. 4.—Simplify $\left(\dfrac{256}{625}\right)^{-\frac{3}{4}}$.

$$\left(\dfrac{256}{625}\right)^{-\frac{3}{4}} = \left(\dfrac{625}{256}\right)^{\frac{3}{4}}$$

$$= \left\{\left(\dfrac{625}{256}\right)^{\frac{1}{4}}\right\}^3 = \left(\dfrac{5}{4}\right)^3 = \dfrac{125}{64}.$$

The student will observe that we have extracted the fourth root of $\dfrac{625}{256}$ before cubing it. We might have raised $\dfrac{625}{256}$ to the third power and then extracted the fourth root, but the largeness of the numbers involved would have made the process cumbrous.

THEORY OF INDICES.

Ex. 5.—Multiply $x^{\frac{2}{3}} - 2x^{\frac{1}{3}} + 1$ by $x^{\frac{1}{3}} - 1$.

From observation we see that
$$x^{\frac{2}{3}} - 2x^{\frac{1}{3}} + 1 = (x^{\frac{1}{3}} - 1)^2;$$
$$\therefore (x^{\frac{2}{3}} - 2x^{\frac{1}{3}} + 1) \times (x^{\frac{1}{3}} - 1) = (x^{\frac{1}{3}} - 1)^3$$
$$= (x^{\frac{1}{3}})^3 - 3(x^{\frac{1}{3}})^2 + 3(x^{\frac{1}{3}}) - 1$$
$$= x - 3x^{\frac{2}{3}} + 3x^{\frac{1}{3}} - 1.$$

Otherwise,

$$x^{\frac{2}{3}} - 2x^{\frac{1}{3}} + 1$$
$$\underline{x^{\frac{1}{3}} - 1}$$
$$x\ \ - 2x^{\frac{2}{3}} +\ \ x^{\frac{1}{3}}$$
$$\underline{\ \ -\ \ x^{\frac{2}{3}} + 2x^{\frac{1}{3}} - 1}$$
$$x\ \ - 3x^{\frac{2}{3}} + 3x^{\frac{1}{3}} - 1$$

Ex. 6.—Divide $x^2y^{-2} + 2 + x^{-2}y^2$ by $xy^{-1} + x^{-1}y$.

$$xy^{-1} + x^{-1}y \,)\ x^2y^{-2} + 2 + x^{-2}y^2\ (xy^{-1} + x^{-1}y$$
$$\underline{x^2y^{-2} + 1 +\phantom{x^{-2}y^2}}$$
$$1 + x^{-2}y^2$$
$$\underline{1 + x^{-2}y^2}$$

We might have performed this division by first changing terms with negative indices into equivalent ones with positive indices, and then proceeding as in ordinary fractions.

Thus, $\quad x^2y^{-2} + 2 + x^{-2}y^2 = \dfrac{x^2}{y^2} + 2 + \dfrac{y^2}{x^2}$

and $\quad xy^{-1} + x^{-1}y = \dfrac{x}{y} + \dfrac{y}{x};$

$\therefore x^2y^{-2} + 2 + x^{-2}y^2 \div xy^{-1} + x^{-1}y$

$$= \dfrac{x^2}{y^2} + 2 + \dfrac{y^2}{x^2} \div \dfrac{x}{y} + \dfrac{y}{x}$$

$$= \dfrac{x}{y} + \dfrac{y}{x} = xy^{-1} + yx^{-1}.$$

THEORY OF INDICES.

Ex. 7.—Extract the square root of

$$9x^{-4} - 18x^{-3}y^{\frac{1}{2}} + 15x^{-2}y - 6x^{-1}y^{\frac{3}{2}} + y^2.$$

$$9x^{-4} - 18x^{-3}y^{\frac{1}{2}} + 15x^{-2}y - 6x^{-1}y^{\frac{3}{2}} + y^2 \; (3x^{-2} - 3x^{-1}y^{\frac{1}{2}} + y$$
$$\underline{9x^{-4}}$$

$6x^{-2} - 3x^{-1}y^{\frac{1}{2}}) -18x^{-3}y^{\frac{1}{2}} + 15x^{-2}y$
$\qquad\qquad\qquad \underline{-18x^{-3}y^{\frac{1}{2}} + \; 9x^{-2}y}$

$6x^{-2} - 6x^{-1}y^{\frac{1}{2}} + y) + \;\; 6x^{-2}y - 6x^{-1}y^{\frac{3}{2}} + y^2$
$\qquad\qquad\qquad\quad \underline{+ \;\; 6x^{-2}y - 6x^{-1}y^{\frac{3}{2}} + y^2}$

214. In solving the following examples let the beginner bear in mind the index laws and the meaning of such expressions as a^0, a^{-n}, and no serious difficulty will be experienced.

EXERCISE LXV.

Express with fractional exponents

1. $\sqrt{x^3}$, $\sqrt[3]{x^2}$, $(\sqrt{x})^5$, $\sqrt[3]{a^4}$, $\sqrt[5]{a^6}$, $(\sqrt[3]{a})^7$, $\sqrt[6]{a^2b^2}$.

2. $\sqrt[3]{xy^2z^3}$, $\sqrt[5]{x^3y^2z^4}$, $\sqrt[7]{a^5b^6c^7}$, $5\sqrt{a^2bc^3x^4}$.

Express with radical signs

3. $a^{\frac{3}{5}}$, $a^{\frac{1}{2}}b^{\frac{1}{3}}$, $4x^{\frac{1}{5}}y^{-\frac{2}{3}}$, $3x^{\frac{3}{2}}y^{-\frac{3}{4}}$.

Express with positive exponents

4. a^{-2}, $3x^{-1}y^{-3}$, $6x^{-3}y$, x^4y^{-5}, $\dfrac{2a^{-1}x}{3^{-1}b^2y^{-3}}$.

Write in the form of integral expressions

5. $\dfrac{3xy}{z^2}$, $\dfrac{z}{x^3y^4}$, $\dfrac{a}{bc}$, $\dfrac{c^5}{a^3b^{-2}}$, $\dfrac{x^{-\frac{1}{3}}}{y^{-\frac{2}{3}}}$, $\dfrac{x^{-2}}{y^{\frac{1}{3}}}$.

THEORY OF INDICES.

Simplify

6. $a^{\frac{1}{2}} \times a^{\frac{1}{3}}$, $b^{\frac{3}{8}} \times b^{\frac{1}{8}}$, $c^{\frac{2}{3}} \times c^{\frac{1}{12}}$, $d^{\frac{3}{8}} \times d^{\frac{1}{16}}$.

7. $m^{\frac{1}{2}} \times m^{-\frac{3}{8}}$, $a^0 \times a^{\frac{1}{2}}$, $a^0 \times a^{\frac{1}{2}} \times a^{-\frac{1}{2}}$.

8. $a^{\frac{1}{2}} \times \sqrt{a}$, $c^{-\frac{1}{2}} \times \sqrt{c}$, $y^{\frac{1}{4}} \times \sqrt[3]{y}$, $x^{\frac{2}{3}} \times \sqrt{x^{-1}}$.

9. $a^{\frac{2}{3}} b^{\frac{1}{2}} c^{-\frac{1}{4}} \times a^{\frac{1}{3}} b^{-\frac{1}{3}} c^{\frac{1}{2}} d$.

10. $x^{\frac{1}{3}} y^{\frac{2}{3}} z^{\frac{1}{6}} \times x^{-\frac{2}{3}} y^{-\frac{1}{2}} z^{-\frac{1}{2}}$, $x^{\frac{5}{6}} y^{\frac{1}{4}} z^{\frac{1}{2}} \times x^{-\frac{1}{3}} y^{-\frac{1}{2}} z^{-\frac{1}{2}}$.

11. $a^{\frac{1}{2}} \times a^{-\frac{1}{8}} \times a^{-\frac{1}{4}} \times a^{-\frac{1}{8}}$, $\left(\dfrac{ay}{x}\right)^{\frac{1}{2}} \times \left(\dfrac{bx}{y^2}\right)^{\frac{1}{3}} \times \left(\dfrac{y^2}{a^2 b^2}\right)^{\frac{1}{4}}$.

12. $\dfrac{a^{\frac{1}{2}}}{a^{\frac{1}{3}}} \cdot \dfrac{c^{\frac{5}{6}}}{c^{\frac{1}{2}}} \cdot \dfrac{n^{\frac{7}{12}}}{n^{\frac{1}{3}}} \cdot \dfrac{a^{\frac{2}{3}}}{\sqrt[3]{a^2}}$.

13. $(a^6)^{\frac{1}{2}} \div (a^6)^{\frac{2}{3}}$, $(c^{-\frac{1}{2}})^{\frac{2}{3}}$, $(m^{-\frac{1}{2}})^4$, $(n^{\frac{1}{3}})^{-3}$.

14. $(p^{-\frac{2}{3}})^{-\frac{2}{3}}$, $(q^{\frac{3}{5}})^{-\frac{1}{2}}$, $(a^{\frac{2}{3}} \times a^{\frac{4}{7}})^{-1\frac{1}{2}}$.

15. $\left(\dfrac{16 a^{-4}}{81 b^3}\right)^{-\frac{3}{4}}$, $\left(\dfrac{9 a^4}{16 b^{-3}}\right)^{-\frac{3}{2}}$, $(3^{\frac{2}{3}} a^{-3})^{-\frac{3}{2}}$.

EXERCISE LXVI.

Multiply

1. $x^{\frac{2}{3}} - x^{\frac{1}{3}} + 1$ by $x^{\frac{1}{3}} + 1$.

2. $a^2 b^{-2} + 2 + a^{-2} b^2$ by $a^2 b^{-2} - 2 + a^{-2} b^2$.

3. $4x^{-3} + 3x^{-2} + 2x^{-1} + 1$ by $x^{-2} - x^{-1} + 1$.

4. $x^{2p} + x^p y^p + y^{2p}$ by $x^{2p} - x^p y^p + y^{2p}$.

5. $x^{\frac{3}{2}} - xy^{\frac{1}{2}} + x^{\frac{1}{2}} y - y^{\frac{3}{2}}$ by $x + x^{\frac{1}{2}} y^{\frac{1}{2}} + y$.

6. $x^2 + x(a^{\frac{1}{2}} - b^{\frac{1}{2}}) - a^{\frac{1}{2}} b^{\frac{1}{2}}$ by $x^2 - x(a^{\frac{1}{2}} - b^{\frac{1}{2}}) - a^{\frac{1}{2}} b^{\frac{1}{2}}$.

7. $x^{(m-1)n} - y^{(n-1)m}$ by $x^n - y^m$.

THEORY OF INDICES.

Divide

8. $x^{\frac{3}{2}} - xy^{\frac{1}{2}} + x^{\frac{1}{2}}y - y^{\frac{3}{2}}$ by $x^{\frac{1}{2}} - y^{\frac{1}{2}}$.

9. $a^{\frac{5}{2}} - a^{\frac{3}{2}}b + ab^{\frac{3}{2}} - 2a^{\frac{1}{2}}b^2 + b^{\frac{5}{2}}$ by $a^{\frac{3}{2}} - ab^{\frac{1}{2}} + a^{\frac{1}{2}}b - b^{\frac{3}{2}}$.

10. $a^{\frac{3n}{2}} - a^{-\frac{3n}{2}}$ by $a^{\frac{n}{2}} - a^{-\frac{n}{2}}$.

11. $x - 2(x^{\frac{2}{3}} - x^{-\frac{2}{3}}) + 2(x^{\frac{4}{3}} - x^{-\frac{4}{3}}) - x^{-1}$ by $x^{\frac{1}{3}} - x^{-\frac{1}{3}}$.

12. $(x^n)^2 - 1$ by $x^n - 1$.

Simplify

13. $(2x^{\frac{1}{3}} + 3y^{\frac{1}{3}})(2x^{\frac{1}{3}} - 3y^{\frac{1}{3}})(4x^{\frac{2}{3}} + 6x^{\frac{1}{3}}y^{\frac{1}{3}} + 9y^{\frac{2}{3}})$
$\quad\quad (4x^{\frac{2}{3}} - 6x^{\frac{1}{3}}y^{\frac{1}{3}} + 9y^{\frac{2}{3}})$.

14. $\sqrt[3]{x-1} \times \sqrt[3]{x+1} \times \sqrt[3]{x^2-x+1} \times \sqrt[3]{x^2+x+1} \times (x^6-1)^{\frac{2}{3}}$.

15. Show that $\dfrac{2^{n+4} - 2 \times 2^n}{2^{n+2} \times 4} = \dfrac{7}{8}$.

16. Simplify $\dfrac{2^n \times (2^{n-1})^n}{2^{n+1} \times 2^{n-1}}$.

Find the square root of

17. $x^{\frac{3}{2}} - 2x^{\frac{3}{2}}y^{\frac{1}{3}} + 2x^{\frac{1}{2}}z^{\frac{1}{3}} + y^{\frac{2}{3}} - 2y^{\frac{1}{3}}z^{\frac{1}{3}} + z^{\frac{2}{3}}$.

18. $x^{\frac{1}{2}} + 4x^{\frac{1}{4}}y^{\frac{1}{4}} - 2x^{\frac{1}{4}}z^{\frac{1}{4}} + 4y^{\frac{1}{2}} - 4y^{\frac{1}{4}}z^{\frac{1}{4}} + z^{\frac{1}{2}}$.

19. $4x^{-4} + 12x^{-3} + 9x^{-2}$.

20. $x^2 + 4x + 2 - 4x^{-1} + x^{-2}$.

21. $x - 2 + 2x^{-\frac{1}{2}} + x^{-1} - 2x^{-\frac{3}{2}} + x^{-2}$.

22. $4a - 12a^{\frac{1}{2}}b^{\frac{3}{2}} + 9b^3 + 16a^{\frac{1}{2}}c^{\frac{1}{2}} - 24b^{\frac{3}{2}}c^{\frac{1}{2}} + 16c^{\frac{3}{2}}$

23. $256x^{\frac{4}{3}} - 512x + 640x^{\frac{2}{3}} - 512x^{\frac{1}{3}} + 304 - 128x^{-\frac{1}{3}} + 40x^{-\frac{2}{3}} - 8x^{-1} + x^{-\frac{4}{3}}$

24. $\dfrac{y^2}{x} + \dfrac{x^2}{4y} + \dfrac{2y^{\frac{3}{2}} - x^{\frac{3}{2}}}{(xy)^{\frac{1}{2}}}$.

25. $(x + x^{-1})^2 - 4(x - x^{-1})$.

26. $4\left\{\sqrt[3]{\dfrac{a^2}{c^2}} - \sqrt[3]{\dfrac{a}{c}} + \sqrt[3]{\dfrac{c^2}{a^2}} - \sqrt[3]{\dfrac{c}{a}}\right\} + 9$.

Simplify

27. $\dfrac{a^x + b^y}{a^{-x} + b^{-y}} \times \dfrac{a^y - b^x}{a^{-y} - b^{-x}}$.

28. $\dfrac{x^3 + x^{-3} + 2(x + x^{-1})}{x^3 - x^{-3} - 2(x - x^{-1})} \times \left(\dfrac{x^2 - 1}{x^2 + 1}\right)^2$.

29. $\dfrac{3a^{-2}x^2 + 5a^{-1}x - 12}{a^{-3}x^3 - 8a^{-2}x^2 - 12a^{-1}x + 63}$.

30. $\dfrac{x - 4 - 3x^{\frac{1}{2}} + 4y^{\frac{1}{2}} - x^{\frac{1}{2}}y^{\frac{1}{2}}}{x - 8 - 2x^{\frac{1}{2}} + 12y^{\frac{1}{2}} - 3x^{\frac{1}{2}}y^{\frac{1}{2}}}$.

31. $\dfrac{e^{x-y} + xy^{-1} + yx^{-1} + e^{y-x}}{xy^{-1}e^{x-y} + 2 + yx^{-1}e^{y-x}}$.

32. $\{(a-b)^2 + 4ab\}^{\frac{1}{2}} \times \{(a+b)^2 - 4ab\}^{\frac{1}{2}} \left\{\dfrac{a^4 - b^4}{a - b} + 2ab(a+b)\right\}^{\frac{3}{2}}$.

33. $3(a^{\frac{1}{2}} + b^{\frac{1}{2}})^2 - 4(a^{\frac{1}{2}} + b^{\frac{1}{2}})(a^{\frac{1}{2}} - b^{\frac{1}{2}}) + (a^{\frac{1}{2}} - b^{\frac{1}{2}})^2$.

34. $\left(\dfrac{x^{p+q}}{x^q}\right)^p \div \left(\dfrac{x^q}{x^{q-p}}\right)^{p-q}$

35. $\dfrac{x^{2p(q-1)} - y^{2q(p-1)}}{x^{p(q-1)} + y^{q(p-1)}}$.

36. $\left\{(a^m)^{m-\frac{1}{m}}\right\}^{\frac{1}{m+1}}$.

THEORY OF INDICES. 183

37. $\{(a+b)^2 - 4ab\}^{\frac{1}{2}} \cdot \left\{\dfrac{a^4-b^4}{a-b} + 2ab(a+b)\right\}^{\frac{1}{3}}$.

38. $\left\{\left(\dfrac{1+x}{1-x}\right)^{\frac{3}{2}} - \left(\dfrac{1-x}{1+x}\right)^{\frac{3}{2}}\right\} \div \left(\dfrac{1+x}{1-x}\right)^{\frac{3}{2}}$.

Find the H. C. M. of

39. $1 + x + x^{\frac{1}{2}} + x^{\frac{3}{2}}$ and $2x + 2x^{\frac{3}{2}} + 3x^2 + 3x^{\frac{5}{2}}$.

Find the L. C. M. of

40. ax^2-1, ax^2+1, $(a^{\frac{1}{2}}x-1)^2$ $a^{\frac{3}{2}}x^3-1$, $a^{\frac{3}{2}}x^3+1$.

41. If $x^{\frac{1}{3}} + y^{\frac{1}{3}} + z^{\frac{1}{3}} = 0$, prove $(x+y+z)^3 = 27xyz$.

Simplify

42. $\dfrac{(x^a)^3}{x^{b+c}} \times \dfrac{(x^b)^3}{x^{c+a}} \times \dfrac{(x^c)^3}{x^{a+b}}$.

43. $(8^{\frac{2}{3}} + 4^{\frac{3}{2}}) \times 16^{-\frac{3}{4}}$.

44. $\dfrac{9^n \times 3^2 \times \dfrac{1}{3^{-n}} - 27^n}{3^{3n} \times 9}$.

45. $27^{\frac{2}{3}} + 16^{\frac{3}{4}} - \dfrac{2}{8^{-\frac{2}{3}}} + \dfrac{\sqrt[4]{2}}{4^{-\frac{2}{3}}}$.

46. $(x+1)(x^2+x+1)^{-1} + (x-1)(x^2-x+1)^{-1} + 2(x^4+x^2+1)^{-1}$.

47. $\dfrac{1}{1+x^{m-n}+x^{m-p}} + \dfrac{1}{1+x^{n-m}+x^{n-p}} + \dfrac{1}{1+x^{p-m}+x^{p-n}}$.

CHAPTER XIII.

SURDS.

215. When a root is indicated but cannot be exactly determined it is called a **Surd**. Thus $\sqrt{2}$, $\sqrt[3]{4}$, $\sqrt{6}$, are *surds*. When the root is indicated but can be exactly determined it is said to have the *form* of a surd, as $\sqrt{4}$, $\sqrt[3]{8}$. Surds are also called *irrational numbers*.

216. A **Quadratic Surd** is one in which the *second* root is required; a **Cubic Surd** one in which the *third* root is required, and so on.

217. The product of a rational factor (*i.e.*, a factor not containing a surd) and a surd factor is called a **Mixed Surd**, as $3\sqrt{2}$, $5\sqrt[3]{3}$, $3\sqrt[4]{6}$.

218. When there is no rational factor outside the radical sign the surd is said to be *entire*, as $\sqrt{b}$, $\sqrt[3]{5}$.

219. A *mixed* surd can be expressed as an *entire* surd. Thus $2\sqrt{3} = \sqrt{4} \times \sqrt{3} = \sqrt{12}$. Generally $a\sqrt[n]{b}$ can be expressed as an entire surd, for in the chapter on Indices it has been proved that
$$a^{\frac{1}{n}} \times b^{\frac{1}{n}} = (ab)^{\frac{1}{n}}.$$
or
$$\sqrt[n]{a} \times \sqrt[n]{b} = \sqrt[n]{ab}.$$
Now,
$$a = \sqrt[n]{a^n}$$
$$\therefore a\sqrt[n]{b} = \sqrt[n]{a^n} \times \sqrt[n]{b} = \sqrt[n]{a^n b}.$$

SURDS.

220. Hence, to reduce a mixed surd to an entire surd, raise the rational factor to the power indicated by the root to be extracted, multiply the result by the factor under the radical sign, and write the radical sign over the product.

221. An *entire* surd can often be expressed as a *mixed* surd.

Thus $\sqrt{8} = 2\sqrt{2}$, $\sqrt{50} = 5\sqrt{2}$.

For $\sqrt{8} = \sqrt{4 \times 2} = (4 \times 2)^{\frac{1}{2}} = 4^{\frac{1}{2}} \times 2^{\frac{1}{2}} = 2 \times 2^{\frac{1}{2}} = 2\sqrt{2}$.

Similarly, $\sqrt{50} = \sqrt{25 \times 2} = (25 \times 2)^{\frac{1}{2}} = (25)^{\frac{1}{2}} \times 2^{\frac{1}{2}} = 5\sqrt{2}$.

Generally, $\sqrt[n]{a^n b} = (a^n b)^{\frac{1}{n}} = (a^n)^{\frac{1}{n}} \times b^{\frac{1}{n}} = a \cdot b^{\frac{1}{n}} = a\sqrt[n]{b}$.

222. Hence, to reduce (when possible) an entire surd to a mixed surd, separate the quantity under the radical sign into two factors, of one of which the required root can be obtained, and set the root outside the radical sign.

223. The expression under the radical sign is called the **Surd-factor** or **Base**. When the *surd-factor* is as *small as possible*, and *integral*, the surd is said to be in its *simplest form*.

224. Similar Surds are those which have, or may be made to have, the *same surd-factor* and *surd-index*. Thus $2\sqrt{2}$, $3\sqrt{2}$, $\sqrt{50}$, are similar surds; so also are $\sqrt{a}$, $\sqrt{b^2 a}$, $3\sqrt{a}$.

The student will notice that before two surds can be said to be similar they must have the *same quantity* under the radical sign, and the same *surd-index* or number which indicates the root to be extracted. Thus $\sqrt[3]{a}$ and $\sqrt{a}$ are not similar surds, for they have not the same *surd-index*. Sometimes the term *radical-index* is used for that which indicates the root to be extracted.

225. When the *surd-factor* is a fraction, and it is required to reduce the surd to its simplest form, it will be necessary to express the fraction in the form of an equivalent fraction, with a

13

denominator whose root can be taken; for the expression under the radical sign must be made *integral*. For instance,

$$\sqrt[3]{\frac{4}{25}} = \sqrt[3]{\frac{20}{125}} = \sqrt[3]{\frac{1}{125} \times 20} = \frac{1}{5}\sqrt[3]{20}.$$

In this example the cube root of 25 cannot be taken; but as $25 = 5^2$ and $25 \times 5 = 5^3$, therefore if we multiply both numerator and denominator by 5 the new denominator will be a perfect cube, and its cube root can be removed outside the radical sign.

Ex. 1.—Express in the form of a cubic surd 2, 3, $\frac{1}{2}$, $(a-b)$, $\frac{a^2}{b^2}$.

$$2 = \sqrt[3]{8}, \quad 3 = \sqrt[3]{27}, \quad \frac{1}{2} = \sqrt[3]{\frac{1}{8}}, \quad a-b = \sqrt[3]{(a-b)^3}, \quad \frac{a^2}{b} = \sqrt[3]{\frac{a^6}{b^3}}.$$

Ex. 2.—Reduce the following *entire* surds to the form of a *mixed* surd:—

$$\sqrt[3]{108}, \quad \sqrt{72}, \quad \sqrt[3]{7x^2y^7}, \quad \sqrt{\frac{7}{12}}, \quad \sqrt[4]{\frac{5a}{2b^3c^2}}, \quad \sqrt[3]{405}.$$

(1) $\sqrt[3]{108} = \sqrt[3]{27 \times 4} = \sqrt[3]{27} \times \sqrt[3]{4} = 3\sqrt[3]{4}.$

(2) $\sqrt{72} = \sqrt{36 \times 2} = \sqrt{36} \times \sqrt{2} = 6\sqrt{2}.$

(3) $\sqrt[3]{7x^2y^7} = \sqrt[3]{7x^2y^2 \times y^5} = \sqrt[3]{7x^2y^2} \times \sqrt[3]{y^5} = y\sqrt[3]{7x^2y^2}.$

(4) $\sqrt{\frac{7}{12}} = \sqrt{\frac{21}{36}} = \sqrt{\frac{1}{36} \times 21} = \frac{1}{6}\sqrt{21}.$

(5) $\sqrt[4]{\frac{5a}{2b^3c^2}} = \sqrt[4]{\frac{40abc^2}{16b^4c^4}} = \sqrt[4]{40abc^2 \times \frac{1}{16b^4c^4}} = \frac{1}{2bc}\sqrt[4]{40abc^2}.$

(6) $\sqrt[3]{405} = \sqrt[3]{27 \times 15} = \sqrt[3]{27} \times \sqrt[3]{15} = 3\sqrt[3]{15}.$

Ex. 3.—Express as *entire* surds the following *mixed* surds:—

$a^2b\sqrt{bc}, \quad 3\sqrt{21}, \quad 5\sqrt{32}, \quad 3c^2\sqrt[3]{abc}, \quad a^3\sqrt[3]{a^3b^2}, \quad 2x\sqrt[3]{xy},$

$$(x+y)\sqrt{\frac{x+y}{x^2+2xy+y^2}}.$$

SURDS. 187

(1) $a^2b \sqrt{bc} = \sqrt{a^4b^2} \times \sqrt{bc} = \sqrt{a^4b^2 \times bc} = \sqrt{a^4b^3c}$.

(2) $3\sqrt{21} = \sqrt{9} \times \sqrt{21} = \sqrt{189}$.

(3) $5\sqrt{32} = \sqrt{25} \times \sqrt{32} = \sqrt{800}$.

(4) $3c^2 \sqrt[3]{abc} = \sqrt[3]{27c^6} \times \sqrt[3]{abc} = \sqrt[3]{27abc^7}$.

(5) $a^3 \sqrt[4]{a^3b^2} = \sqrt[4]{a^{12}} \times \sqrt[4]{a^3b^2} = \sqrt[4]{a^{15}b^2}$.

(6) $2x \sqrt[5]{xy} = \sqrt[5]{32x^5} \times \sqrt[5]{xy} = \sqrt[5]{32x^6y}$.

(7) $(x+y)\sqrt{\dfrac{x+y}{x^2+2xy+y^2}} = \sqrt{(x+y)^2} \times \sqrt{\dfrac{x+y}{x^2+2xy+y^2}}$

$= \sqrt{\dfrac{(x+y)^2(x+y)}{x^2+2xy+y^2}} = \sqrt{x+y}$.

226. Surds are *simple* or *compound* according as they contain one or more terms.

EXERCISE LXVII.

Express as mixed surds

1. $\sqrt{x^2y^4z}$, $\sqrt{8a^3b}$, $\sqrt[3]{54a^4x^2y^3}$, $\sqrt{24}$, $\sqrt{125a^4d^3}$.

2. $\sqrt[3]{1000a}$, $\sqrt[3]{160x^4y^7}$, $\sqrt[3]{108m^9n^{10}}$, $\sqrt[3]{1372a^{15}b^{16}}$.

3. $\sqrt[3]{a^4 - 3a^3b + 3a^2b^2 - ab^3}$, $\sqrt{50a^2 - 100ab + 50b^2}$.

4. $\sqrt{\dfrac{5}{9}}$, $\sqrt{\dfrac{27}{16}}$, $\sqrt{\dfrac{25}{8}}$, $\sqrt[3]{\dfrac{1}{2}}$, $\sqrt[3]{\dfrac{1}{5}}$.

5. $\sqrt{\dfrac{a+b}{a-b}}$, $\sqrt[3]{\dfrac{(a-b)^2}{a+b}}$, $\sqrt[4]{\dfrac{b}{2a}}$, $\sqrt{\dfrac{3a^2bx}{4cy^3}}$.

Express as entire surds

6. $3\sqrt{2}$, $2\sqrt[3]{7}$, $3\sqrt[3]{5}$, $5\sqrt[3]{9}$, $2\sqrt[4]{6}$.

7. $3\sqrt{a}$, $4a\sqrt{3x}$, $3\sqrt[3]{\dfrac{3}{7}}$, $2ax\sqrt{\dfrac{3a}{4x}}$,

8. $(m+n)\sqrt{\dfrac{m-n}{m+n}}$, $(a+b)\sqrt{\dfrac{1}{a^2-b^2}}$, $\left(\dfrac{x-y}{x+y}\right)\sqrt{\dfrac{x^2+xy}{(x-y)^2}}$.

Simplify

9. $2\sqrt[4]{80a^5b^2c^6}$, $9\sqrt[3]{81x^2y^3z}$, $5\sqrt{726}$.

10. $\sqrt[3]{\dfrac{2xy^2}{z}}$, $\dfrac{a}{b}\sqrt[4]{\dfrac{b}{2a^3}}$, $\dfrac{x^3y^2}{z^2}\sqrt[6]{\dfrac{z^5}{x^5y^5}}$.

11. $ax\sqrt{b^2x}$, $2a^2b^4\sqrt[4]{b^2x^3}$.

12. Show that $\sqrt{20}$, $\sqrt{45}$, $\sqrt{\dfrac{4}{5}}$ are similar surds.

13. Show that $2\sqrt[3]{a^3b^2}$, $\sqrt[3]{8b^5}$, $\dfrac{1}{2}\sqrt[3]{\dfrac{a^6}{b}}$ are similar surds.

14. Show that $\sqrt{50}$, $\sqrt{72}$, $\sqrt{32}$, $\sqrt{\dfrac{1}{2}}$ are similar surds.

227. Surds having the same radical sign are said to be of the same order. Thus $\sqrt[3]{a}$, $2\sqrt[3]{b^2}$, $\dfrac{4}{5}\sqrt[3]{3}$, are each of the third order.

228. It is often necessary to bring surds of different orders to the same order. This can be done as follows:—

Let $a^{\frac{1}{n}}$, $b^{\frac{1}{m}}$ be surds of different orders. Then $a^{\frac{1}{n}} = a^{\frac{m}{mn}}$ and $b^{\frac{1}{m}} = b^{\frac{n}{mn}}$; but $a^{\frac{m}{mn}} = (a^m)^{\frac{1}{mn}}$ and $b^{\frac{n}{mn}} = (b^n)^{\frac{1}{mn}}$, therefore $a^{\frac{1}{n}} = \sqrt[mn]{a^m}$ and $b^{\frac{1}{m}} = \sqrt[mn]{b^n}$. It is evident that $\sqrt[mn]{a^m}$ and $\sqrt[mn]{b^n}$ are of the same order.

Hence, to reduce surds to the same order, find the L. C. M. of their radical indices for a common radical index. Raise each expression under the radical sign to the power indicated by the number obtained by dividing its radical index into the common radical index, and write over the result the common radical index.

SURDS. 189

Or, express the different surds in the form of quantities having fractional exponents with a common denominator. The common denominator will be the common radical index. Thus $\sqrt{2}$ and $\sqrt[3]{3}$ may be expressed in this form: $2^{\frac{3}{6}}$ and $3^{\frac{2}{6}}$ or $\sqrt[6]{2^3}$ and $\sqrt[6]{3^2}$.

Ex. 1.—Reduce to a common radical index $\sqrt{3}$ and $\sqrt[3]{4}$.

$$\sqrt{3} = 3^{\frac{1}{2}} = 3^{\frac{3}{6}} = (3^3)^{\frac{1}{6}} = \sqrt[6]{3^3}.$$

Ex. 2.—Reduce to common index $\sqrt[5]{8}$ and $\sqrt{2}$.

$$\sqrt[5]{8} = 8^{\frac{1}{5}} = 8^{\frac{2}{10}} = (8^2)^{\frac{1}{10}} = \sqrt[10]{8^2}$$

$$\sqrt{2} = 2^{\frac{1}{2}} = 2^{\frac{5}{10}} = (2^5)^{\frac{1}{10}} = \sqrt[10]{2^5}.$$

EXERCISE LXVIII.

Reduce to a common radical index

1. $\sqrt{5}$, $\sqrt[3]{4}$.
2. $\sqrt[3]{6}$, $\sqrt{7}$.
3. $\sqrt[4]{2}$, $\sqrt[3]{3}$.
4. $\sqrt[3]{4}$, $\sqrt[4]{8}$.
5. $\sqrt{2}$, $\sqrt[3]{5}$, $\sqrt[4]{6}$.
6. $\sqrt[3]{7}$, $\sqrt[4]{5}$, $\sqrt[5]{120}$.
7. $\sqrt[3]{3}$, $\sqrt[4]{6}$, $\sqrt[5]{10}$.
8. $\sqrt[3]{a^2 - ab + b^2}$, $\sqrt[4]{a+b}$.
9. $\sqrt[4]{\dfrac{a}{b}}$, $\sqrt[5]{\dfrac{b}{a}}$.

229. Surds can be compared with respect to magnitude by reducing them to the same order, and then expressing them as entire surds.

Ex. 1.—Which is the greater, $\sqrt{2}$ or $\sqrt[3]{3}$?

$$\sqrt{2} = \sqrt[6]{2^3} = \sqrt[6]{8}.$$

$$\sqrt[3]{3} = \sqrt[6]{3^2} = \sqrt[6]{9}.$$

$\therefore \sqrt[3]{3}$ is greater than $\sqrt{2}$.

Ex. 2.—Which is the greater, $\sqrt[4]{3}$ or $\sqrt[3]{5}$?

$$\sqrt[4]{3} = \sqrt[12]{3^4} = \sqrt[12]{81}$$

$$\sqrt[3]{5} = \sqrt[12]{5^3} = \sqrt[12]{125}.$$

$\therefore \sqrt[3]{5}$ is greater than $\sqrt[4]{3}$

EXERCISE LXIX.

1. Which is the greater, $3\sqrt{7}$ or $2\sqrt{15}$?
2. Arrange in order of magnitude $9\sqrt{3}$, $6\sqrt{7}$, $5\sqrt{10}$.
3. Arrange in order of magnitude $4\sqrt[3]{4}$, $3\sqrt[3]{5}$, $5\sqrt[3]{3}$.
4. Arrange in order of magnitude $6\sqrt[3]{7}$, $4\sqrt[3]{9}$, $3\sqrt[3]{18}$.

230. *Similar* surds, in their simplest forms, may be added or subtracted by adding or subtracting their rational factors and then affixing the common radical factor.

Ex. 1.—Find the algebraic sum of $\sqrt{50} + \sqrt{72} - 3\sqrt{2} + \dfrac{1}{\sqrt{2}}$.

$$\sqrt{50} = 5\sqrt{2}, \quad \sqrt{72} = 6\sqrt{2}, \quad \dfrac{1}{\sqrt{2}} = \dfrac{1}{2}\sqrt{2}.$$

$\therefore$ Algebraic sum $= \left(5 + 6 - 3 + \dfrac{1}{2}\right)\sqrt{2} = \dfrac{17}{2}\sqrt{2}.$

Ex. 2.—Simplify $2\sqrt[3]{320} - 2\sqrt[3]{40}$.

$$\sqrt[3]{320} = \sqrt[3]{64 \times 5} = 4\sqrt[3]{5},$$

$\therefore 2\sqrt[3]{320} = 8\sqrt[3]{5}.$

$$\sqrt[3]{40} = \sqrt[3]{8 \times 5} = 2\sqrt[3]{5},$$

$\therefore 3\sqrt[3]{40} = 6\sqrt[3]{5}.$

$\therefore$ Result $= 8\sqrt[3]{5} - 6\sqrt[3]{5} = (8-6)\sqrt[3]{5} = 2\sqrt[3]{5}.$

SURDS. 191

231. Surds having the *same* radical index can be multiplied together by finding the product of their rational factors or coefficients, and affixing to the result the product of their *bases* with the common radical index written over it.

Ex. 1.—Find the product of $\sqrt{a}$ and $\sqrt{b^3}$.

$$\sqrt{a} \times \sqrt{b^3} = \sqrt{ab^3}.$$

For $\sqrt{a} = a^{\frac{1}{2}}$ and $\sqrt{b^3} = b^{\frac{3}{2}},$

$$\therefore \sqrt{a} \times \sqrt{b^3} = a^{\frac{1}{2}} \times b^{\frac{3}{2}} = (ab^3)^{\frac{1}{2}} = \sqrt{ab^3}.$$

Ex. 2.—Find the product of $a\sqrt[n]{c}$ and $b\sqrt[n]{d}$.

$$a\sqrt[n]{c} \times b\sqrt[n]{d} = ab\sqrt[n]{cd}.$$

For $\sqrt[n]{c} = c^{\frac{1}{n}}$ and $\sqrt[n]{d} = d^{\frac{1}{n}},$

$$\therefore \sqrt[n]{c} \times \sqrt[n]{d} = c^{\frac{1}{n}} \times d^{\frac{1}{n}} = (cd)^{\frac{1}{n}} = \sqrt[n]{cd}.$$

$$\therefore a\sqrt[n]{c} \times b\sqrt[n]{d} = ab\sqrt[n]{cd}.$$

Hence the rule.

232. Surds having *different* radical indices can be multiplied together by first reducing them to surds of the same order, or common radical index, and then finding their product by the preceding rule.

Ex. 1.—Find the product of $\sqrt{2}$, $\sqrt[3]{3}$ and $\sqrt[5]{4}$.

The L. C. M. of 2, 3 and 5 is 30, therefore the common radical index is 30. Hence

$$\sqrt{2} = \sqrt[30]{2^{15}},$$
$$\sqrt[3]{3} = \sqrt[30]{3^{10}},$$
$$\sqrt[5]{4} = \sqrt[30]{4^6};$$
$$\therefore \sqrt{2} \times \sqrt[3]{3} \times \sqrt[5]{4} = \sqrt[30]{2^{15} \times 3^{10} \times 4^6}$$
$$= \sqrt[30]{2^{15} \times 3^{10} \times 2^{12}} = \sqrt[30]{2^{27} \times 3^{10}}.$$

192 SURDS.

Ex. 2.—Find the product of $\sqrt[n]{a}$ and $\sqrt[m]{b}$.

$$\sqrt[n]{a} = a^{\frac{1}{n}} = a^{\frac{m}{mn}},$$

$$\sqrt[m]{b} = b^{\frac{1}{m}} = b^{\frac{m}{mn}},$$

$$\therefore \sqrt[n]{a} \times \sqrt[m]{b} = a^{\frac{m}{mn}} \times b^{\frac{n}{mn}} = (a^m b^n)^{\frac{1}{mn}} = \sqrt[mn]{a^m b^n}$$

Hence the rule.

233. If two surds are of the *same* order the quotient of the one by the other can be obtained by dividing the rational factor of the dividend by the rational factor of the divisor, and to the result affixing the quotient, with the common radical index written over it, of the one base by the other. If the surds are of *different* orders they can be reduced to the same order, and the preceding rule applied.

Ex. 1.—Divide $\sqrt[3]{9}$ by $\sqrt[3]{15}$.

$$\sqrt[3]{9} \div \sqrt[3]{15} = 9^{\frac{1}{3}} \div 15^{\frac{1}{3}} = \left(\frac{9}{15}\right)^{\frac{1}{3}} = \left(\frac{3}{5}\right)^{\frac{1}{3}} = \sqrt[3]{\frac{3}{5}}.$$

Ex. 2.—Divide $\sqrt[2]{3}$ by $\sqrt[3]{4}$.

$$\sqrt[2]{3} \div \sqrt[3]{4} = 3^{\frac{1}{2}} \div 4^{\frac{1}{3}} = 3^{\frac{3}{6}} \div 4^{\frac{2}{6}}$$

$$= (3^3)^{\frac{1}{6}} \div (4^2)^{\frac{1}{6}}$$

$$= \left(\frac{3^3}{4^2}\right)^{\frac{1}{6}} = \left(\frac{27}{16}\right)^{\frac{1}{6}} = \sqrt[6]{\frac{27}{16}}$$

Ex. 3.—Divide $\sqrt[3]{5}$ by $\sqrt[5]{7}$.

$$\sqrt[3]{5} = \sqrt[15]{5^5},$$

$$\sqrt[5]{7} = \sqrt[15]{7^3};$$

$$\therefore \sqrt[3]{5} \div \sqrt[5]{7} = \sqrt[15]{5^5} \div \sqrt[15]{7^3} = \sqrt[15]{\frac{5^5}{7^3}}.$$

234. The multiplication and division of compound surds differ in no respect from the multiplication and division of polynomial

SURDS. 193

rational expressions, except that surd terms are used instead of rational terms. An example or two will illustrate this.

Ex. 1.—Multiply together $\sqrt{3} + \sqrt{2}$ and $\sqrt{3} - \sqrt{2}$.

Here we have the sum of two expressions multiplied by their difference, therefore the product = the difference of their squares;

$$\therefore (\sqrt{3} + \sqrt{2})(\sqrt{3} - \sqrt{2}) = 3 - 2 = 1.$$

Ex. 2.—Find the product of $2\sqrt{3} + \sqrt[3]{2}$ and $\sqrt[3]{4} + \sqrt{5}$.

$$2\sqrt{3} + \sqrt[3]{2}$$
$$\sqrt[3]{4} + \sqrt{5}$$
$$\overline{2\sqrt[3]{3^3 \times 4^3} + \sqrt[3]{8}}$$
$$+ \quad 2\sqrt{15} + \sqrt[3]{2^2 \times 5^3}$$
$$\overline{2\sqrt[3]{27 \times 16} + 2 + 2\sqrt{15} + \sqrt[3]{4 \times 125}}$$

Ex. 3.—Square $\sqrt{2} + \sqrt[3]{3} + 1$.

$(\sqrt{2} + \sqrt[3]{3} + 1)^2$

$= (\sqrt{2})^2 + (\sqrt[3]{3})^2 + 1 + 2\sqrt{2} \times \sqrt[3]{3} + 2\sqrt{2} + 2\sqrt[3]{3}$

$= 2 + \sqrt[3]{9} + 1 + 2\sqrt[3]{8 \times 9} + 2\sqrt{2} + 2\sqrt[3]{3}$

$= 3 + \sqrt[3]{9} + 2\sqrt[3]{72} + 2\sqrt{2} + 2\sqrt[3]{3}$

Ex. 4.—Divide $\sqrt{8} + 3\sqrt{24} + 5\sqrt{56}$ by $\sqrt{12}$.

$$\sqrt{8} + 3\sqrt{24} + 5\sqrt{56} \div \sqrt{12}$$

$$= \sqrt{\frac{8}{12}} + 3\sqrt{2} + 5\sqrt{\frac{56}{12}}$$

$$= \sqrt{\frac{2}{3}} + 3\sqrt{2} + 5\sqrt{\frac{14}{3}}.$$

235. If the divisor consists of two or three terms it is better to rationalize it before proceeding to division. The method of rationalizing a divisor or denominator will be explained hereafter.

EXERCISE LXX.

Simplify

1. $2\sqrt{3} + 3\sqrt{3}, \quad 11\sqrt{13} - 4\sqrt{13}.$

2. $5\sqrt{7} + 3\sqrt{7} + 2\sqrt{7}, \quad 8\sqrt{10} + 3\sqrt{10} + 2\sqrt{10}.$

3. $4\sqrt{11} + 3\sqrt{11} - 5\sqrt{11}, \quad 2\sqrt{3} - 5\sqrt{3} + 9\sqrt{3}.$

4. $4\sqrt[3]{3} - 3\sqrt[3]{3} + 2\sqrt[3]{3}, \quad 7\sqrt{2} - \sqrt{18}.$

5. $5\sqrt[3]{4} + 2\sqrt[3]{32} - \sqrt[3]{108}, \quad 3\sqrt[3]{2} + 4\sqrt[3]{2} - \sqrt[3]{64}.$

6. $2\sqrt{175} - 3\sqrt{63} + 5\sqrt{28}, \quad \dfrac{1}{2}\sqrt[3]{5} + \dfrac{5}{2}\sqrt[3]{5} + \dfrac{1}{4}\sqrt[3]{40}.$

7. $\sqrt{2} + 3\sqrt{32} + \dfrac{1}{2}\sqrt{128} - 6\sqrt{18}.$

8. $\sqrt{75} + \sqrt{48} - \sqrt{147} + \sqrt{300}.$

9. $20\sqrt{245} - \sqrt{5} + \sqrt{125} - \dfrac{5}{2}\sqrt{180}.$

10. $3\sqrt[3]{3} - 5\sqrt[3]{48} + \sqrt[3]{243}.$

11. $\sqrt[3]{27c^4} - \sqrt[3]{8c^4} + \sqrt[3]{125c}.$

12. $\sqrt[3]{a^5b} - \sqrt[3]{b^6} + \sqrt[3]{32b}.$

13. $\sqrt{a^4x} + \sqrt{b^4x} - \sqrt{4a^2b^2x}.$

14. $6a\sqrt{63ab^3} - 3\sqrt{112a^3b^3} + 2ab\sqrt{343ab} - 5b\sqrt{28a^3b}.$

15. $\sqrt{32a^4b^5} + 6\sqrt{72b} + 3\sqrt{128a^2b^3} - 4\sqrt{288a^2b^3}.$

16. $\sqrt[3]{\dfrac{4}{27}} - \sqrt[3]{\dfrac{1}{2}} - \sqrt[3]{256} + \sqrt[3]{\dfrac{32}{27}} - \sqrt[3]{\dfrac{125}{54}}.$

SURDS. 195

17. $\sqrt{9} + \sqrt{20} - \sqrt{\dfrac{1}{16}} - \sqrt{\dfrac{9}{5}} + \sqrt{\dfrac{9}{16}} + \sqrt{\dfrac{25}{81}} + \sqrt{\dfrac{20}{9}}.$

18. $2\sqrt{\dfrac{5}{3}} + \sqrt{60} - \sqrt{15} + \sqrt{\dfrac{3}{5}} + \sqrt{\dfrac{4}{15}}.$

19. $2\sqrt{\dfrac{2}{3}} + 3\sqrt{\dfrac{3}{2}} - 3\sqrt{\dfrac{1}{6}} + 2\sqrt{6}.$

20. $(a+x)\sqrt{a^2-x^2} + (a^2+x^2)\sqrt{\dfrac{a+x}{a-x}}.$

21. $\sqrt{a+x} + \sqrt{ax^2+x^3} + \sqrt{(a+x)^3}.$

EXERCISE LXXI.

Multiply together

1. $\sqrt{3}$ and $\sqrt{27}$, $\sqrt{5}$ and $\sqrt{20}$, $\sqrt{2}$ and $\sqrt{18}$.

2. $\sqrt[3]{3}$ and $\sqrt[3]{9}$, $\sqrt[3]{2}$ and $\sqrt[3]{32}$, $\sqrt[3]{16}$ and $\sqrt[3]{32}$.

3. $\sqrt[3]{27}$ and $\sqrt[3]{3}$, $\sqrt[3]{108}$ and $\sqrt[3]{12}$, $\sqrt[3]{4}$ and $\sqrt[3]{8}$.

4. $\sqrt[3]{1331}$, $\sqrt[3]{11}$ and $\sqrt[3]{121}$; $\sqrt[3]{9}$, $\sqrt[3]{81}$ and $\sqrt[3]{729}$.

5. $\sqrt{6}$, $\sqrt{12}$ and $\sqrt{72}$; $\sqrt[3]{5^3}$, $\sqrt[3]{5^7}$ and $\sqrt[3]{5^6}$.

6. $3\sqrt[3]{16}$ and $5\sqrt[3]{4}$; $\dfrac{1}{2}\sqrt[3]{8}$ and $\dfrac{3}{4}\sqrt[3]{2}$.

7. $(\sqrt{18} + 2\sqrt{72} - 3\sqrt{8} - \dfrac{1}{2}\sqrt{128})$ and $\sqrt{2}$.

8. $(\sqrt[3]{32} - \dfrac{1}{2}\sqrt[3]{864} + 3\sqrt[3]{4})$ and $\sqrt[3]{2}$

9. $3 + 2\sqrt{5}$ and $2 - \sqrt{5}$.

10. $8 + 3\sqrt{2}$ and $2 - \sqrt{2}$.

11. $5 + 2\sqrt{3}$ and $3 - 5\sqrt{3}$.

12. $3 - \sqrt{6}$ and $6 - 3\sqrt{6}$.

13. $2\sqrt{6} - 3\sqrt{5}$ and $\sqrt{3} + 2\sqrt{2}$.

14. $\sqrt[3]{9} - 2\sqrt[3]{4}$ and $4\sqrt[3]{3} + \sqrt[3]{2}$.

15. $\sqrt{5 - \sqrt{4}}$ and $\sqrt{5 + \sqrt{4}}$.

16. $\sqrt{3 + \sqrt{2}}$ and $\sqrt{3 - \sqrt{2}}$.

17. $\sqrt{11 - \sqrt{10}}$ and $\sqrt{11 + \sqrt{10}}$.

18. $\sqrt{6 + \sqrt{11}}$ and $\sqrt{6 - \sqrt{11}}$.

19. $\sqrt[3]{10 - \sqrt{36}}$ and $\sqrt[3]{10 + \sqrt{36}}$.

20. $\sqrt[3]{9 + \sqrt{17}}$ and $\sqrt[3]{9 - \sqrt{17}}$.

21. $\sqrt[3]{12 + \sqrt{19}}$ and $\sqrt[3]{12 - \sqrt{19}}$.

22. $2\sqrt{30} - 3\sqrt{5} + 5\sqrt{3}$ and $\sqrt{8} + \sqrt{3} - \sqrt{5}$.

23. $3\sqrt{3} - 4\sqrt{8} + 5\sqrt{18}$ and $5\sqrt{2} + \sqrt{3} + \sqrt{\dfrac{1}{2}}$.

24. $x^2 - x\sqrt{3} + 1$ and $x^2 + x\sqrt{3} + 1$.

25. $\sqrt{a} + \sqrt{b} + \sqrt{c}$, $\sqrt{a} + \sqrt{b} - \sqrt{c}$, $\sqrt{a} - \sqrt{b} + \sqrt{c}$, and $\sqrt{b} + \sqrt{c} - \sqrt{a}$.

26. $\sqrt{5}$ and $\sqrt[3]{4}$; $\sqrt{125}$ and $\sqrt[3]{36}$.

27. $\sqrt[3]{16}$ and $\sqrt[3]{250}$; $\sqrt[3]{7^5}$ and $\sqrt{5^6}$.

28. $\sqrt[5]{\dfrac{3}{8}}$ and $\sqrt[3]{\dfrac{2}{3}}$; $\sqrt[4]{\dfrac{7}{9}}$ and $\sqrt[6]{\dfrac{15}{7}}$.

29. $\sqrt[9]{\dfrac{9}{28}}$ and $\sqrt[4]{\dfrac{2}{3}}$; $\sqrt[10]{\dfrac{125}{16}}$ and $\sqrt[4]{\dfrac{2}{3}}$.

SURDS.

EXERCISE LXXII.

Perform the following divisions:—

1. $\sqrt{243} \div \sqrt{3}$; $\sqrt[4]{80} \div \sqrt[4]{5}$; $\sqrt[3]{81} \div \sqrt[3]{3}$.

2. $\sqrt[3]{243} \div \sqrt[3]{3}$; $\sqrt[3]{56} \div \sqrt[3]{7}$; $\sqrt[3]{500} \div \sqrt[3]{4}$.

3. $\sqrt{\dfrac{7}{12}} \div \sqrt{\dfrac{3}{7}}$; $\sqrt{\dfrac{2}{9}} \div \sqrt{\dfrac{1}{2}}$; $\sqrt{\dfrac{12}{35}} \div \sqrt{\dfrac{21}{5}}$.

4. $3\sqrt{6} + 45\sqrt{2} \div 3\sqrt{3}$.

5. $42\sqrt{5} - 30\sqrt{3} \div 2\sqrt{15}$.

6. $84\sqrt{15} + 168\sqrt{6} \div 3\sqrt{21}$.

7. $56\sqrt{30} - 84\sqrt{10} + 100\sqrt{14} \div 4\sqrt{35}$.

8. $30\sqrt[3]{4} - 36\sqrt[3]{10} + 30\sqrt[3]{90} \div 3\sqrt[3]{20}$.

9. $50\sqrt[3]{18} + 18\sqrt[3]{20} - 48\sqrt[3]{5} \div 2\sqrt[3]{30}$

10. $\sqrt{\dfrac{a^3}{b^3}} \div \sqrt{\dfrac{a}{b^3}}$; $\sqrt[3]{2y} \div \sqrt[3]{\dfrac{1}{4y^2}}$.

11. $\dfrac{b-1}{\sqrt{b}-1}$; $\dfrac{a-b}{\sqrt{a}-\sqrt{b}}$; $\dfrac{1-a}{1-\sqrt{a}}$.

12. $\dfrac{a^2+ab+b^2}{a+\sqrt{ab}+b}$; $\dfrac{acx^2\sqrt{y}-bcy^2\sqrt{x}}{c\sqrt{xy}}$.

13. $\sqrt{54} \div \sqrt[3]{36}$; $\sqrt[3]{9} \div \sqrt{3}$.

14. $\sqrt{\dfrac{8}{45}} \div \sqrt[3]{6\tfrac{1}{4}}$; $\sqrt[3]{12} \div \sqrt{6}$.

15. $\sqrt[3]{2x} \div \sqrt{x^3}$; $\sqrt[3]{8x^3y^2} \div 2x^2y^3$.

SURDS.

236. It is evident from its definition that a *surd* cannot be equal to a *rational* quantity. From this fact the following important theorems are deduced:—

I. *A surd and a rational quantity cannot be equal to a rational quantity.* For, if possible, let $a + \sqrt{b} = c$, where a and c are rational quantities and $\sqrt{b}$ a surd; therefore $\sqrt{b} = c - a$. But $c - a$ is a rational quantity, being the difference between two rational quantities; therefore $\sqrt{b} = $ a rational quantity, that is, a surd equals a rational quantity, which is impossible. Therefore $a + \sqrt{b}$ does not $= c$.

II. *A surd and a rational quantity cannot equal a surd.*

Let $\qquad a + \sqrt{b} = \sqrt{c}$.

Squaring, $\qquad a^2 + b + 2a\sqrt{b} = c$

or $\qquad 2a\sqrt{b} = c - a^2 - b;$

$\qquad \therefore \sqrt{b} = \dfrac{c - a^2 - b}{2a}.$

But $\dfrac{c - a^2 - b}{2a}$ is a rational quantity, therefore $\sqrt{b} = $ a rational quantity, which is impossible. Therefore $a + \sqrt{b}$ does not $= \sqrt{c}$.

III. *If a surd and a rational quantity be together equal to a surd and a rational quantity, then the rational parts of the equation will be equal, and therefore the surds parts equal.* In brief, if

$$a + \sqrt{b} = c + \sqrt{d},$$

then $\qquad a = c$

and $\qquad \sqrt{b} = \sqrt{d}.$

For, if possible, let $\quad a + \sqrt{b} = c + \sqrt{d};$

then, transposing, $(a - c) + \sqrt{b} = \sqrt{d}.$

Therefore the rational quantity $(a - c)$ and the surd $\sqrt{b} = $ the surd $\sqrt{d}$, which is impossible unless $a - c = 0$, and therefore $a = c$. If $a = c$ it follows that $\sqrt{b} = \sqrt{d}.$

NOTE.—*This is a very important theorem.*

SURDS.

IV. If $a + \sqrt{b} = c + \sqrt{d}$,
then $a - \sqrt{b} = c - \sqrt{d}$.
For, from III., $a = c$
and $\sqrt{b} = \sqrt{d}$.
∴ subtracting, $a - \sqrt{b} = c - \sqrt{d}$.

V. If $\sqrt{a + \sqrt{b}} = \sqrt{x} + \sqrt{y}$, (1)
then $\sqrt{a - \sqrt{b}} = \sqrt{x} - \sqrt{y}$. (2)
Squaring (1), $a + \sqrt{b} = x + y + 2\sqrt{xy}$;
∴ $a = x + y$
and $\sqrt{b} = 2\sqrt{xy}$.
Subtracting, $a - \sqrt{b} = x + y - 2\sqrt{xy}$.
Extracting sq. root, $\sqrt{a - \sqrt{b}} = \sqrt{x} - \sqrt{y}$.

237. We are now in a position to extract the square root of a binomial surd.

Ex. 1.—Extract the square root of $7 + 4\sqrt{3}$.

Assume $\sqrt{7 + 4\sqrt{3}} = \sqrt{x} + \sqrt{y}$, (1)

∴ $\sqrt{7 - 4\sqrt{3}} = \sqrt{x} - \sqrt{y}$. (2)

Multiplying together (1) and (2),

∴ $\sqrt{49 - 48} = x - y$ (3)

or $1 = x - y$.

Squaring (1), $7 + 4\sqrt{3} = x + y + 2\sqrt{xy}$;

∴ $7 = x + y$.

But $1 = x - y$.

Solving, ∴ $x = 4$ and $y = 3$.

∴ $\sqrt{7 + 4\sqrt{3}} = \sqrt{4} + \sqrt{3} = 2 + \sqrt{3}$.

SURDS.

Ex. 2.—Extract the square root of $a + \sqrt{b}$.

Assume $\qquad \sqrt{a + \sqrt{b}} = \sqrt{x} + \sqrt{y},\qquad$ (1)

Then $\qquad \sqrt{a - \sqrt{b}} = \sqrt{x} - \sqrt{y}.\qquad$ (2)

Multiplying together (1) and (2),

$$\therefore \sqrt{a^2 - b} = x - y.$$

Squaring (1) and equating rational terms,

$$a = x + y.$$

Hence, to find x and y, we have the equations,

$$a = x + y, \qquad (3)$$
$$\sqrt{a^2 - b} = x - y. \qquad (4)$$

Therefore, by addition and subtraction of (3) and (4), we obtain

$$2x = a + \sqrt{a^2 - b},$$
and
$$2y = a - \sqrt{a^2 - b},$$
or
$$x = \frac{a + \sqrt{a^2 - b}}{2}$$
and
$$y = \frac{a - \sqrt{a^2 - b}}{2}.$$

Hence $\qquad \sqrt{a + \sqrt{b}} = \sqrt{\dfrac{a + \sqrt{a^2 - b}}{2}} + \sqrt{\dfrac{a - \sqrt{a^2 - b}}{2}}$

and $\qquad \sqrt{a - \sqrt{b}} = \sqrt{\dfrac{a + \sqrt{a^2 - b}}{2}} - \sqrt{\dfrac{a - \sqrt{a^2 - b}}{2}}.$

If $a^2 - b$ is not an exact square these results are more complicated than the original surds.

SURDS.

EXERCISE LXXIII.

Extract the square root of

1. $7 - 4\sqrt{3}$.
2. $4 + 2\sqrt{3}$.
3. $27 + 10\sqrt{2}$.
4. $27 - 10\sqrt{2}$.
5. $6 + 2\sqrt{5}$.
6. $6 - 2\sqrt{5}$.
7. $10 + 4\sqrt{6}$.
8. $21 - 8\sqrt{5}$.
9. $a + b + 2\sqrt{ab}$.
10. $a^2 + b + 2a\sqrt{b}$.
11. $2a + 2\sqrt{a^2 - b^2}$.
12. $2(a^2 + b^2) + 2\sqrt{a^4 + a^2b^2 + b^4}$.
13. $\dfrac{7}{9} + \dfrac{4}{3}\sqrt{\dfrac{1}{3}}$.
14. $\dfrac{26}{100} - 4\sqrt{\dfrac{1}{250}}$.
15. $\dfrac{11}{3} + \dfrac{4}{3}\sqrt{7}$.
16. $\dfrac{7}{2} + \dfrac{3}{2}\sqrt{5}$.
17. $a^2 - 2b\sqrt{a^2 - b^2}$.
18. $(a + b)^2 - 4(a - b)\sqrt{ab}$.
19. $a^2 + x^2 + \sqrt{-(3a^4 - 2a^2x^2 - x^4)}$.
20. $3b^2 + a^2 + \sqrt{6a^2b^2 + a^4}$.

238. The square root of a binomial surd can often be obtained by *inspection*.

Since $\qquad (\sqrt{a} + \sqrt{b})^2 = a + b + 2\sqrt{ab},$

the square root of $\sqrt{a + b + 2\sqrt{ab}} = \sqrt{a} + \sqrt{b},$

from which we see that when the binomial surd is written with 2 for the coefficient of its surd term the terms in its square root are found by obtaining two quantities which, when added together, give $(a + b)$, and multiplied together give ab.

Let us take the example $7 + 4\sqrt{3}$.

$$7 + 4\sqrt{3} = 4 + 3 + 4\sqrt{3}$$
$$= 4 + 3 + 2\sqrt{12}.$$

202 SURDS.

The numbers which, when multiplied together, give 12, and added together, 7, are 4 and 3,

$$\therefore \sqrt{7 + 4\sqrt{3}} = \sqrt{4} + \sqrt{3}$$
$$= 2 + \sqrt{3}.$$

EXERCISE LXXIV.

Obtain by inspection

1. $\sqrt{10 + 2\sqrt{21}}, \quad \sqrt{16 + 2\sqrt{55}}, \quad \sqrt{14 - 4\sqrt{6}}.$

2. $\sqrt{13 - 2\sqrt{30}}, \quad \sqrt{38 - 12\sqrt{10}}, \quad \sqrt{57 - 12\sqrt{15}}.$

3. $\sqrt{14 + 6\sqrt{5}}, \quad \sqrt{20 - 8\sqrt{6}}, \quad \sqrt{11 - 6\sqrt{2}}.$

4. $\sqrt{9 - 2\sqrt{14}}, \quad \sqrt{87 - 12\sqrt{42}}, \quad \sqrt{17 + 4\sqrt{15}}.$

239. When the denominator of a fraction consists of one or more surd terms it is sometimes necessary to reduce the fraction to the form of one with a rational denominator. In simple examples this process, called *rationalizing the denominator*, consists of multiplying both the numerator and the denominator of the fraction by its denominator, with one sign changed.

Ex. 1.—Rationalize the denominator of $\dfrac{2}{\sqrt{3} + \sqrt{2}}$.

Multiplying both numerator and denominator by $\sqrt{3} - \sqrt{2}$,

$$\therefore \frac{2}{\sqrt{3} + \sqrt{2}} = \frac{2(\sqrt{3} - \sqrt{2})}{(\sqrt{3} + \sqrt{2})(\sqrt{3} - \sqrt{2})} = \frac{2\sqrt{3} - 2\sqrt{2}}{3 - 2}$$
$$= \frac{2\sqrt{3} - 2\sqrt{2}}{1}.$$

SURDS. 203

Ex. 2.—Rationalize the denominator of $\dfrac{1}{\sqrt{3}+\sqrt{2}+1}$.

Multiplying both numerator and denominator by $\sqrt{3}-\sqrt{2}+1$,

$$\therefore \frac{1}{\sqrt{3}+\sqrt{2}+1} = \frac{\sqrt{3}-\sqrt{2}+1}{(\sqrt{3}+1+\sqrt{2})(\sqrt{3}+1-\sqrt{2})}$$

$$= \frac{\sqrt{3}-\sqrt{2}+1}{(\sqrt{3}+1)^2-2} = \frac{\sqrt{3}-\sqrt{2}+1}{4+2\sqrt{3}-2}$$

$$= \frac{\sqrt{3}-\sqrt{2}+1}{2\sqrt{3}+2} = \frac{(\sqrt{3}-\sqrt{2}+1)(\sqrt{3}-1)}{2(\sqrt{3}-1)(\sqrt{3}+1)}$$

$$= \frac{2-\sqrt{6}+\sqrt{2}}{4}.$$

240. It is not often required in elementary algebra to rationalize denominators with radical indices higher than the second order. When such denominators occur their rationalization may be accomplished by finding a factor which, taken with the given denominator, will give as a product a rational expression. We give a simple example, as the more difficult problems are not suitable for classes in elementary work.

Ex.—Rationalize the denominator of $\dfrac{1}{\sqrt{x}+\sqrt[3]{y}}$.

$$\frac{1}{\sqrt{x}+\sqrt[3]{y}} = \frac{1}{x^{\frac{1}{2}}+y^{\frac{1}{3}}}.$$

Now, the question arises, of what rational quantity is $x^{\frac{1}{2}}+y^{\frac{1}{3}}$ a factor? As the sum of two quantities will divide the difference of two even powers of these quantities, therefore $x^{\frac{1}{2}}+y^{\frac{1}{3}}$ is a factor of $(x^{\frac{1}{2}})^n - (y^{\frac{1}{3}})^n$ if n is an *even* number; and this expression is *rational* if $n=6$. Therefore if we multiply both numerator and denominator by $\dfrac{(x^{\frac{1}{2}})^6-(y^{\frac{1}{3}})^6}{x^{\frac{1}{2}}+y^{\frac{1}{3}}}$ or $\dfrac{x^3-y^2}{x^{\frac{1}{2}}+y^{\frac{1}{3}}}$, that is, by $x^{\frac{5}{2}}-x^2y^{\frac{1}{3}}+$

$x^{\frac{2}{3}}y^{\frac{2}{3}} - xy + x^{\frac{1}{3}}y^{\frac{4}{3}} - x^{\frac{5}{3}}$, the denominator of the fraction $\dfrac{1}{\sqrt{x} + \sqrt[3]{y}}$ will be a rational quantity.

EXERCISE LXXV.

Rationalize the denominators of the following:—

1. $\dfrac{2\sqrt{2}}{3\sqrt{2} - 2\sqrt{3}}$, $\dfrac{\sqrt{11} + \sqrt{5}}{\sqrt{11} - \sqrt{5}}$, $\dfrac{7 - 3\sqrt{10}}{5 + 4\sqrt{5}}$.

2. $\dfrac{21\sqrt{3}}{4\sqrt{3} - 3\sqrt{2}}$, $\dfrac{120\sqrt{2}}{5\sqrt{6} + 3\sqrt{10}}$, $\dfrac{75\sqrt{14}}{8\sqrt{2} + 2\sqrt{7}}$.

3. $\dfrac{\sqrt{5} - \sqrt{3}}{\sqrt{5} + \sqrt{3}}$, $\dfrac{\sqrt{13} - \sqrt{10}}{\sqrt{13} + \sqrt{10}}$, $\dfrac{2 + 2\sqrt{2}}{6 + 5\sqrt{6}}$.

4. $\dfrac{\sqrt{8} + \sqrt{7}}{\sqrt{7} - \sqrt{2}}$, $\dfrac{\sqrt{6} + \sqrt{10}}{\sqrt{10} - \sqrt{6}}$, $\dfrac{3 + 5\sqrt{3}}{4 - 3\sqrt{2}}$.

5. $\dfrac{1 - \sqrt{2} + \sqrt{3}}{1 + \sqrt{2} - \sqrt{3}}$, $\dfrac{2 + \sqrt{6} - \sqrt{2}}{2 - \sqrt{6} + \sqrt{2}}$, $\dfrac{3 - \sqrt{5} - \sqrt{2}}{3 + \sqrt{5} + \sqrt{2}}$.

6. $\dfrac{7 - 2\sqrt{3} + 3\sqrt{2}}{3 + 3\sqrt{3} - 2\sqrt{2}}$, $\dfrac{2\sqrt{3} - 3\sqrt{6}}{\sqrt{2} - \sqrt{3} + \sqrt{5} - \sqrt{6}}$.

7. $\dfrac{\sqrt{a+x} + \sqrt{a-x}}{\sqrt{a+x} - \sqrt{a-x}}$, $\dfrac{\sqrt{a^2+b^2} + \sqrt{a^2-b^2}}{\sqrt{a^2+b^2} - \sqrt{a^2-b^2}}$.

8. $\dfrac{\sqrt{x+3a} + \sqrt{x-3a}}{\sqrt{x+3a} - \sqrt{x-3a}}$, $\dfrac{x + \sqrt{4a^2 - x^2}}{x - \sqrt{4a^2 - x^2}}$.

9. $\dfrac{1}{\sqrt{2} + \sqrt[3]{3}}$, $\dfrac{2\sqrt{3}}{\sqrt[3]{4} + \sqrt[3]{3}}$, $\dfrac{2\sqrt{2} - \sqrt{3}}{\sqrt{3} - \sqrt[3]{5}}$.

10. $\dfrac{7 - 2\sqrt{5}}{4 - \sqrt{5}} - \dfrac{15 + 6\sqrt{5}}{2 + \sqrt{5}}$.

11. $\dfrac{\sqrt{3} + 1}{\sqrt{3} - 1} + \dfrac{\sqrt{3} - 1}{\sqrt{3} + 1}$.

SURDS.

12. $\dfrac{20 + 30\sqrt{2}}{3 + \sqrt{3}} - \dfrac{5 - 2\sqrt{2}}{2 - \sqrt{3}}.$

13. $\dfrac{5 + \sqrt{5}}{1 + \sqrt{5}} \times \dfrac{5 + 2\sqrt{5}}{2 + \sqrt{5}}.$

14. $\dfrac{a + x + \sqrt{a + x}}{a + x + \sqrt{(a + x)^3}}.$

15. $\dfrac{2\sqrt{a + b} + 3\sqrt{a - b}}{2\sqrt{a + b} - 3\sqrt{a - b}}.$

241. It is often necessary to find the value of a surd expression when a particular value is given to the unknown quantity involved.

Ex.—Find the value of $x^2 + xy + y^2$ when $x = \dfrac{\sqrt{3} + 1}{\sqrt{3} - 1}$ and $y = \dfrac{\sqrt{3} - 1}{\sqrt{3} + 1}.$

$x^2 + xy + y^2 = (x + y)^2 - xy$

$= \left(\dfrac{\sqrt{3} + 1}{\sqrt{3} - 1} + \dfrac{\sqrt{3} - 1}{\sqrt{3} + 1}\right)^2 - \dfrac{\sqrt{3} + 1}{\sqrt{3} - 1} \times \dfrac{\sqrt{3} - 1}{\sqrt{3} + 1}.$

$= \left(\dfrac{4 + 2\sqrt{3} + 4 - 2\sqrt{3}}{3 - 1}\right)^2 - 1$

$= (4)^2 - 1 = 15.$

This example might have been solved by rationalizing the denominators of $\dfrac{\sqrt{3} + 1}{\sqrt{3} - 1}$ and $\dfrac{\sqrt{3} - 1}{\sqrt{3} + 1}$ before substituting.

Thus, $\dfrac{\sqrt{3} + 1}{\sqrt{3} - 1} = \dfrac{(\sqrt{3} + 1)^2}{3 - 1} = \dfrac{4 + 2\sqrt{3}}{2} = 2 + \sqrt{3}$

and $\dfrac{\sqrt{3} - 1}{\sqrt{3} + 1} = \dfrac{(\sqrt{3} - 1)^2}{3 - 1} = \dfrac{4 - 2\sqrt{3}}{2} = 2 - \sqrt{3},$

$\therefore x^2 = (2 + \sqrt{3})^2 = 7 + 4\sqrt{3}$

and $y^2 = (2 - \sqrt{3})^2 = 7 - 4\sqrt{3};$

also, $xy = 1,$

$\therefore x^2 + xy + y^2 = 7 + 4\sqrt{3} + 1 + 7 - 4\sqrt{3} = 15.$

242. In the following example instead of directly substituting the problem is reduced to a simpler form by means of a well-known theorem in fractions. (Art. 169, Theorem I.)

Ex.—Find the value of $\dfrac{\sqrt{a+x}+\sqrt{a-x}}{\sqrt{a+x}-\sqrt{a-x}}$ when $x = \dfrac{2ab}{1+b^2}$.

Let $\dfrac{\sqrt{a+x}+\sqrt{a-x}}{\sqrt{a+x}-\sqrt{a-x}} = \dfrac{k}{1}$.

Adding numerator and denominator of each fraction, and dividing by their difference,

$$\dfrac{2\sqrt{a+x}}{2\sqrt{a-x}} = \dfrac{k+1}{k-1}$$

or $\dfrac{\sqrt{a+x}}{\sqrt{a-x}} = \dfrac{k+1}{k-1}$.

Squaring, $\dfrac{a+x}{a-x} = \dfrac{k^2+2k+1}{k^2-2k+1}$.

Again, adding numerator and denominator of each fraction, and dividing by their difference,

$$\dfrac{2a}{2x} = \dfrac{2k^2+2}{4k},$$

or $\dfrac{a}{x} = \dfrac{k^2+1}{2k}$ or $\dfrac{x}{a} = \dfrac{2k}{k^2+1}$.

Substituting now for x its value $\dfrac{2ab}{1+b^2}$ we obtain

$$\dfrac{2b}{1+b^2} = \dfrac{2k}{k^2+1}$$

or $\dfrac{1+2b+b^2}{1-2b+b^2} = \dfrac{1+2k+k^2}{1-2k+k^2}$.

Extracting square root of each side,

$$\dfrac{1+k}{1-k} = \pm\dfrac{1+b}{1-b}.$$

SURDS. 207

If we take the *positive* sign we obtain for k the value b; if we take the *negative* sign we find $k = \dfrac{1}{b}$. Therefore the value of the fraction $= b$ or $\dfrac{1}{b}$.

Had this problem been solved by direct substitution the process would have been somewhat as follows:—

$$\frac{\sqrt{a+x}+\sqrt{a-x}}{\sqrt{a+x}-\sqrt{a-x}} = \frac{\sqrt{a+\dfrac{2ab}{1+b^2}}+\sqrt{a-\dfrac{2ab}{1+b^2}}}{\sqrt{a+\dfrac{2ab}{1+b^2}}-\sqrt{a-\dfrac{2ab}{1+b^2}}}$$

$$= \frac{\sqrt{a}\left\{\sqrt{1+\dfrac{2b}{1+b^2}}+\sqrt{1-\dfrac{2b}{1+b^2}}\right\}}{\sqrt{a}\left\{\sqrt{1+\dfrac{2b}{1+b^2}}-\sqrt{1-\dfrac{2b}{1+b^2}}\right\}}$$

$$= \frac{\sqrt{\dfrac{1+2b+b^2}{1+b^2}}+\sqrt{\dfrac{1-2b+b^2}{1+b^2}}}{\sqrt{\dfrac{1+2b+b^2}{1+b^2}}-\sqrt{\dfrac{1-2b+b^2}{1+b^2}}}$$

$$= \frac{\dfrac{1+b}{\sqrt{1+b^2}} \pm \dfrac{1-b}{\sqrt{1+b^2}}}{\dfrac{1+b}{\sqrt{1+b^2}} \mp \dfrac{1-b}{\sqrt{1+b^2}}}$$

$$= \frac{(1+b)\pm(1-b)}{(1+b)\mp(1-b)}$$

$$= \frac{2}{2b} \text{ or } \frac{2b}{2} = \frac{1}{b} \text{ or } b.$$

N.B.—The double value, it will be noticed, is obtained by using the double sign before the square root of the second terms of numerator and denominator. Had the double sign been used in all the terms the result would not have been changed.

SURDS.

EXERCISE LXXVI.

1. Find the value of $\dfrac{\sqrt{a^2+x^2}+x}{\sqrt{a^2+x^2}-x}$ when $x=\dfrac{a(b-c)}{2\sqrt{bc}}$.

2. Find the value of $\dfrac{\sqrt{x^2+a^2}+\sqrt{x^2-a^2}}{\sqrt{x^2+a^2}-\sqrt{x^2-a^2}}$ when $x=\sqrt{\dfrac{a^4+1}{2}}$.

3. Find the value of $\dfrac{ax+1+\sqrt{a^2x^2-1}}{ax+1-\sqrt{a^2x^2-1}}$ when $x=\dfrac{2}{\sqrt{4ab-b^2}}$.

4. Find the value of $x^2+\dfrac{1}{x^2}+x+\dfrac{1}{x}$ when $x=\dfrac{3+\sqrt{5}}{2}$.

5. Find the value of $\sqrt{\dfrac{x+4}{x-4}}$ when $x=4\sqrt{2}$.

6. Given $\sqrt{3}=1.7320508$, find the value of $\dfrac{1}{2+\sqrt{3}}$.

7. Find the value of $\dfrac{1+x}{1+\sqrt{1+x}}+\dfrac{1-x}{1+\sqrt{1-x}}$ when $x=\dfrac{\sqrt{3}}{2}$.

8. Find the value of $\dfrac{1+x}{1+\sqrt{1+x}}+\dfrac{1-x}{1-\sqrt{1-x}}$ when $x=\dfrac{\sqrt{3}}{2}$.

9. Find the value of $\dfrac{\sqrt{a^2+x^2}+\sqrt{a^2-x^2}}{\sqrt{a^2+x^2}-\sqrt{a^2-x^2}}$ when $x=\dfrac{2a}{\sqrt{5}}$.

10. If $2m=x+\dfrac{1}{x}$, $2n=y+\dfrac{1}{y}$, find the value of

    $mn+\sqrt{(m^2-1)(n^2-1)}$ in terms of x and y.

11. If $x=\dfrac{2ac}{b(1+c^2)}$ find the value of $\dfrac{\sqrt{a+bx}+\sqrt{a-bx}}{\sqrt{a+bx}-\sqrt{a-bx}}$.

IMAGINARY EXPRESSIONS.

243. The square root of a positive quantity may be positive or negative. For $(+a) \times (+a) = a^2$; also $(-a) \times (-a) = a^2$; therefore $\sqrt{a^2} = \pm a$. As both $(+a)^2$ and $(-a)^2$ are positive it is easily seen that the square root of $-a^2$ cannot be either $+a$ or $-a$. Such an expression as $\sqrt{-a^2}$ is therefore called an *impossible*, or *imaginary*, expression. We mean by an *imaginary* expression one which, if raised to an *even* power, will give a *negative* expression. All other expressions are said to be *real*.

244. All imaginary square roots may be made to take the same algebraic form.

Thus $\sqrt{-a^2} = \sqrt{a^2 \times (-1)} = a\sqrt{-1}$;

$\sqrt{-x} = \sqrt{x \times (-1)} = \sqrt{x}\sqrt{-1}$.

245. The beginner should note carefully such results as the following:—

(1) $\sqrt{-1} \times \sqrt{-1} = -1$, *i.e.*, $(\sqrt{-1})^2 = -1$.

(2) $(\sqrt{-1})^3 = (\sqrt{-1})^2 \times (\sqrt{-1}) = (-1) \times \sqrt{-1} = -\sqrt{-1}$.

(3) $(\sqrt{-1})^4 = (\sqrt{-1})^2 \times (\sqrt{-1})^2 = (-1) \times (-1) = +1$

Generally, $\sqrt{-1}$ raised to an *even* power is rational, to an odd power, *imaginary*.

If the power is *even* and a multiple of 4 the result is positive; if *even*, and not a multiple of 4, it is negative.

Ex. 1.—Express $\sqrt{-4}$ in the form of $a\sqrt{-1}$.

$\sqrt{-4} = \sqrt{4 \times -1} = 2\sqrt{-1}$.

IMAGINARY EXPRESSIONS.

Ex. 2.—Simplify $\dfrac{\sqrt{-9}}{\sqrt{-3}}$.

Since $\sqrt{-9} = 3\sqrt{-1}$ and $\sqrt{-3} = \sqrt{3} \cdot \sqrt{-1}$,

$$\therefore \dfrac{\sqrt{-9}}{\sqrt{-3}} = \dfrac{3\sqrt{-1}}{\sqrt{3}\sqrt{-1}} = \dfrac{3}{\sqrt{3}} = \sqrt{3}.$$

Ex. 3.—Multiply $\sqrt{5} + \sqrt{-6}$ by $\sqrt{6} - \sqrt{-8}$.

$$(\sqrt{5} + \sqrt{-6}) \times (\sqrt{6} - \sqrt{-8})$$
$$= (\sqrt{5} + \sqrt{6} \cdot \sqrt{-1}) \times (\sqrt{6} - \sqrt{8} \cdot \sqrt{-1})$$
$$= \sqrt{30} + 6\sqrt{-1} - \sqrt{40} \cdot \sqrt{-1} + \sqrt{48}$$
$$= \sqrt{30} + (6 - \sqrt{40})\sqrt{-1} + \sqrt{48}$$
$$= \sqrt{30} + (6 - 2\sqrt{10})\sqrt{-1} + 4\sqrt{3}.$$

Ex. 4.—Multiply $a + \sqrt{-1}$ by $a - \sqrt{-1}$.

$$(a + \sqrt{-1})(a - \sqrt{-1}) = a^2 - (\sqrt{-1})^2 = a^2 - (-1) = a^2 + 1.$$

EXERCISE LXXVII.

Express in the form of $a\sqrt{-1}$,

1. $\sqrt{-625}$, $\sqrt{-25}$, $\sqrt{-81}$, $\sqrt{-144}$, $\sqrt{-b^4}$.
2. $\sqrt{-16}$, $\sqrt{-64}$, $\sqrt{-9x^6}$, $\sqrt{-81m^6}$.

Find the value of

3. $(\sqrt{-1})^6$, $(\sqrt{-1})^8$, $(\sqrt{-1})^7$, $(\sqrt{-1})^{16}$, $(\sqrt{-1})^{21}$.

Add together

4. $\sqrt{-25}$, $\sqrt{-49}$, $\sqrt{-121}$, $-\sqrt{-64}$.

IMAGINARY EXPRESSIONS.

5. $5+\sqrt{-16}$, $3-\sqrt{-4}$, $8+\sqrt{-4}$.

6. $3+2\sqrt{-1}$, $4-2\sqrt{-1}$, $7+3\sqrt{-1}$, $4+2b\sqrt{-1}$.

Multiply

7. $\sqrt{-16} \times \sqrt{-9}$, $\sqrt{-25} \times \sqrt{-49}$, $\sqrt{-9y^2} \times \sqrt{-16z^6}$.

8. $(3+5\sqrt{-1}) \times (4-7\sqrt{-1})$, $(3x-8\sqrt{-1}) \times (3x+3\sqrt{-1})$.

9. $(2a+3b\sqrt{-1}) \times (c-d\sqrt{-1})$, $(m+b\sqrt{-1}) \times (n-b\sqrt{-1})$.

Divide

10. $\dfrac{a}{\sqrt{-1}}$, $\dfrac{x}{\sqrt{-x}}$, $\dfrac{\sqrt{-x^2}}{\sqrt{-x}}$, $\dfrac{b}{\sqrt{-b^2}}$, $\dfrac{\sqrt{-10x^3}}{\sqrt{-5x}}$.

11. If $x = \dfrac{-1+\sqrt{-3}}{2}$ find the value of x^2+x+1.

12. If $x = \dfrac{1-\sqrt{-3}}{2}$ find the value of x^2-x+1.

13. Simplify $\dfrac{7+\sqrt{-3}}{2-\sqrt{-3}} + \dfrac{8+3\sqrt{-3}}{2+\sqrt{-3}} - \dfrac{4(2-\sqrt{-3})}{1-\sqrt{-3}}$.

14. If $x = \dfrac{-1+\sqrt{-3}}{2}$ and $y = \dfrac{-1-\sqrt{-3}}{2}$

prove $x^2 = y$ and $y^2 = x$.

CHAPTER XIV.

SIMPLE EQUATIONS INVOLVING SURDS.

246. When an equation involves one or more surd expressions it may frequently be solved by a process called *clearing of surds*.

247. If an equation contains but *one* surd the radical sign can be removed by bringing the surd to one side of the equation and the remaining terms to the other, and then raising both sides to the power indicated by the radical index.

Ex. 1.—Solve $\sqrt{4x^2 + 3x - 16} = 2x + 2$.

Squaring both sides,
$$4x^2 + 3x - 16 = 4x^2 + 8x + 4;$$
$$\therefore -5x = -20 \text{ or } x = 4.$$

Ex. 2.—Solve $\sqrt{x^2 - 9} + x = 9$.

Transposing, $\quad \sqrt{x^2 - 9} = 9 - x$

Squaring, $\quad x^2 - 9 = 81 - 18x + x^2$.

$$\therefore 18x = 90 \text{ or } x = 5.$$

248. When two quadratic surds occur in the equation the clearing of surds may be accomplished by bringing one surd to one side of the equation and all the other terms to the other, and then squaring both sides. There may still remain one surd, which can be removed by the method of the preceding article.

SIMPLE EQUATIONS INVOLVING SURDS. 213

Ex. 1.—Solve $\sqrt{x+4} + \sqrt{x+15} = 11$.

Transposing, $\sqrt{x+15} = 11 - \sqrt{x+4}$

Squaring both sides, $x + 15 = 121 - 22\sqrt{x+4} + x + 4$.

Transposing, $22\sqrt{x+4} = 110$;

$\therefore \sqrt{x+4} = 5$.

Squaring again, $x + 4 = 25$;

$\therefore x = 21$.

Sometimes it is more convenient to keep both surds on one side and take all the other terms to the other side.

Ex. 2.—Solve $\sqrt{x+4} - \sqrt{x-4} = 4$.

Squaring, $x + 4 + x - 4 - 2\sqrt{x^2 - 16} = 16$.

Simplifying, $-\sqrt{x^2 - 16} = 8 - x$.

Squaring again, $x^2 - 16 = 64 - 16x + x^2$.

$\therefore 16x = 80$ and $x = 5$.

Ex. 3.—Solve $\dfrac{\sqrt{12x+1} + \sqrt{12x}}{\sqrt{12x+1} - \sqrt{12x}} = 18$.

$\dfrac{\sqrt{12x+1} + \sqrt{12x}}{\sqrt{12x+1} - \sqrt{12x}} = \dfrac{18}{1}$.

Therefore, adding numerator and denominator of each fraction, and dividing by difference,

$\dfrac{2\sqrt{12x+1}}{2\sqrt{12x}} = \dfrac{19}{17}$

or

$\sqrt{\dfrac{12x+1}{12x}} = \dfrac{19}{17}$.

Squaring both sides, $\dfrac{12x+1}{12x} = \dfrac{361}{289}$.

Dividing numerators by denominators,

$$1 + \frac{1}{12x} = 1 + \frac{72}{289}.$$

Simplifying,
$$\frac{1}{12x} = \frac{72}{289}$$

and
$$x = \frac{289}{864}.$$

Ex. 4.—Solve $\sqrt{x} + \sqrt{x+3} = \dfrac{5}{\sqrt{x+3}}$.

Multiplying through by the denominator $\sqrt{x+3}$,

$$\sqrt{x(x+3)} + x + 3 = 5.$$

Transposing, $\sqrt{x(x+3)} = 2 - x.$

Squaring both sides, $x^2 + 3x = 4 - 4x + x^2.$

$$\therefore 7x = 4 \text{ or } x = \frac{4}{7}.$$

249. Sometimes when the equation contains *three* quadratic surds it assumes the form of a simple equation after clearing of surds. In such cases to clear of radicals we employ the same methods as for equations containing one or two surd expressions.

Ex. 1.—Solve $\sqrt{x-5} + \sqrt{x+5} = \sqrt{4x-6}.$

Squaring both sides,

$$x - 5 + x + 5 + 2\sqrt{x^2 - 25} = 4x - 6.$$

Simplifying and transposing,

$$2\sqrt{x^2 - 25} = 2x - 6$$

or $\sqrt{x^2 - 25} = x - 3.$

Squaring again, $x^2 - 25 = x^2 - 6x + 9.$

$$\therefore 6x = 34 \text{ or } x = 5\tfrac{2}{3}.$$

SIMPLE EQUATIONS INVOLVING SURDS.

Ex. 2.—Solve $\sqrt{x+3}+\sqrt{x+8}=2\sqrt{x}$.

Squaring both sides,

$$x+3+x+8+2\sqrt{x^2+11x+24}=4x.$$

Simplifying and transposing,

$$2\sqrt{x^2+11x+24}=2x-11.$$

Squaring again, $\quad 4x^2+44x+96=4x^2-44x+121.$

$$\therefore 88x=25 \text{ or } x=\frac{25}{88}.$$

EXERCISE LXXVIII.

Solve the following equations:—

1. $2\sqrt{x+5}=\sqrt{28}$.
2. $3\sqrt{4x-8}=\sqrt{13x-3}$.
3. $4=2\sqrt{x}-3$.
4. $5-3\sqrt{x}=4$.
5. $9+4\sqrt{x}=11$.
6. $6-3\sqrt{x}=4$.
7. $11-4\sqrt{5x}=9$.
8. $7+2\sqrt[3]{3x}=5$.
9. $\sqrt{x+9}=5\sqrt{x-3}$.
10. $2x-8=\sqrt{4x^2-12x+32}$.
11. $\sqrt{x-5}-\sqrt{x-4}=1$.
12. $\sqrt{x+15}+\sqrt{x}=15$.
13. $\sqrt{x-16}+\sqrt{x}=8$.
14. $\sqrt{x-5}+\sqrt{x+7}=6$.
15. $\sqrt{x-1}+\sqrt{x-4}-3=0$.
16. $\sqrt{x-1}=3-\sqrt{x+4}$.
17. $\sqrt{x}+\sqrt{x-9}=\dfrac{36}{\sqrt{x-9}}$.
18. $\sqrt{x+7}+\sqrt{x}=\dfrac{28}{\sqrt{x+7}}$.
19. $\sqrt{x-15}+\sqrt{x}=\dfrac{105}{\sqrt{x-15}}$.
20. $\sqrt{x}+\sqrt{x-4}=\dfrac{8}{\sqrt{x-4}}$.

21. $\sqrt{x} + \sqrt{x-21} = \dfrac{35}{\sqrt{x}}$.

22. $\dfrac{\sqrt{x}-3}{\sqrt{x}+3} = \dfrac{\sqrt{x}+1}{\sqrt{x}-2}$.

23. $\dfrac{\sqrt{x}-8}{\sqrt{x}-6} = \dfrac{\sqrt{x}-4}{\sqrt{x}+2}$.

24. $\dfrac{\sqrt{x}+16}{\sqrt{x}+4} = \dfrac{\sqrt{x}+32}{\sqrt{x}+12}$.

25. $\dfrac{1+\sqrt{1-x}}{1-\sqrt{1-x}} = 3$.

26. $\dfrac{3x+\sqrt{4x+x^2}}{3x-\sqrt{4x+x^2}} = 2$.

27. $\sqrt{x+1} + \sqrt{x+16} = 2\sqrt{x+25}$.

28. $\sqrt{4a+x} = 2\sqrt{b+x} - \sqrt{x}$.   29. $\sqrt{4+x} + \sqrt{x+1} = 2\sqrt{x-2}$.

30. $\sqrt{4x+5} - \sqrt{x} = \sqrt{x+3}$.   31. $\dfrac{1}{\sqrt{x+1}} + \dfrac{1}{\sqrt{x-1}} = \dfrac{1}{\sqrt{x^2-1}}$.

32. $\sqrt{(x-a)^2 + 2ab + b^2} = x - a + b$.

33. $\sqrt{(x+a)^2 + 2ab + b^2} = b - a - x$.

34. $x^{\frac{1}{2}} + (x-9)^{\frac{1}{2}} = 36(x-9)^{-\frac{1}{2}}$.   35. $(12+x)^{\frac{1}{2}} + x^{\frac{1}{2}} = 6$.

36. $(x-3)^{\frac{1}{2}} + x^{\frac{1}{2}} = \dfrac{3}{(x-3)^{\frac{1}{2}}}$.   37. $x^{\frac{1}{2}} - \left(\dfrac{a}{x}\right)^{\frac{1}{2}} = (a+x)^{\frac{1}{2}}$.

38. $\dfrac{\sqrt{1-x}}{2-\sqrt{1+x}} = \dfrac{\sqrt{1+x}}{2+\sqrt{1-x}}$.   39. $\dfrac{5x-1}{\sqrt{5x}+1} = 1 + \dfrac{\sqrt{5x}-1}{2}$.

40. $\dfrac{1}{1+\sqrt{1-x}} + \dfrac{1}{1-\sqrt{1-x}} = \dfrac{2x}{9}$.

41. $\dfrac{2}{x+\sqrt{2+x^2}} + \dfrac{2}{x-\sqrt{2+x^2}} = x + 5$.

42. $\dfrac{5}{x+\sqrt{5+x^2}} + \dfrac{5}{x-\sqrt{5+x^2}} = x + 8$.

43. $\sqrt{4a+x} + \sqrt{a+x} = 2\sqrt{x+2a}$.

CHAPTER XV.

SYMMETRICAL EXPRESSIONS.

250. An expression is said to be symmetrical with regard to any number of letters when its value is not altered by interchanging any two of the given letters. Thus, if we interchange a and b in ab, $a+b$, a^2+ab+b^2, $(x-a)(x-b)$, we get ba, $b+a$, b^2+ba+a^2, $(x-b)(x-a)$. But these are the same expressions as before; they are therefore said to be symmetrical with regard to a and b. If in the last expression we interchange a and x we get $(a-x)(a-b)$, which is not the same as before; it is therefore *not* symmetrical with regard to a and x.

Again, $a+b+c$, $a^2+b^2+c^2-ab-bc-ca$, $(x-a)(x-b)(x-c)$ are unaltered by interchanging any two of the letters a, b, c; these expressions are therefore symmetrical with regard to the three letters a, b, c.

Similarly, expressions may be formed which are symmetrical with regard to four or more letters, but we shall confine our attention chiefly to those formed from three.

251. The simplest form of a symmetrical expression is the sum or the product of the letters involved, thus $a+b+c$ or abc. If one term, involving one or more of the letters, be given, the remaining terms necessary to form a symmetrical group can be at once written down by using all the letters in the same way; thus if a^2 be the given term, $a^2+b^2+c^2$ is the group required. Similarly from $-ab$ we write $-ab-bc-ca$; from a^2b we write $a^2(b+c)+b^2(c+a)+c^2(a+b)$ to complete the symmetry. In the latter example we are given the square of one letter into the first power of another; therefore to complete the symmetry we must have the *square of each into the first power of each of the others*. If two or more terms be given we can complete the symmetry by writing a group as before from each term; thus from a^2-bc we write $a^2+b^2+c^2-ab-bc-ca$. But from $ab+bc$ we should only write *one* group, viz., $ab+bc+ca$, because the two given terms belong to the same symmetrical group.

A symmetrical group of terms is evidently homogeneous.

252. If two or more symmetrical expressions be combined in any way by addition, subtraction, multiplication or division, or if a root be extracted, the result in any case will be symmetrical; for if two letters be interchanged in the given expressions the same letters will be interchanged in the result, and since the given expressions are unchanged the result will be unchanged, *i.e.*, it will be symmetrical.

253. The product of two homogeneous expressions will evidently be homogeneous and of a number of dimensions equal to the sum of the dimensions of the factors, since each term of the product is formed by multiplying two terms, one being taken from each factor. Similarly, the quotient of two homogeneous expressions will be homogeneous, and of a number of dimensions equal to the difference between the dimensions of the dividend and the divisor.

This Art. and the preceding are of great assistance in testing the accuracy of algebraical work and in remembering the factors of algebraical expressions.

254. Many expressions which have not the perfect symmetry described in Art. 250 are still unaffected by particular interchanges of the letters involved. For example, the expression $(ax-by)^2-(ay-bx)^2$ is unaltered by interchanging a and x, providing we at the same time interchange b and y; or we may interchange a and b providing we also interchange x and y. Again, the expressions $a^2b+b^2c+c^2a$ and $(a-b)(b-c)(c-a)$ remain the

SYMMETRICAL EXPRESSIONS. 219

same if we change a into b, b into c, and c into a; but if we interchange any *two* letters the latter expression has the same numerical value as before, but of the opposite sign, while the former expression is entirely different. Expressions like the latter are sometimes called **Alternating Expressions.** Expressions which are unaffected by a series of changes similar to that just given are of the greatest importance, and the greater portion of this chapter is devoted to their consideration.

255. We shall now explain the formation and investigate the properties of the particular class of symmetrical expressions referred to in the preceding Art.

Place any three letters, a, b, c, on the circumference of a circle; then in passing around the circle they will be found in the order of one of the three groups, abc, bca, cab. No other order is possible so long as we pass around the circle in the same direction, and each group is derived from the preceding by changing a into b, b into c, and c into a. Similarly, if any expression whatever containing one or more of these letters be written down, two others may be derived from it by the same series of changes.

Thus from $x + a$ we get $x + b$ and $x + c$.
" $a - b$ we get $b - c$ and $c - a$.
" $a^2(b - c)$ we get $b^2(c - a)$ and $c^2(a - b)$.
" $(b - c)(x - a)$ we get $(c - a)(x - b)$ and $(a - b)(x - c)$.

256. Similarly, from $x + b$ we get $x + c$ and $x + a$; from $c^2(a - b)$ we get $a^2(b - c)$ and $b^2(c - a)$, etc. Thus from any one of the three expressions the other two may be obtained by the same series of changes, the expressions following each other in the same way as a, b, c follow each other on the circle. If now we form a compound expression by taking the sum or the product of the three expressions so obtained, it will remain unchanged when a is changed into b, b into c, and c into a, just as $a + b + c = b + c + a$ or $abc = bca$. For example, $(a - b)(b - c)(c - a) = (b - c)(c - a)(a - b)$; $a^2(b - c) + b^2(c - a) + c^2(a - b) = b^2(c - a) + c^2(a - b) + a^2(b - c)$, etc. In these examples the letters are said to be written in circular order,

SYMMETRICAL EXPRESSIONS.

and the expressions so formed are said to be symmetrical in circular order, to distinguish them from those which are perfectly symmetrical.

257. The orderly manner in which the letters of symmetrical expressions follow each other enables us to transform or simplify such expressions without writing down all the terms which compose them. This will be understood from the following examples:

Ex. 1.—Arrange in powers of x the expression
$$c(x-a)(x-b) + a(x-b)(x-c) + b(x-c)(x-a).$$

We observe that the letters a, b, c in the successive terms follow each other in circular order. Expanding the first term we get
$c(x-a)(x-b) = cx^2 - c(a+b)x + abc.$

The coefficient of x^2 is c; in the next term it will be a, and in the next b; hence the coefficient of x^2 in the result is $a+b+c$.

The coefficient of x is $-(bc+ca)$; in the following terms it will be $-(ca+ab)$ and $-(ab+bc)$; hence the coefficient of x in the result is $-2(ab+bc+ca)$.

From abc we know we shall get bca and cab in the following terms.

The whole result is therefore $(a+b+c)x^2 - 2(ab+bc+ca)x + 3abc.$

Ex. 2.—Simplify $(a-b)(b-c) + (b-c)(c-a) + (c-a)(a-b)$.
$$(a-b)(b-c) = ab + bc - ca - b^2.$$

From $ab+bc-ca$ we learn that if all the terms were expanded we should have all the products ab, bc, ca three times over, twice with the positive sign and once with the negative sign.

From $-b^2$ we learn that we should have the square of each letter taken with the negative sign.

The result is therefore $ab + bc + ca - a^2 - b^2 - c^2.$

Ex. 3.—Find the value of $x^2 + y^2 + z^2 - xy - yz - zx$ in terms of a, b, c, when $x = a+b$, $y = b+c$, $z = c+a$.

This expression consists of two parts, each symmetrical in itself; the value of each must be found separately.

SYMMETRICAL EXPRESSIONS. 221

Now, $x^2 = (a+b)^2 = a^2+b^2+2ab$, from which we see that we shall have all the squares twice over, and also twice all the products, viz., $2(a^2+b^2+c^2+ab+bc+ca)$;

And $xy = (a+b)(b+c) = b^2+ab+bc+ca$, from which we see that we shall have all the squares once and the products three times, viz., $a^2+b^2+c^2+3(ab+bc+ca)$.

Subtracting this result from the former we get the final result, $a^2+b^2+c^2-ab-bc-ca$.

An easy solution may also be obtained by writing the given expression in the form $x(x-y)+y(y-z)+z(z-x)$.

Ex. 4.—Simplify

$(ax+by-cz)(ax-by+cz) + (bx+cy-az)(bx-cy+az)$
$\qquad + (cx+ay-bz)(cx-ay+bz).$

$(ax+by-cz)(ax-by+cz) = \{ax+(by-cz)\}\{ax-(by-cz)\}$
$\qquad = a^2x^2 - b^2y^2 - c^2z^2 + 2bcyz;$

$\therefore (a^2+b^2+c^2)(x^2-y^2-z^2) + 2(bc+ca+ab)yz$ is the result required.

The student should write the terms in full a few times and carefully observe the order in which the letters follow each other until he can write the required result from examining a single term, as in the preceding example.

258. The three following combinations of letters occur so frequently that they should be carefully remembered:—

1. $(a-b)+(b-c)+(c-a) = 0$.
2. $c(a-b)+a(b-c)+b(c-a) = 0$.
3. $(x-a)(x-b)(x-c) = x^3 - (a+b+c)x^2 + (ab+bc+ca)x - abc$.

EXERCISE LXXIX.

Simplify

1. $(a+b)^2+(b+c)^2+(c+a)^2$.   2. $(a-b)^2+(b-c)^2+(c-a)^2$.
3. $(a+b)(b+c)+(b+c)(c+a)+(c+a)(a+b)$.
4. $(a-b)(b-c)+(b-c)(c-a)+(c-a)(a-b)$.

SYMMETRICAL EXPRESSIONS.

5. $(a-b)(b+c)+(b-c)(c+a)+(c-a)(a+b)$.
6. $(a+b)(b-c)+(b+c)(c-a)+(c+a)(a-b)$.
7. $(a+b-c)^2+(b+c-a)^2+(c+a-b)^2$.
8. $(a-b-c)^2+(b-c-a)^2+(c-a-b)^2$.
9. $(a-b)(a+b-c)+(b-c)(b+c-a)+(c-a)(c+a-b)$.
10. $(a+b)(a-b+c)+(b+c)(b-c+a)+(c+a)(c-a+b)$.
11. $(a-b)(ma+mb-nc)+(b-c)(mb+mc-na)+(c-a)(mc+ma-nb)$.
12. $\dfrac{1}{(a-b)(b-c)}+\dfrac{1}{(b-c)(c-a)}+\dfrac{1}{(c-a)(a-b)}$.
13. $\dfrac{a}{(a-b)(a-c)}+\dfrac{b}{(b-c)(b-a)}+\dfrac{c}{(c-a)(c-b)}$.
14. $\dfrac{a+b}{(b-c)(c-a)}+\dfrac{b+c}{(c-a)(a-b)}+\dfrac{c+a}{(a-b)(b-c)}$.
15. $\dfrac{(2x-y-z)^2+(2y-z-x)^2+(2z-x-y)^2}{(y-z)^2+(z-x)^2+(x-y)^2}$.
16. $\dfrac{x+y-z}{(y-z)(z-x)}+\dfrac{y+z-x}{(z-x)(x-y)}+\dfrac{z+x-y}{(x-y)(y-z)}$.
17. $\dfrac{x+y}{(x^2-yz)(y^2-zx)}+\dfrac{y+z}{(y^2-zx)(z^2-xy)}+\dfrac{z+x}{(z^2-xy)(x^2-yz)}$.
18. $\dfrac{x^2(y-z)}{(z+x)(x+y)}+\dfrac{y^2(z-x)}{(x+y)(y+z)}+\dfrac{z^2(x-y)}{(y+z)(z+x)}$.
19. $\dfrac{\{(ax+b)^2-(bx+a)(cx+a)\}+\text{two similar terms}}{\{(ax+b)^2-(cx+a)(cx+b)\}+\text{two similar terms}}$.

Arrange in powers of x

20. $(x+a)(x+b)+(x+b)(x+c)+(x+c)(x+a)$.
21. $(x-a)(x-b)+(x-b)(x-c)+(x-c)(x-a)$.
22. $c(x-a)(x-b)+a(x-b)(x-c)+b(x-c)(x-a)$.
23. $c(x-a)(x+b)+a(x-b)(x+c)+b(x-c)(x+a)$.

SYMMETRICAL EXPRESSIONS.

24. $(a-b)(x-c) + (b-c)(x-a) + (c-a)(x-b)$.
25. $(a-b)(x-c)^2 + (b-c)(x-a)^2 + (c-a)(x-b)^2$.
26. $(a-b)(x-c)^3 + (b-c)(x-a)^3 + (c-a)(x-b)^3$.
27. $(a-b)(x-a)(x-b) + (b-c)(x-b)(x-c) + (c-a)(x-c)(x-a)$.
28. $(a+b)(x+a)(x-b) + (b+c)(x+b)(x-c) + (c+a)(x+c)(x-a)$.
29. $a(b-c)(x-bc) + b(c-a)(x-ca) + c(a-b)(x-ab)$.
30. $a(b-c)(x-bc)^2 + b(c-a)(x-ca)^2 + c(a-b)(x-ab)^2$.

If $2s = a+b+c$ simplify the following:—

31. $s(s-a) + (s-b)(s-c)$.
32. $(s-a)^2 + (s-b)^2 + (s-c)^2 + s^2$.
33. $(s-a)(s-b) + (s-b)(s-c) + (s-c)(s-a) + s^2$.
34. $a(s-a) + b(s-b) + c(s-c) + 2s^2$.
35. $a(s-b)(s-c) + b(s-c)(s-a) + c(s-a)(s-b) + 2(s-a)(s-b)(s-c)$.
36. $(s-a)^2 + (s-b)^2 + (s-c)^2 + 2(s-a)(s-b) + 2(s-b)(s-c) + 2(s-c)(s-a)$.
37. $(s-a)^3 + (s-b)^3 + 3(s-a)(s-b)c$.
38. $(s-a)^3 + (s-b)^3 + (s-c)^3 + 3abc$.
39. $s\{(s-2b)(s-2c) + (s-2c)(s-2a) + (s-2a)(s-2b) + 2s^2\}$
$\qquad - (s-2a)(s-2b)(s-2c)$.

If $2s = a+b+c$ and $2S^2 = a^2 + b^2 + c^2$, show that

40. $a^2(S^2-a^2) + b^2(S^2-b^2) + c^2(S^2-c^2) = 8s(s-a)(s-b)(s-c)$.
41. $(S^2-a^2)(S^2-b^2) + (S^2-b^2)(S^2-c^2) + (S^2-c^2)(S^2-a^2)$
$\qquad = 4s(s-a)(s-b)(s-c)$.

259. Factor $a(b^2-c^2) + b(c^2-a^2) + c(a^2-b^2)$.

Removing the brackets, adding and subtracting b^3, and rearranging the terms, we get

$$a(b^2-c^2) + b(c^2-a^2) + c(a^2-b^2) = (a-b)(b^2-c^2) - (a^2-b^2)(b-c)$$
$$= (a-b)(b-c)\{b+c-(a+b)\}$$
$$= (a-b)(b-c)(c-a).$$

By changing the signs and re-arranging the terms the given expression may be reduced to either of the following forms,

$$a^2(b-c) + b^2(c-a) + c^2(a-b),$$
$$ab(a-b) + bc(b-c) + ca(c-a),$$

whose factors are, therefore, known to be $-(a-b)(b-c)(c-a)$. The three forms of this example should be carefully noted and used in working the following exercise. By their use a large number of the examples may be solved mentally. Similar remarks apply to Exercises LXXXI.–LXXXV.

Ex.—Simplify $\dfrac{(x+a)^2}{(a-b)(a-c)} + \dfrac{(x+b)^2}{(b-c)(b-a)} + \dfrac{(x+c)^2}{(c-a)(c-b)}$.

We observe that the terms of one factor in each denominator are the reverse of the regular circular order; we therefore take $-(a-b)(b-c)(c-a)$ for the L. C. M. of the denominators. Reducing the fractions to the L. C. D. we get $(b-c)(x+a)^2$ for the first numerator, and we know that the other numerators may be obtained from this by the usual interchange of letters. Then $(b-c)(x^2 + 2ax + a^2) = (b-c)x^2 + 2a(b-c)x + a^2(b-c)$, from which we see that when the sum of all the numerators is taken the terms involving x^2 and x will vanish. The sum of the given fractions, therefore, is $\dfrac{a^2(b-c) + b^2(c-a) + c^2(a-b)}{-(a-b)(b-c)(c-a)} = 1$, the result required.

EXERCISE LXXX.

Simplify

1. $\dfrac{a(b+c)}{(c-a)(a-b)} + \dfrac{b(c+a)}{(a-b)(b-c)} + \dfrac{c(a+b)}{(b-c)(c-a)}$.

2. $\dfrac{a^2}{(a-b)(a-c)} + \dfrac{b^2}{(b-c)(b-a)} + \dfrac{c^2}{(c-a)(c-b)}$.

3. $\dfrac{ab}{(a-c)(b-c)} + \dfrac{bc}{(b-a)(c-a)} + \dfrac{ca}{(c-b)(a-b)}$.

4. $\dfrac{a^2-bc}{(c-a)(a-b)} + \dfrac{b^2-ca}{(a-b)(b-c)} + \dfrac{c^2-ab}{(b-c)(c-a)}$.

SYMMETRICAL EXPRESSIONS. 225

5. $\dfrac{x-a}{a(a-b)(a-c)} + \dfrac{x-b}{b(b-c)(b-a)} + \dfrac{x-c}{c(c-a)(c-b)}.$

6. $\dfrac{(x-a)^2}{(a-b)(a-c)} + \dfrac{(x-b)^2}{(b-c)(b-a)} + \dfrac{(x-c)^2}{(c-a)(c-b)}.$

7. $\dfrac{(x-a)(x-b)}{(a-c)(b-c)} + \dfrac{(x-b)(x-c)}{(b-a)(c-a)} + \dfrac{(x-c)(x-a)}{(c-b)(a-b)}.$

8. $\dfrac{a^2(x-b)(x-c)}{(a-b)(a-c)} + \dfrac{b^2(x-c)(x-a)}{(b-c)(b-a)} + \dfrac{c^2(x-a)(x-b)}{(c-a)(c-b)}.$

9. $\dfrac{bc(x-a)^2}{(a-b)(a-c)} + \dfrac{ca(x-b)^2}{(b-c)(b-a)} + \dfrac{ab(x-c)^2}{(c-a)(c-b)}.$

10. $\dfrac{(1+ab)(1+bc)}{(a-b)(b-c)} + \dfrac{(1+bc)(1+ca)}{(b-c)(c-a)} + \dfrac{(1+ca)(1+ab)}{(c-a)(a-b)}.$

11. $\dfrac{a-b}{a+b} + \dfrac{b-c}{b+c} + \dfrac{c-a}{c+a} - \dfrac{(a-b)(b-c)(c-a)}{(a+b)(b+c)(c+a)}.$

12. $\dfrac{(ax+b)(bx+c)(cx+a) - (ax+c)(bx+a)(cx+b)}{(ax+b)^3 + (bx+c)^3 + (cx+a)^3 - (ax+c)^3 - (bx+a)^3 - (cx+b)^3}.$

13. $\dfrac{a}{(a-b)(a-c)(x-a)} + \dfrac{b}{(b-c)(b-a)(x-b)} + \dfrac{c}{(c-a)(c-b)(x-c)}.$

14. $\dfrac{a^2}{(a-b)(a-c)(x-a)} + \dfrac{b^2}{(b-c)(b-a)(x-b)} + \dfrac{c^2}{(c-a)(c-b)(x-c)}.$

260. *Ex. 1.* $a(b^3-c^3) + b(c^3-a^3) + c(a^3-b^3)$
$= (a-b)(b^3-c^3) - (a^3-b^3)(b-c)$
$= (a-b)(b-c)\{b^2+bc+c^2-(a^2+ab+b^2)\}$
$= (a-b)(b-c)(c-a)(a+b+c).$

Similarly, by rearranging the terms as in Art. 259, the factors of the first two examples in the following exercise may be obtained.

Ex. 2.—Simplify

$$\dfrac{(a+b)(a+c)}{bc(a-b)(a-c)} + \dfrac{(b+c)(b+a)}{ca(b-c)(b-a)} + \dfrac{(c+a)(c+b)}{ab(c-a)(c-b)}.$$

226 SYMMETRICAL EXPRESSIONS.

The L. C. M. of the denominators is $-abc(a-b)(b-c)(c-a)$. Reducing to the L. C. D., the first numerator is

$$a(b-c)(a+b)(a+c) = a(b-c)(a^2+ab+bc+ca)$$
$$= a^3(b-c) + (ab+bc+ca)\{a(b-c)\}.$$

Now, the numerators of the other fractions may be obtained from this by the usual change of letters; and by separating the expression into two parts as above we see that the second part will vanish, because $ab+bc+ca$ remains constant and $a(b-c) + b(c-a) + c(a-b) = 0$. The numerator, therefore, reduces to $a^3(b-c) + b^3(c-a) + c^3(a-b)$, and the whole of the given expression reduces to $\dfrac{a+b+c}{abc}$.

EXERCISE LXXXI.

1. Factor $a^3(b-c) + b^3(c-a) + c^3(a-b)$.

2. Factor $ab(a^2-b^2) + bc(b^2-c^2) + ca(c^2-a^2)$.

Simplify

3. $\dfrac{a(b^2+bc+c^2)}{(c-a)(a-b)} + \dfrac{b(c^2+ca+a^2)}{(a-b)(b-c)} + \dfrac{c(a^2+ab+b^2)}{(b-c)(c-a)}$.

4. $\dfrac{ab(a+b)}{(a-c)(b-c)} + \dfrac{bc(b+c)}{(b-a)(c-a)} + \dfrac{ca(c+a)}{(c-b)(a-b)}$.

5. $\dfrac{a^2}{bc(a-b)(a-c)} + \dfrac{b^2}{ca(b-c)(b-a)} + \dfrac{c^2}{ab(c-a)(c-b)}$.

6. $\dfrac{a^3}{(c-a)(a-b)} + \dfrac{b^3}{(a-b)(b-c)} + \dfrac{c^3}{(b-c)(c-a)}$.

7. $\dfrac{a(a+b)(a+c)}{(a-b)(a-c)} + \dfrac{b(b+c)(b+a)}{(b-c)(b-a)} + \dfrac{c(c+a)(c+b)}{(c-a)(c-b)}$.

8. $\dfrac{(a+b)(x-a)(x-b)}{(a-c)(b-c)} + \dfrac{(b+c)(x-b)(x-c)}{(b-a)(c-a)} + \dfrac{(c+a)(x-c)(x-a)}{(c-b)(a-b)}$.

9. $\dfrac{(x-a)^3}{(a-b)(a-c)} + \dfrac{(x-b)^3}{(b-c)(b-a)} + \dfrac{(x-c)^3}{(c-a)(c-b)}$.

SYMMETRICAL EXPRESSIONS.

10. $\dfrac{(1+a^2b)(1+a^2c)}{(a-b)(a-c)} + \dfrac{(1+b^2c)(1+b^2a)}{(b-c)(b-a)} + \dfrac{(1+c^2a)(1+c^2b)}{(c-a)(c-b)}$.

11. $\dfrac{b+c-a}{(b+c)(c-a)(a-b)} + \dfrac{c+a-b}{(c+a)(a-b)(b-c)} + \dfrac{a+b-c}{(a+b)(b-c)(c-a)}$.

12. $\dfrac{(a-b)^3}{(b+c-2a)(c+a-2b)} + \dfrac{(b-c)^3}{(c+a-2b)(a+b-2c)}$
$\qquad\qquad\qquad + \dfrac{(c-a)^3}{(a+b-2c)(b+c-2a)}$.

13. $\dfrac{(ax+b)^4+(bx+c)^4+(cx+a)^4-(ax+c)^4-(bx+a)^4-(cx+b)^4}{(a^2x+b^2)(b^2x+c^2)(c^2x+a^2)-(a^2x+c^2)(b^2x+a^2)(c^2x+b^2)}$.

14. $\dfrac{(a-x)(a-y)(a-z)}{(a-b)(a-c)} + \dfrac{(b-x)(b-y)(b-z)}{(b-c)(b-a)} + \dfrac{(c-x)(c-y)(c-z)}{(c-a)(c-b)}$.

15. Solve the equation, $\dfrac{a^2-b^2}{x-c} + \dfrac{b^2-c^2}{x-a} + \dfrac{c^2-a^2}{x-b} = 0$.

261. *Ex. 1.* $a(b^4-c^4)+b(c^4-a^4)+c(a^4-b^4)$
$= (a-b)(b^4-c^4) - (a^4-b^4)(b-c)$
$= (a-b)(b-c)\{b^3+b^2c+bc^2+c^3-(a^3+a^2b+ab^2+b^3)\}$
$= (a-b)(b-c)(c-a)(a^2+b^2+c^2+ab+bc+ca)$.

Ex. 2.—Solve the equation,

$(a-b)(x-c)^4+(b-c)(x-a)^4+(c-a)(x-b)^4$
$\qquad = 6\{(a-b)(cx+y)^2+(b-c)(ax+y)^2+(c-a)(bx+y)^2\}$.

$(a-b)(x^4-4cx^3+6c^2x^2-4c^3x+c^4) + \&c.$
$\qquad\qquad = 6\{(a-b)(c^2x^2+2cxy+y^2) + \&c.\};$

$-4x\{c^3(a-b)+\&c.\} + \{c^4(a-b)+\&c.\} = 0;$

$x = \dfrac{c^4(a-b)+a^4(b-c)+b^4(c-a)}{4\{c^3(a-b)+a^3(b-c)+b^3(c-a)\}} = \dfrac{a^2+b^2+c^2+ab+bc+ca}{4(a+b+c)}$.

In the above the coefficients of x^4 and of x^3 vanish, those of x^2 on the two sides are equal, and those of xy and y^2 on the second side vanish, leaving the result as given.

SYMMETRICAL EXPRESSIONS.

EXERCISE LXXXII.

1. Factor $a^4(b-c)+b^4(c-a)+c^4(a-b)$.
2. Factor $ab(a^3-b^3)+bc(b^3-c^3)+ca(c^3-a^3)$.

Simplify

3. $\dfrac{a^4}{(a-b)(a-c)}+\dfrac{b^4}{(b-c)(b-a)}+\dfrac{c^4}{(c-a)(c-b)}$.

4. $\dfrac{a(b^2+c^2)}{(c^2-a^2)(a^2-b^2)}+\dfrac{b(c^2+a^2)}{(a^2-b^2)(b^2-c^2)}+\dfrac{c(a^2+b^2)}{(b^2-c^2)(c^2-a^2)}$.

5. $\dfrac{a^2+ab+b^2}{c(c-a)(c-b)}+\dfrac{b^2+bc+c^2}{a(a-b)(a-c)}+\dfrac{c^2+ca+a^2}{b(b-c)(b-a)}$.

6. $\dfrac{a^2(a+b)(a+c)}{(a-b)(a-c)}+\dfrac{b^2(b+c)(b+a)}{(b-c)(b-a)}+\dfrac{c^2(c+a)(c+b)}{(c-a)(c-b)}$.

7. $\dfrac{a^2}{(a-b)(a-c)(b+c)}+\dfrac{b^2}{(b-c)(b-a)(c+a)}+\dfrac{c^2}{(c-a)(c-b)(a+b)}$.

8. $\dfrac{a^2(x-a)^2}{(a-b)(a-c)}+\dfrac{b^2(x-b)^2}{(b-c)(b-a)}+\dfrac{c^2(x-c)^2}{(c-a)(c-b)}$.

9. $\dfrac{a(x-a)^3}{(a-b)(a-c)}+\dfrac{b(x-b)^3}{(b-c)(b-a)}+\dfrac{c(x-c)^3}{(c-a)(c-b)}$.

10. $\dfrac{(a-b)^4}{(b+c-2a)(c+a-2b)}+\dfrac{(b-c)^4}{(c+a-2b)(a+b-2c)}$
$+\dfrac{(c-a)^4}{(a+b-2c)(b+c-2a)}$.

11. $\dfrac{(a^2x+b)^3+(b^2x+c)^3+(c^2x+a)^3-(a^2x+c)^3-(b^2x+a)^3-(c^2x+b)^3}{(a^2x+b)^2+(b^2x+c)^2+(c^2x+a)^2-(a^2x+c)^2-(b^2x+a)^2-(c^2x+b)^2}$.

12. $\dfrac{(a^2x-b)^3+(b^2x-c)^3+(c^2x-a)^3-(a^2x-c)^3-(b^2x-a)^3-(c^2x-b)^3}{(a^4x+c)^2+(b^4x+a)^2+(c^4x+b)^2-(a^4x+b)^2-(b^4x+c)^2-(c^4x+a)^2}$.

Solve

13. $(a-b)(x-c^2)^2+(b-c)(x-a^2)^2+(c-a)(x-b^2)^2=0$.
14. $(a-b)(x-c)(x-c^3)+(b-c)(x-a)(x-a^3)+(c-a)(x-b)(x-b^3)=0$.
15. $c^2(a-b)(x-c^2)+a^2(b-c)(x-a^2)+b^2(c-a)(x-b^2)=0$.

SYMMETRICAL EXPRESSIONS. 229

262.—*Ex.* $a^2(b^3 - c^3) + b^2(c^3 - a^3) + c^2(a^3 - b^3)$
$= (a^2 - b^2)(b^3 - c^3) - (a^3 - b^3)(b^2 - c^2)$
$= (a - b)(b - c)\{(a + b)(b^2 + bc + c^2) - (a^2 + ab + b^2)(b + c)\}$
$= (a - b)(b - c)(c - a)(ab + bc + ca).$

Similarly the first two examples in the following exercise may be factored.

EXERCISE LXXXIII.

1. Factor $a^3(b^2 - c^2) + b^3(c^2 - a^2) + c^3(a^2 - b^2).$

2. Factor $a^2b^2(a - b) + b^2c^2(b - c) + c^2a^2(c - a).$

Simplify

3. $\dfrac{a^3(b+c)}{(a-b)(a-c)} + \dfrac{b^3(c+a)}{(b-c)(b-a)} + \dfrac{c^3(a+b)}{(c-a)(c-b)}.$

4. $\dfrac{a(b^2 + bc + c^2)}{bc(c-a)(a-b)} + \dfrac{b(c^2 + ca + a^2)}{ca(a-b)(b-c)} + \dfrac{c(a^2 + ab + b^2)}{ab(b-c)(c-a)}.$

5. $\dfrac{ab}{c(a-c)(b-c)} + \dfrac{bc}{a(b-a)(c-a)} + \dfrac{ca}{b(c-b)(a-b)}.$

6. $\dfrac{a^3}{(a-b)(a-c)(a-x)} + \dfrac{b^3}{(b-c)(b-a)(b-x)} + \dfrac{c^3}{(c-a)(c-b)(c-x)}$
$\qquad + \dfrac{x^3}{(x-a)(x-b)(x-c)}.$

7. $\dfrac{ab(a+b)}{(b-c)(c-a)(x-c)} + \dfrac{bc(b+c)}{(c-a)(a-b)(x-a)} + \dfrac{ca(c+a)}{(a-b)(b-c)(x-b)}.$

8. $\dfrac{ab(a-b)(a^2+b^2) + bc(b-c)(b^2+c^2) + ca(c-a)(c^2+a^2)}{a^2b^2(a-b) + b^2c^2(b-c) + c^2a^2(c-a)}.$

9. $\dfrac{ab(a-b)(a^3+b^3-2c^3) + bc(b-c)(b^3+c^3-2a^3) + ca(c-a)(c^3+a^3-2b^3)}{\{(a-b)^2+(b-c)^2+(c-a)^2\}\{(a-b)^3+(b-c)^3+(c-a)^3\}}.$

Solve

10. $(a - b)(x + a^2)(x + b^2) + (b - c)(x + b^2)(x + c^2)$
$\qquad + (c - a)(x + c^2)(x + a^2) = 0.$

11. $c(a-b)(x+a^2)(x+b^2) + a(b-c)(x+b^2)(x+c^2)$
$\qquad + b(c-a)(x+c^2)(x+a^2) = 0.$

12. $(a-b)(x+bc)(x+ca) + (b-c)(x+ca)(x+ab)$
$\qquad + (c-a)(x+ab)(x+bc) = 0.$

13. $(a^2x+b)(b^2x+c)(c^2x+a) = (a^2x+c)(b^2x+a)(c^2x+b).$

263. Various artifices may be employed to obtain the factors of complicated expressions from those of simpler expressions whose factors are known. We shall give two examples which illustrate some of the most important methods.

Ex. 1.—Factor $(a+b+c)^5 - a^5 - b^5 - c^5.$

Writing the expression in the form $\{(a+b+c)^5 - a^5\} - (b^5 + c^5)$ we see that $b+c$ is a factor. Similarly $c+a$ and $a+b$ are factors. Now, it may easily be shown that $(a+b)^5 - a^5 - b^5 = 5ab(a+b)(a^2+ab+b^2)$; and since the expression on the left is what the given expression becomes when for c we write 0, the factors on the right are what the factors of the given expression become when for c we write 0. The given expression is perfectly symmetrical with regard to a, b and c, therefore its factors are also symmetrical.

$\therefore (a+b+c)^5 - a^5 - b^5 - c^5$
$\qquad = 5(a+b)(b+c)(c+a)(a^2+b^2+c^2+ab+bc+ca).$

Ex. 2.—Factor $(x-y)^5 + (y-z)^5 + (z-x)^5.$

Let $x-y=a$, $y-z=b$, then $z-x=-(a+b)$ and
$(x-y)^5 + (y-z)^5 + (z-x)^5$
$\qquad = a^5 + b^5 - (a+b)^5$
$\qquad = -5ab(a+b)(a^2+ab+b^2)$
$\qquad = 5(x-y)(y-z)(z-x)(x^2+y^2+z^2-xy-yz-zx).$

To express a^2+ab+b^2 in terms of x, y and z, it is best to write it in the form $(a+b)^2 - ab$, when the substitutions can be easily effected.

SYMMETRICAL EXPRESSIONS. 231

EXERCISE LXXXIV.

Factor

1. $(a+b)^3 - a^3 - b^3$. 2. $(a+b)^5 - a^5 - b^5$. 3. $(a+b)^7 - a^7 - b^7$.
4. $(a-b)^5 - a^5 + b^5$. 5. $(a-b)^7 - a^7 + b^7$. 6. $a^7 - (a-b)^7 - b^7$.
7. $(x-y)^3 + (y-z)^3 + (z-x)^3$. 8. $(x-y)^7 + (y-z)^7 + (z-x)^7$.
9. $a(b-c)^3 + b(c-a)^3 + c(a-b)^3$. 10. $a^2(b-c)^3 + b^2(c-a)^3 + c^2(a-b)^3$.
11. $(x+y)^4 + x^4 + y^4$. 12. $(x-y)^4 + (y-z)^4 + (z-x)^4$.
13. $(a+b+c)^3 - a^3 - b^3 - c^3$. 14. $(a+b-c)^5 - a^5 - b^5 + c^5$.
15. $(x+y+z)^3 - (x+y-z)^3 - (x-y+z)^3 - (-x+y+z)^3$.
16. $8(x+y+z)^3 - (x+y)^3 - (y+z)^3 - (z+x)^3$.
17. $(a-b)^3 + (b-c)^3 + (c-d)^3 + (d-a)^3$.
18. $(a-b)^5 + (b-c)^5 + (c-d)^5 + (d-a)^5$.

Prove

19. $\{(x-y)^2 + (y-z)^2 + (z-x)^2\}^2 = 2\{(x-y)^4 + (y-z)^4 + (z-x)^4\}$.
20. $25\{(x-y)^7 + (y-z)^7 + (z-x)^7\}\{(x-y)^3 + (y-z)^3 + (z-x)^3\}$
$= 21\{(x-y)^5 + (y-z)^5 + (z-x)^5\}^2$.

If $a+b+c=0$ prove

21. $\dfrac{a^5 + b^5 + c^5}{5} = \dfrac{a^3 + b^3 + c^3}{3} \cdot \dfrac{a^2 + b^2 + c^2}{2}$.

22. $\dfrac{a^7 + b^7 + c^7}{7} = \dfrac{a^5 + b^5 + c^5}{5} \cdot \dfrac{a^2 + b^2 + c^2}{2}$

$= \dfrac{a^3 + b^3 + c^3}{3} \cdot \dfrac{a^4 + b^4 + c^4}{2}$.

23. If $a+b+c+d=0$ prove

$\dfrac{1}{5}(a^5 + b^5 + c^5 + d^5) = \dfrac{1}{6}(a^3 + b^3 + c^3 + d^3)(a^2 + b^2 + c^2 + d^2)$.

24. Factor $(b+c-a-x)^4(b-c)(a-x) + (c+a-b-x)^4(c-a)(b-x)$
$+ (a+b-c-x)^4(a-b)(c-x)$.

SYMMETRICAL EXPRESSIONS.

264. It is frequently convenient to express the cube of a trinomial in one of the following forms:—

1. $(a+b+c)^3 = (a+b)^3 + c^3 + 3c(a+b)(a+b+c)$
$= a^3 + b^3 + c^3 + 3ab(a+b) + 3c(a+b)(a+b+c)$
$= a^3 + b^3 + c^3 + 3(a+b)(ab+ac+bc+c^2)$
$= a^3 + b^3 + c^3 + 3(a+b)(b+c)(c+a).$

2. $(a+b+c)^3 - (a^3+b^3+c^3-3abc)$
$= (a+b+c)\{(a+b+c)^2 - (a^2+b^2+c^2-ab-bc-ca)\}$
$= 3(a+b+c)(ab+bc+ca);$
$\therefore (a+b+c)^3 = a^3+b^3+c^3 + 3\{(a+b+c)(ab+bc+ca) - abc\}.$

265. The following identities may easily be verified, and should be committed to memory:—

1. $a(b^2+c^2) + b(c^2+a^2) + c(a^2+b^2) + 2abc = (a+b)(b+c)(c+a).$

2. $a(b^2+c^2) + b(c^2+a^2) + c(a^2+b^2) + 3abc = (a+b+c)(ab+bc+ca).$

3. $a(b^2+c^2) + b(c^2+a^2) + c(a^2+b^2) = a^2(b+c) + b^2(c+a) + c^2(a+b)$
$= ab(a+b) + bc(b+c) + ca(c+a).$

EXERCISE LXXXV.

Factor

1. $a(b+c)^2 + b(c+a)^2 + c(a+b)^2 - 4abc.$

2. $ab(a+b) + bc(b+c) + ca(c+a) + 3abc.$

3. $a^2(b+c) + b^2(c+a) + c^2(a+b) + 2abc.$

4. $(a+b)(c-a)(c-b) + (b+c)(a-b)(a-c) + (c+a)(b-c)(b-a) + 8abc$

5. $a(b+c)(b+c-a) + b(c+a)(c+a-b) + c(a+b)(a+b-c).$

6. $a(1+b^2)(1+c^2) + b(1+c^2)(1+a^2) + c(1+a^2)(1+b^2) + 4abc.$

7. $a(1-b^2)(1-c^2) + b(1-c^2)(1-a^2) + c(1-a^2)(1-b^2) - 4abc.$

8. $(a+b)(1-bc)(1-ca) + (b+c)(1-ca)(1-ab) + (c+a)(1-ab)(1-bc)$
$- (a+b)(b+c)(c+a).$

9. $(a+b)^3 + (b+c)^3 + (c+a)^3 - 3(a+b)(b+c)(c+a).$

10. $(ax+b)^3 + (bx+c)^3 + (cx+a)^3 - 3(ax+b)(bx+c)(cx+a).$

SYMMETRICAL EXPRESSIONS.

Prove the following identities:—

11. $\dfrac{1}{2}(a+b+c)\{(a-b)^2+(b-c)^2+(c-a)^2\} = a^3+b^3+c^3-3abc$.

12. $(a+b+c)^3-3\{a(b-c)^2+b(c-a)^2+c(a-b)^2\} = a^3+b^3+c^3+24abc$.

13. $\dfrac{1}{2}\{(a+b+7c)(a-b)^2+(b+c+7a)(b-c)^2+(c+a+7b)(c-a)^2\}$
$= (a+b+c)^3 - 27abc$.

14. $(4a+4b+c)(a-b)^2+(4b+4c+a)(b-c)^2+(4c+4a+b)(c-a)^2$
$= 9(a^3+b^3+c^3)-(a+b+c)^3$.

15. $a^2(b+c)^2+b^2(c+a)^2+c^2(a+b)^2+2abc(a+b+c) = 2(ab+bc+ca)^2$.

16. $(b+c)^2(c+a)^2(a+b)^2+2a^2b^2c^2-a^4(b+c)^2-b^4(c+a)^2-c^4(a+b)^2$
$= 2(bc+ca+ab)^3$.

17. $(b+c)^2(c+a)^2(a+b)^2+2a^2b^2c^2-a^2b^2(a+b)^2-b^2c^2(b+c)^2-c^2a^2(c+a)^2$
$= 2abc(a+b+c)^3$.

18. $2(a+b+c)^3+(a+b+c)(ab+bc+ca)+abc$
$= (2a+b+c)(a+2b+c)(a+b+2c)$.

19. $4(a+b+c)^3+2(a+b+c)(ab+bc+ca)-abc$
$= (a+2b+2c)(2a+b+2c)(2a+2b+c)$.

20. $(3a-2b-c)^2(3b-2c-a)^2+(3b-2c-a)^2(3c-2a-b)^2$
$+(3c-2a-b)^2(3a-2b-c)^2 = 49(ab+bc+ca-a^2-b^2-c^2)^2$.

21. $(a+b-c)^3(b+c-a)^3+(b+c-a)^3(c+a-b)^3+(c+a-b)^3(a+b-c)^3$
$+24abc(a+b-c)(b+c-a)(c+a-b)$
$= (2ab+2bc+2ca-a^2-b^2-c^2)^3$.

CHAPTER XVI.

THEORY OF DIVISORS AND COMPLETE SQUARES.

266. Before proceeding to study the subject to which this chapter is devoted it will be necessary for the student to clearly comprehend the exact meaning of the word "condition" in Algebra. A simple example will make its meaning clear.

267. Is $x-2$ a factor of x^2-ax+b? A little consideration will show that the answer depends upon the values of a and b. If $a=5$ and $b=6$ then $x-2$ is a factor of x^2-ax+b. Here, then, are two conditions which, being fulfilled, render $x-2$ a factor of the other expression. But if $a=1$ and $b=-2$, or if $a=3$ and $b=2$, it is also a factor. Hence we conclude that the conditions $a=5$ and $b=6$ were *sufficient* but not *necessary*. If we give a and b any values such that $b-2a+4=0$, we shall find that $x-2$ is a factor of x^2-ax+b; but if such values be given that $b-2a+4$ is not $=0$, then it is not a factor. Therefore $b-2a+4=0$ is the necessary and sufficient condition that $x-2$ may be a factor of x^2-ax+b.

268. To find the conditions that x^2+px+q may be a factor of x^3+ax^2+bx+c for all values of x.

Proceed as in ordinary division as follows:—

$$x^2+px+q \overline{\smash{\big)}\, x^3+ax^2+bx+c} \, (x+(a-p)$$
$$\underline{x^3+px^2+qx}$$
$$(a-p)x^2+\ (b-q)x+c$$
$$(a-p)x^2+p(a-p)x+q(a-p)$$

Now, if $b-q=p(a-p)$ and $c=q(a-p)$, the remainder is zero for all values of x, and the first expression is a factor of the second. These are the necessary and sufficient conditions required. If, however, we divide the former equation by the latter we obtain $\dfrac{b-q}{c} = \dfrac{p}{q}$ or $q(b-q) = pc$, a necessary condition that the first expression may be a factor of the second, but which in itself is not sufficient.

269. One expression may be a factor of another for *particular* values of their leading letter though not for *all* values. Thus, if $x = 10$ both $x - 2$ and $x - 3$ are factors of $x^2 - 6x + 16$; for

$$x^2 - 6x + 16 = (x-2)(x-3) - (x-10)$$
$$= (x-2)(x-3) \text{ when } x = 10.$$

Such values may be found by dividing one expression by the other and equating any remainder which contains x to zero; for if the remainder is zero the division is complete.

Ex.—Find values of x which will render $x^2 - 2x + 3$ a factor of $x^3 - x^2 + 5x - 21$.

Dividing in the usual way we obtain $x + 1$ for quotient with $4x - 24$ remainder, which vanishes when $x = 6$. If we place -1 in the quotient instead of $+1$, as before, the remainder is $2x^2 - 18$, which vanishes when $x = \pm 3$. Again, we may take -7 for the last term, giving $8x^2 - 12x$ remainder, which is zero when $x = 0$ or $\dfrac{3}{2}$. Each of these values of x renders the former expression an exact factor of the latter. This, however, is using the word "factor" in a restricted sense. Properly speaking, one expression is a factor of another only when the remainder *is always zero*.

270. To find the condition that $ax^2 + bx + c$ may be a complete square for all values of x.

If the given expression is a complete square it must be the square of $x\sqrt{a} + \sqrt{c}$, since no other expression when squared could give the terms ax^2 and c; therefore

$$ax^2 + bx + c = (x\sqrt{a} + \sqrt{c})^2 = ax^2 + 2\sqrt{ac} \cdot x + c,$$

and therefore $b = 2\sqrt{ac}$ or $b^2 = 4ac$ is the condition required.

THEORY OF DIVISORS AND COMPLETE SQUARES.

It should be observed that the preceding condition merely enables us to write the given expression in the *form* of the square of a binomial; but it does not ensure that the numerical value of the expression will be an exact square when for the various letters we substitute numbers which satisfy the given condition. For example, let $a=2$, $b=4$, $c=2$, and $x=10$, then $ax^2+bx+c = 242$, which is not an exact square.

Again, let $a=-1$, $b=2$, $c=-1$ and $x=10$, then $ax^2+bx+c = -81$, a negative number, whilst all *square* numbers are positive. Since $b^2 = 4ac$ we have

$$ax^2+bx+c = a\left(x^2+\frac{b}{a}x+\frac{b^2}{4a^2}\right) = a\left(x+\frac{b}{2a}\right)^2,$$

which shows that if a is an exact square and $b^2 = 4ac$ then the whole expression is also an exact square for all values of the letters which satisfy these conditions.

EXERCISE LXXXVI.

1. What values of a and b will render x^2-2x+3 a factor of $2x^4-4x^3+9x^2+ax+b$?

2. What values of a and b will render x^2-3x+a a factor of $x^4+x^3-5x^2+7x+6b$?

3. Find the values of a and b in order that $x^2+3xy+4y^2$ may exactly divide $x^6+7x^5y+6x^4y^2+5x^3y^3+ax^2y^4+bxy^5+12y^6$.

4. What value of x will render $x^4+6x^3+11x^2+3x+31$ a perfect square?

5. What values of x will render $x^6+3x^5+6x^4+7x^3+7x^2-7x+25$ a perfect cube?

6. What value of m will make $16x^4-4mx^3+20x^2+2mx+4$ an exact square for all values of x?

7. What value of x will render $m^2x^2+px+pq+q^2$ an exact square for all values of the other letters involved?

THEORY OF DIVISORS AND COMPLETE SQUARES. 237

8. What values of x will render $x^4 + 2ax^3 + 3b^2x^2 - 4a^3x + 4b^4$ an exact square for all values of a and b?

9. What value of x will make $x^2 + px + q$ a factor of
$$x^3 + ax^2 + bx + c\,?$$

10. Find the condition that $ax^2 + 2bx + c$ may be an exact square for all values of x.

11. Find the condition that $ax^3 + 3bx^2 + 3cx + d$ may be a complete cube for all values of x.

12. If $ax^2 + 2bx + c$ is a factor of $ax^3 + 3bx^2 + 3cx + d$ then the former is a complete square and the latter a complete cube.

13. If $x^2 + ax + b$ is a factor of $x^3 + 2ax^2 + 3bx + 4c$ then $ax^2 + 2bx + 2c$ is an exact square and $x^3 + 3ax^2 + 6bx + 8c$ is an exact cube.

14. If $x^2 + max + a^2$ is a factor of $x^4 - ax^3 + a^2x^2 - a^3x + a^4$ then $m^2 + m = 1$.

15. If $(x-1)^2$ is a factor of $x^3 + ax^2 + bx + c$ then $b + 2c = 1$. Is the converse true?

16. If $x^3 + qx + r$ contains a square factor then $4q^3 + 27r^2 = 0$. Is the converse true?

17. Find the conditions that $x^4 + ax^3 + bx^2 + cx + d$ shall be a perfect square for all values of x.

18. If $(a-b)x^2 + (a+b)^2x + (a^2 - b^2)(a+b)$ is a complete square then $a = 3b$ or $b = 3a$.

19. Find a value of x which will render $(n^2+n+1)x^2 + (n^4+n^2)x - (n^2 - n + 1)$ a complete square for all values of n.

20. Find values for a and b which render the fraction
$$\frac{2x^2 + (a-b)x + 2a^2 - 3b^2}{3x^2 + (a-7)x + 3(a^2 + 2ab + 3b^2)}$$ the same for all values of x.

21. Find values for x and y which render the fraction
$$\frac{2z^2 + (x-a)z + 2b(x - 2c)}{3z^2 + (y-b)z + 3a(y-3c)}$$ independent of the value of z.

271. A Function of x is any algebraical expression whose value depends, in whole or in part, upon the value of x.

Thus $2x - 3$, $\dfrac{2}{3x}$ and $\sqrt{x^2+5}$ are functions of x; but $\dfrac{2x}{3x}$, x^0 and $(a-b)(x-c) + (b-c)(x-a) + (c-a)(x-b)$ are not functions of x, because when reduced to their simplest forms their values are independent of the value of x.

272. A Rational Integral Function of x is a function in which x does not appear in the denominator of a fraction, and is not affected by any root sign.

Thus $x^3 - 4x^2 + \dfrac{2}{3}x + \sqrt{7}$ is a rational integral function of x, but $x^2 - 2x + \dfrac{3}{2-x}$ and $x + \sqrt{bx+c}$ are not; the former is not integral and the latter is not rational with regard to x.

273. *If any rational integral function of x vanishes when $x = m$ then the function is divisible by $x - m$.*

Let $a + bx + cx^2 + dx^3 +$ &c., be any rational integral function of x which vanishes when $x = m$, that is, let $a + bm + cm^2 + dm^3 +$ &c. $= 0$, then $x - m$ is a factor of the proposed expression.

For $a + bx + cx^2 + dx^3 +$ &c.
$= a + bx + cx^2 + dx^3 +$ &c. $- (a + bm + cm^2 + dm^3 +$ &c.$)$
$= b(x - m) + c(x^2 - m^2) + d(x^3 - m^3) +$ &c.
$= (x - m)\{b + c(x + m) + d(x^2 + mx + m^2) +$ &c.$\}$,

which shows that $x - m$ is a factor.

Cor.—If $a + bx + cx^2 + dx^3 +$ &c. be divided by $x - m$ the remainder will be $a + bm + cm^2 + dm^3 +$ &c.; for this expression does not contain x, and when it is subtracted from the given expression the remainder is divisible by $x - m$.

274. The symbol $f(x)$ is frequently used to denote any function of x. The value of such a function when $x = m$ is then conveniently expressed by $f(m)$; thus, if $f(x)$ denotes $ax^2 + bx + c$, $f(m)$

denotes $am^2 + bm + c$. This notation enables us to give the result of Art. 273 very neatly as follows:—

275. *If $f(x)$, any rational integral function of x, be divided by $x - m$, the remainder will be $f(m)$; and if $f(m)$ be zero then $x - m$ is a factor of $f(x)$.*

Divide $f(x)$ by $x - m$ until the remainder no longer contains x. Let Q be the quotient and R the remainder; then $f(x) = Q(x-m) + R$ is an identity, and therefore true for all values of x. Let $x = m$, then $Q(x - m) = 0$ and we get $f(m) = R$; but R is independent of x, therefore $R = f(m)$ for all values of x; and if $f(m) = 0$, $R = 0$, which proves the proposition.

Ex. 1.—Factor $x^3 - 5x^2 - 46x - 40$.

The expression vanishes when for x we substitute 10, -1 or -4; therefore $x - 10$, $x + 1$ and $x + 4$ are factors. There can be no other factor containing x, since the given expression is of only three dimensions. There can be no numerical factor, since the first term of the product of these factors, viz., x^3, is the first term of the given expression.

Two special cases are worthy of note in connection with examples like the preceding:—

(1) When the sum of the positive coefficients is equal to the sum of the negative coefficients, $x - 1$ is a factor.

(2) When the coefficients are all positive, and the sum of the coefficients of the *odd* powers is equal to the sum of the coefficients of the even powers, $x + 1$ is a factor.

Ex. 2.—Factor $(a+b+c)^4 - (b+c)^4 - (c+a)^4 - (a+b)^4 + a^4 + b^4 + c^4$.

For a substitute 0 and the expression vanishes; therefore $a - 0$, that is, a, is a factor.

The expression is symmetrical with regard to a, b and c, therefore b and c are also factors.

The expression is of 4 dimensions, therefore there is another factor of one dimension.

240   THEORY OF DIVISORS AND COMPLETE SQUARES.

The factors a, b, c by themselves form a symmetrical group, therefore the other factor must also be symmetrical in itself.

The only symmetrical expression of one dimension is $a+b+c$,
$$\therefore (a+b+c)^4 - (b+c)^4 - (c+a)^4 - (a+b)^4 + a^4 + b^4 + c^4 = Nabc(a+b+c)$$
where N is a number independent of a, b, c.

To find N, give a, b, c any values which will *not* make the expression vanish; in this case $a=b=c=1$, then
$$3^4 - 2^4 - 2^4 - 2^4 + 1 + 1 + 1 = 3N \text{ or } N = 12;$$
therefore the given expression $= 12abc(a+b+c)$.

Ex. 3.—Factor $(a-b)^5 + (b-c)^5 + (c-a)^5$.

For a substitute b and the expression vanishes, therefore $a-b$ is a factor; then $(a-b)(b-c)(c-a)$ is a factor by symmetry.

The expression is symmetrical and of 5 dimensions, therefore there is another symmetrical factor of 2 dimensions.

Let this factor be $m(a^2+b^2+c^2)+n(ab+bc+ca)$, in which m and n are not functions of $a, b,$ and c, and will therefore be the same for all values of these letters. Then
$$(a-b)^5 + (b-c)^5 + (c-a)^5$$
$$= (a-b)(b-c)(c-a)\{m(a^2+b^2+c^2) + n(ab+bc+ca)\}.$$

We have now two unknown quantities, m and n. It will therefore be necessary to have two independent equations to determine them. These may be obtained by giving two sets of values to a, b, c.

First, let $a=0, b=1, c=2$;
Second, let $a=0, b=1, c=-1$, and we get
$$\left.\begin{array}{r}5m+2n=15\\2m-n=15\end{array}\right\} \text{from which } m=5, n=-5.$$
$$\therefore (a-b)^5 + (b-c)^5 + (c-a)^5 = 5(a-b)(b-c)(c-a)(a^2+b^2+c^2-ab-bc-ca).$$

276. *If $x-m$ is a factor of any rational integral function of x, when m is substituted for x the result will vanish.*

Let $a+bx+cx^2+dx^3+$ &c. be the function, $x-m$ a factor of it, then $a+bm+cm^2+dm^3+$ &c. $=0$.

THEORY OF DIVISORS AND COMPLETE SQUARES. 241

Since $x - m$ is a factor of $a + bx + cx^2 + dx^3 + $ &c. by hypothesis, and of $a + bx + cx^2 + dx^3 + $ &c. $- (a + bm + cm^2 + dm^3 + $ &c.) by Art. 273, therefore $x - m$ is a factor of their difference, $a + bm + cm^2 + dm^3 + $ &c.; but this is impossible, since the latter expression does not contain x; therefore $a + bm + cm^2 + dm^3 + $ &c. must be zero. This result is also evident from the fact that when for x we substitute m one factor of the expression, viz., $x - m$, becomes zero, and therefore the whole expression is zero.

Ex. 1.—Find the condition that $ax^2 + bx + c$ and $a'x^2 + b'x + c'$ may have a common factor, and find that factor.

Let $x - m$ be the common factor, then

$$am^2 + bm + c = 0, \quad \text{Art. 276} \quad (1)$$
$$a'm^2 + b'm + c' = 0. \quad (2)$$

Eliminating first m^2, and second c and c', we get

$$(a'b - ab')m + a'c - ac' = 0, \quad (3)$$
$$(ac' - a'c)m^2 + (bc' - b'c)m = 0. \quad (4)$$

Equating the values of m obtained from (3) and (4) we get

$$\frac{a'c - ac'}{ab' - a'b} = \frac{bc' - b'c}{a'c - ac'} \quad \text{or} \quad (a'c - ac')^2 = (bc' - b'c)(ab' - a'b),$$

the condition required,

and $x - \dfrac{a'c - ac'}{ab' - a'b}$ or $x - \dfrac{bc' - b'c}{a'c - ac'}$ is the common factor required.

Again, if in (1) and (2) we substitute the values of m from (3) and (4), we get

$$a(a'c - ac')^2 + b(a'c - ac')(ab' - a'b) + c(ab' - a'b)^2 = 0;$$
$$a(bc' - b'c)^2 + b(bc' - b'c)(a'c - ac') + c(a'c - ac')^2 = 0;$$
$$a'(a'c - ac')^2 + b'(a'c - ac')(ab' - a'b) + c'(ab' - a'b)^2 = 0;$$
$$a'(bc' - b'c)^2 + b'(bc' - b'c)(a'c - ac') + c'(a'c - ac')^2 = 0.$$

All these results are merely different forms of the result previously given, and may be deduced from it. This is a simple example of an indirect method of proving the identity of different forms of an algebraical expression.

Ex. 2.—Find the value of $3x^7 - 8x^6 - 11x^4 - 3x^3 + 28x^2 - 9x + 7$ when $x^2 = 2x + 3$.

Dividing the given expression by $x^2 - 2x - 3$ we obtain the quotient $3x^5 - 2x^4 + 5x^3 - 7x^2 - 2x + 3$, with remainder $-9x + 16$.

Now, since the dividend is equal to the product of the divisor and quotient, plus the remainder, we have

$3x^7 - 8x^6 - \&c. = (x^2 - 2x - 3)(3x^5 - 2x^4 + 5x^3 - 7x^2 - 2x + 3) - 9x + 16$.

But since, in this particular example, $x^2 = 2x + 3$, the first factor is zero, therefore the product is zero, and the given expression reduces to $-9x + 16$.

Again, $x^2 - 2x - 3 = (x - 3)(x + 1) = 0$, therefore $x = 3$ or -1, and $-9x + 16 = -11$ or 25, which are the required results.

The division is most conveniently performed by Horners' Method, Art. 86.

Ex. 3.—If $a + b = c$, $(a^2 - b^2)^2 + (b^2 - c^2)^2 + (c^2 - a^2)^2 = a^4 + b^4 + c^4$

For $(a^2 - b^2)^2 + (b^2 - c^2)^2 + (c^2 - a^2)^2 - (a^4 + b^4 + c^4)$

$= a^4 + b^4 + c^4 - 2a^2b^2 - 2b^2c^2 - 2c^2a^2$

$= (a^2 + b^2 - c^2)^2 - 4a^2b^2$

$= (a^2 + b^2 - c^2 + 2ab)(a^2 + b^2 - c^2 - 2ab)$

$= (a + b + c)(a + b - c)(a - b + c)(a - b - c)$

$= 0$, since $a + b - c = 0$.

$\therefore (a^2 - b^2)^2 + (b^2 - c^2)^2 + (c^2 - a^2)^2 = a^4 + b^4 + c^4$.

The same result would evidently follow if any of the other factors were zero.

EXERCISE LXXXVII.

1. Show, without actual division, that $x - a$ is a factor of $x^3 - (a - b)x^2 + (d - ab)x - ad$.

2. Show that $a - b$, $b - c$ and $c - a$ are each factors of $a^n(b - c) + b^n(c - a) + c^n(a - b)$ for all positive integral values of n.

THEORY OF DIVISORS AND COMPLETE SQUARES.   243

3. Find the remainder when $x^3 + a^3 + b^3 - 3abx$ is divided by $x - a + b$, $x + a - b$, and by $x - a - b$.

4. Show that a, $a - x$ and $a - 2x$ are each factors of
$$(a - b)(a - b - x)(a + 2b - 2x) + b(b - x)(3a - 2b - 2x).$$

5. Show that
$$(ab - xy)^2 - (a + b - x - y)\{ab(x + y) - xy(a + b)\}$$
$$= (x - a)(x - b)(y - a)(y - b).$$

6. Find the value of a for which the fraction
$$\frac{x^3 - ax^2 + 19x - a - 4}{x^3 - (a + 1)x^2 + 23x - a - 7}$$
admits of reduction, and reduce it to its lowest terms.

Find the value of

7. $x^5 + 290x^4 + 279x^3 - 2892x^2 - 586x - 312$ when $x = -289$.

8. $3x^6 - 11x^5 + 19x^4 - 13x^3 - x^2 + 10$ when $x^2 = 2x - 3$.

9. $4x^5 - 12x^4 + 5x^3 + 5x^2 - 6x + 3$ when $x = \pm\frac{1}{2}\sqrt{3}$.

10. $2x^5 + 803x^4 - 398x^3 + 1605x^2 - 1204x + 422$
when $x^2 + 401x = 402$.

If $a + b + c = 0$ prove Exs. 11-19.

11. $a^2 - bc = b^2 - ca = c^2 - ab$ and $a^3 + b^3 + c^3 = 3abc$.

12. $(a + b)(b + c)(c + a) + abc = 0$.

13. $a(b^2 + bc + c^2) + b(c^2 + ca + a^2) + c(a^2 + ab + b^2) = 0$.

14. $(a^2 + b^2)^2 + (b^2 + c^2)^2 + (c^2 + a^2)^2 = 3(a^4 + b^4 + c^4)$.

15. $(a + b)(b + c) + (b + c)(c + a) + (c + a)(a + b) = ab + bc + ca$.

16. $(a - b)(b - c) + (b - c)(c - a) + (c - a)(a - b) = 3(ab + bc + ca)$.

17. $(ab + bc + ca)^2 = a^2b^2 + b^2c^2 + c^2a^2$.

18. $(a^2 - 2b^2)^2 + (b^2 - 2c^2)^2 + (c^2 - 2a^2)^2 = 3(a^4 + b^4 + c^4)$.

19. $\dfrac{a^2}{2a^2 + bc} + \dfrac{b^2}{2b^2 + ca} + \dfrac{c^2}{2c^2 + ab} = 1$.

If $x+y+z = xyz$ prove Exs. 20-22.

20. $\dfrac{x}{1-x^2} + \dfrac{y}{1-y^2} + \dfrac{z}{1-z^2} = \dfrac{4xyz}{(1-x^2)(1-y^2)(1-z^2)}$.

21. $\dfrac{x+y}{1-xy} + \dfrac{y+z}{1-yz} + \dfrac{z+x}{1-zx} = \dfrac{(x+y)(y+z)(z+x)}{(1-xy)(1-yz)(1-zx)}$.

22. $\dfrac{x-y}{1+xy} + \dfrac{y-z}{1+yz} + \dfrac{z-x}{1+zx} = \dfrac{(x-y)(y-z)(z-x)}{(1+xy)(1+yz)(1+zx)}$.

23. If $x^2 + y^2 + z^2 + 2xyz = 1$ then
$$x\sqrt{(1-y^2)(1-z^2)} + y\sqrt{(1-z^2)(1-x^2)} + z\sqrt{(1-x^2)(1-y^2)} = 1 + xyz.$$

24. If $x^2 + qx + 1$ and $x^3 + px^2 + qx + 1$ have a common factor of the form $x + a$, then $(p-1)^2 - q(p-1) + 1 = 0$.

25. The expressions $ax^2 + bx + c$ and $ax^2 + mbx + m^2 c$ have a common factor if $(m+1)^2 ac = mb^2$.

26. If $ax^3 + bx + c$ and $mx^3 + nx + p$ have a common factor of the first degree in x, then $(pa - mc)^3 = (nc - pb)(na - mb)^2$.

27. If $x^2 + mx + n$ and $x^2 + px + q$ have a common factor, then $(n-q)^2 - m(n-q)(m-p) + n(m-p)^2 = 0$.

28. If $x^3 + px^2 + qx + r$ is divisible by $x^2 + mx + n$, then the quotient is a factor of $mx^2 + qx + r$.

29. If $x + a$ and $x - a$ are both factors of $x^3 + px^2 + qx + r$, then $pq = r$.

30. If $x^2 - 1$ is a factor of $x^4 + px^2 + qx + a^2$, the other factor is $x^2 - a^2$.

31. If $x-a$ and $x-b$ are each factors of $x^2 + x + 1$, then $a^3 - b^3 = 0$.

32. If $x + m$ and $x + n$ are each factors of $x^3 + ax^2 + b$, then
$$a = \dfrac{m^2 + mn + n^2}{m + n}.$$

33. If $x^3 + ax^2 + b$ and $x^3 + px + q$ have a common factor of the second degree in x, then $a^3 bq = (b - q)^3$.

34. If $ax^3 + bx^2 + c$ and $cx^3 + bx + a$ have a common factor which is a complete square, then $a^2 b^2 = 4bc(c^2 - a^2)$.

THEORY OF DIVISORS AND COMPLETE SQUARES.   245

35. If $a+b+c=0$ then abc is the H. C. F. of $a^3+b^3+c^3$ and $a^5+b^5+c^5$, and $15abc(c^2-ab)$ is their L. C. M.

36. If $x+c$ is the H. C. F. of x^2+ax+b and x^2+mx+a, their L. C. M. is $x^3+(a+m-c)x^2+(am-c^2)x+(a-c)(m-c)c$.

277. From the principle that the square of any real number is positive we are able to deduce various important conclusions with regard to the numerical value of expressions when any numbers whatsoever are substituted for the various letters involved. We give three examples.

Ex. 1.—Show that the sum of the squares of any two real quantities is greater than twice their product.

Let a and b represent the quantities.
Then $(a-b)^2$ is positive;
$\therefore a^2 - 2ab + b^2$ is positive;
$\therefore a^2 + b^2 > 2ab$, which proves the proposition.

Ex. 2.—Prove that $4x^2 - 24x + 41$ is positive for all real values of x, and find what value of x will give the expression the least value possible.

$$4x^2 - 24x + 41 = 4\{x^2 - 6x + 10\tfrac{1}{4}\} = 4\{(x-3)^2 + 1\tfrac{1}{4}\}.$$

Now, $(x-3)^2$ is positive for all values of x; $1\tfrac{1}{4}$ is positive and 4 is positive, therefore the whole expression is always positive. The value of the expression will be least when $x-3=0$, *i.e.*, when $x=3$, and the expression itself then becomes equal to 5.

Ex. 3.—If $a^2+b^2+c^2 = ab+bc+ca$ then $a=b=c$.
For $\qquad a^2+b^2+c^2-ab-bc-ca=0;$
$\therefore 2(a^2+b^2+c^2-ab-bc-ca)=0,$
$\therefore (a-b)^2+(b-c)^2+(c-a)^2=0.$

Now, since the sum of any number of positive quantities cannot be zero, each of these three terms must be zero, and therefore $a=b=c$.

246 THEORY OF DIVISORS AND COMPLETE SQUARES.

EXERCISE LXXXVIII.

1. Prove that the sum of any positive number (except unity) and its reciprocal is greater than 2.

2. Prove that the square of the sum of any two numbers is greater than four times their product.

3. Prove $a^4b^2 + a^2b^4 > 2a^3b^3$ for all real values of a and b.

4. Prove $a^2 + b^2 + c^2 > ab + bc + ca$ for all real values of a, b and c, except when $a = b = c$.

5. Prove $x^3 + 1 > x^2 + x$ if $x + 1$ is positive; and $x^3 - 1 > x^2 - x$ if $x > 1$.

6. Prove $bc(b + c) + ca(c + a) + ab(a + b) > 6abc$ when a, b and c are real, unequal, and positive.

7. Prove that $\dfrac{a+b}{2}$, $\sqrt{ab}$ and $\dfrac{2ab}{a+b}$ are in order of magnitude.

8. Show that $\dfrac{a^3+b^3}{a^2+b^2} > \dfrac{a^2+b^2}{a+b}$.

9. If x is real prove that $x^2 - 8x + 22$ can never be less than 6.

10. Show that the least value of $x^2 - x + 1$ is obtained by making $x = \dfrac{1}{2}$.

11. Show that the greatest value of $24x - x^2$ is 144.

12. Show that the area of a square is greater than that of a rectangle of the same perimeter.

13. If a straight line be divided into two equal and also into two unequal parts, the squares on the unequal parts are together greater than four times the rectangle contained by half the line and the line between the points of section.

14. Show that the area of the largest rectangle which can be enclosed on three sides by a line 60 feet in length is 450 square feet.

THEORY OF DIVISORS AND COMPLETE SQUARES. 247

15. If $a^2 + 2b^2 + c^2 = 2b(a+c)$ then $a = b = c$.

16. If $a^4 + b^4 + b^2c^2 + c^2a^2 = 2abc(a+b)$ then $a = b = c$.

17. If $a^3 + b^3 + c^3 = 3abc$ and $a+b+c$ is not zero then $a = b = c$.

18. If $(1 + a^2 + b^2)(1 + x^2 + y^2) = (1 + ax + by)^2$ then $x = a$ and $y = b$.

19. If $(a^2 + b^2 + c^2)(x^2 + y^2 + z^2) = (ax + by + cz)^2$ then $\dfrac{x}{a} = \dfrac{y}{b} = \dfrac{z}{c}$.

20. If $x^2 + y^2 - 4x - 2y + 5 = 0$ then $x = 2$ and $y = 1$.

21. Prove $a^3 + b^3 + c^3 >$, $=$ or $<$ $3abc$, according as $a+b+c$ is positive, zero or negative.

22. If a, b, c are real numbers, not all equal, then $(a-b)(b-c) + (b-c)(c-a) + (c-a)(a-b)$ is negative.

23. If a, b, c are real, unequal and positive, then
$(a+b+c)(ab+bc+ca) > 9abc$ and $(a+b+c)^3 > 27abc$.

24. If x and y are real and positive, $x^5 + y^5 - x^4y - xy^4$ is positive

25. If a, b, c, d are real, unequal and positive, prove
$ac + bd > 2\sqrt{abcd}$, $ab + cd > 2\sqrt{abcd}$, and thence
$a^2bc + ab^2d + ac^2d + bcd^2 > 4abcd$.

CHAPTER XVII.

QUADRATIC EQUATIONS OF ONE UNKNOWN.

278. Equations involving unknown quantities of two dimensions, and no higher, are called *quadratic* equations.

Thus $x^2 + 6x + 8 = 0$, $x^2 - 9 = 0$, $\left.\begin{array}{r}x^2 + xy + y^2 = 8 \\ x^2 + y^2 = 2\end{array}\right\}$, are quadratic equations.

279. Quadratics of one unknown may be either *pure* or *adfected*. A *pure* quadratic contains the square of the unknown quantity, and no other power; whereas an *adfected* quadratic contains the *first* power as well as the *second* of the unknown quantity. For instance,

$x^2 - 16 = 0$ is a *pure* quadratic,

but $x^2 + 8x + 16 = 0$ is an *adfected* quadratic.

280. A quadratic equation is the statement that the product of two factors, each of one dimension, is equal to zero. These factors may be *rational*, *real* and *irrational*, or *imaginary*.

Thus, in the equation $x^2 + 8x + 15 = 0$, the factors are $(x + 3)$ and $(x + 5)$, and therefore are *rational*. The factors of $x^2 - 2 = 0$ are $x - \sqrt{2}$ and $x + \sqrt{2}$, and therefore are *real* and *irrational*. The equation $x^2 + 4 = 0$ has for factors $(x + 2\sqrt{-1})$ and $(x - 2\sqrt{-1})$, and therefore consists of the product of two *imaginary* quantities.

281. When the *factors* of a quadratic expression are obtained we can at once write down the *roots* of the corresponding equation. This has already been explained in Arts. 120–122.

QUADRATIC EQUATIONS OF ONE UNKNOWN. 249

We have already given examples to be solved by factoring; we here append a few more.

Ex. 1.—Solve $x^2 - 9 = 0$.
Factoring, $(x-3)(x+3) = 0$; $\therefore x = 3$ or -3.

Ex. 2.—Solve $x^2 - (m-n)x - mn = 0$.
Factoring, $(x-m)(x+n) = 0$; $\therefore x = m$ or $-n$.

EXERCISE LXXXIX.

Solve by factoring

1. $x^2 + 9x + 14 = 0$.
2. $x^2 - 8x + 15 = 0$.
3. $x^2 - x - 12 = 0$.
4. $6x^2 - 5x + 1 = 0$.
5. $30x^2 - x - 1 = 0$.
6. $x^2 + x - 20 = 0$.
7. $12x^2 + x - 1 = 0$.
8. $2x^2 - 27x = 14$.
9. $5x^2 - 3x = 2$.
10. $3x^2 - 5x = 2$.
11. $x^2 + ax - a - x = 0$.
12. $5x^2 - 12x + 2 = 11$.
13. $3x^2 - 53x + 34 = 0$.
14. $110x^2 - 21x + 1 = 0$.
15. $780x^2 - 73x + 1 = 0$.
16. $x + 2 - \dfrac{6}{x+2} = 1$.
17. $\dfrac{x^4 + 3x^3 + 6}{x^2 + x - 4} = x^2 + 2x + 15$.
18. $\dfrac{2x-1}{2x+1} + \dfrac{13}{11} = \dfrac{3x+5}{3x-5}$.
19. $\dfrac{x+2}{x-1} - \dfrac{4-x}{2x} = 2\frac{1}{3}$.
20. $\dfrac{12}{5-x} + \dfrac{8}{4-x} = \dfrac{32}{x+2}$.
21. $\dfrac{x^2}{a^2} + \dfrac{x}{b} = \dfrac{2a^2}{b^2}$.
22. $\dfrac{2x(a-x)}{3a-2x} = \dfrac{a}{4}$.
23. $x^2 - 2ax + 4ab = 2bx$.
24. $x^2 - 2ax + 8x = 16a$.
25. $\dfrac{3x-2}{2x-3} = \dfrac{5x}{x+4} - 2$.
26. $\dfrac{3x}{11} + \dfrac{23}{x+4} = \dfrac{x+5}{3}$.

17

27. $\dfrac{2x+1}{x+3} - \dfrac{x-1}{x^2-9} = \dfrac{x+3}{3-x} - \dfrac{4+x}{3+x}$. 28. $\dfrac{x-1}{2x+3} - \dfrac{x+1}{2x-3} = \dfrac{x+7}{4x^2-9} - \dfrac{x-3}{2x-3}$.

29. $\dfrac{1}{x-3} + \dfrac{7}{x+3} = \dfrac{14}{x^2-9} - \dfrac{x-4}{x+3}$. 30. $\dfrac{4b}{a} + \dfrac{a-4b}{x-2b} = \dfrac{a+4b}{x+2b}$.

282. It frequently happens that a quadratic expression cannot readily be factored. In such cases the following method of finding the roots is usually adopted:—

Ex. 1.—Let $x^2 + px + q = 0$, find x.

Transposing, $\qquad x^2 + px = -q$.

Complete the square by adding the square of one-half the coefficient of x to both sides of the equations.

$$\therefore x^2 + px + \left(\frac{p}{2}\right)^2 = \left(\frac{p}{2}\right)^2 - q = \frac{p^2}{4} - q = \frac{p^2 - 4q}{4},$$

$$i.e., \left(x + \frac{p}{2}\right)^2 = \frac{p^2 - 4q}{4}.$$

Extracting square root of both sides,

$$x + \frac{p}{2} = \pm \frac{\sqrt{p^2 - 4q}}{2};$$

$$\therefore x = -\frac{p}{2} \pm \frac{\sqrt{p^2 - 4q}}{2};$$

Hence $\qquad x = -\dfrac{p}{2} + \dfrac{\sqrt{p^2 - 4q}}{2}$

or $\qquad x = -\dfrac{p}{2} - \dfrac{\sqrt{p^2 - 4q}}{2}$.

The student will observe that the double sign $\pm$ is prefixed to $\dfrac{\sqrt{p^2 - 4q}}{2}$. The reason is that the square root of a quantity may be either positive or negative. (See Art. 191.)

QUADRATIC EQUATIONS OF ONE UNKNOWN. 251

Ex. 2.—Solve $ax^2 + bx + c = 0$.

Divide through by a, so as to make the coefficient of x^2 unity and positive;

$$\therefore x^2 + \frac{b}{a}x + \frac{c}{a} = 0.$$

Transposing, $\qquad x^2 + \frac{b}{a}x = -\frac{c}{a}.$

Completing the square,

$$x^2 + \frac{b}{a}x + \left(\frac{b}{2a}\right)^2 = \left(\frac{b}{2a}\right)^2 - \frac{c}{a} = \frac{b^2 - 4ac}{4a^2};$$

Extracting square root,

$$\left(x + \frac{b}{2a}\right) = \pm \frac{\sqrt{b^2 - 4ac}}{2a};$$

$$\therefore x = -\frac{b}{2a} \pm \frac{\sqrt{b^2 - 4ac}}{2a}.$$

Therefore roots are $-\dfrac{b}{2a} + \dfrac{\sqrt{b^2 - 4ac}}{2a}$ and $-\dfrac{b}{2a} - \dfrac{\sqrt{b^2 - 4ac}}{2a}$.

The equation $x^2 + px + q = 0$ is the *form* of all quadratic equations which have the coefficient of x^2, *unity* and positive; and $ax^2 + bx + c = 0$ is the *form* of all quadratics in which the coefficient of x^2 is not *unity*.

Ex. 3.—Solve $x^2 - 12x + 16 = 0$.

Transposing, $\qquad x^2 - 12x = -16.$

Completing the square,

$$x^2 - 12x + (6)^2 = 36 - 16 = 20.$$

Extracting square root, $\quad x - 6 = \pm \sqrt{20};$

$$\therefore x = 6 \pm \sqrt{20}.$$

Hence $\qquad\qquad x = 6 + 2\sqrt{5}$ or $6 - 2\sqrt{5}.$

252 QUADRATIC EQUATIONS OF ONE UNKNOWN.

Ex. 4.—Solve $3x^2 - 8x + 9 = 0$.

Dividing by coefficient of x^2,

$$x^2 - \frac{8}{3}x + 3 = 0$$

or

$$x^2 - \frac{8}{3}x = -3.$$

Completing the square,

$$x^2 - \frac{8}{3}x + \left(\frac{4}{3}\right)^2 = \left(\frac{4}{3}\right)^2 - 3 = \frac{16}{9} - 3$$

or

$$\left(x - \frac{4}{3}\right)^2 = -\frac{11}{9}.$$

Extracting square root, $x - \dfrac{4}{3} = \pm \dfrac{\sqrt{-11}}{3}$;

$$\therefore x = \frac{4}{3} \pm \frac{\sqrt{-11}}{3}.$$

283. This method of solving a quadratic equation is called the *Italian* method, and is the one generally adopted in practice. There is another method, called the *Hindoo* method, two examples of which we will give.

Ex. 1.—Solve $ax^2 + bx + c = 0$.

Multiply the equation by four times the coefficient of x^2,

$$\therefore 4a^2x^2 + 4abx + 4ac = 0.$$

Transposing, $4a^2x^2 + 4abx = -4ac.$

Adding b^2 to both sides,

$$\therefore 4a^2x^2 + 4abx + b^2 = b^2 - 4ac.$$

Extracting square root, $2ax + b = \pm \sqrt{b^2 - 4ac}$

or $2ax = -b \pm \sqrt{b^2 - 4ac};$

$$\therefore x = \frac{-b \pm \sqrt{b^2 - 4ac}}{2a}.$$

QUADRATIC EQUATIONS OF ONE UNKNOWN. 253

Ex. 2.—Solve $2x^2 + 8x + 5 = 0$.

Multiplying by 8, $16x^2 + 64x = -40$.

Adding $(8)^2$ to both sides,

$$16x^2 + 64x + 64 = 24.$$

Extracting square root, $4x + 8 = \pm \sqrt{24}$;

$$\therefore x = \frac{-8 \pm \sqrt{24}}{4}.$$

284. It is often necessary to reduce a given quadratic to one of the forms $ax^2 + bx + c = 0$, $x^2 + px + q = 0$. The methods for the reduction and simplification of quadratic equations are the same as those given for simple equations. An example will suffice:

Ex.—Solve $\dfrac{x+a}{x-a} + \dfrac{x+b}{x-b} + \dfrac{x+c}{x-c} = 3$.

Dividing each numerator by its corresponding denominator we obtain

$$1 + \frac{2a}{x-a} + 1 + \frac{2b}{x-b} + 1 + \frac{2c}{x-c} = 3;$$

$$\therefore \frac{2a}{x-a} + \frac{2b}{x-b} + \frac{2c}{x-c} = 0.$$

Dividing by 2, $\dfrac{a}{x-a} + \dfrac{b}{x-b} + \dfrac{c}{x-c} = 0.$

Clearing of fractions,

$$a(x-b)(x-c) + b(x-a)(x-c) + c(x-a)(x-b) = 0.$$

Multiplying out and collecting coefficients,

$$x^2(a + b + c) - x(2ab + 2ac + 2bc) + 3abc = 0.$$

Now, we know the roots of $ax^2 + bx + c = 0$ are $-\dfrac{b}{2a} \pm \dfrac{\sqrt{b^2 - 4ac}}{2a}$, and as a, b, c stand for any coefficients whatever we can at once write down the roots of a given quadratic by substituting for a, b and c their particular values in the given equation. In this

QUADRATIC EQUATIONS OF ONE UNKNOWN.

example we must write for a, $a+b+c$; for b, $-2(ab+bc+ca)$; and for c, $3abc$. Therefore roots are

$$\frac{2(ab+bc+ca)}{2(a+b+c)} \pm \frac{\sqrt{4(ab+bc+ca)^2 - 12(abc)(a+b+c)}}{2(a+b+c)};$$

or simplifying,

$$\frac{ab+bc+ca}{a+b+c} \pm \frac{\sqrt{a^2b^2 + b^2c^2 + c^2a^2 - abc(a+b+c)}}{a+b+c}.$$

EXERCISE XC.

Solve

1. $2x^2 - 7x + 3 = 0$.
2. $3x^2 - 53x + 34 = 0$.
3. $2x^2 - 2x - \dfrac{3}{2} = 0$.
4. $14x - x^2 = 33$.
5. $(x-1)(x-2) = 6$.
6. $(2x+1)(x+2) = 3x^2 - 4$.
7. $(x+1)(2x+3) = 4x^2 - 22$.
8. $(x+1)^2 + (x+2)^2 = (x+3)^2$.
9. $(x-1)(x-2) + (x-2)(x-4) = 6(2x-5)$.
10. $(x-7)(x-4) + (2x-3)(x-5) = 103$.
11. $\left(x - \dfrac{1}{2}\right)\left(x - \dfrac{1}{3}\right) + \left(x - \dfrac{1}{3}\right)\left(x - \dfrac{1}{4}\right) = \left(x - \dfrac{1}{4}\right)\left(x - \dfrac{1}{5}\right)$.
12. $\dfrac{x}{2} + \dfrac{2}{x} = \dfrac{x}{3} + \dfrac{3}{x}$.
13. $\dfrac{x^2 - 5x}{x+3} = x - 3 + \dfrac{1}{x}$.
14. $\dfrac{x+2}{x-1} - \dfrac{4-x}{2x} = \dfrac{7}{3}$.
15. $\dfrac{x+3}{x+2} + \dfrac{x-3}{x-2} = \dfrac{2x-3}{x-1}$.
16. $\dfrac{x-2}{x+2} + \dfrac{x+2}{x-2} = \dfrac{2(x+3)}{x-3}$.
17. $\dfrac{4}{x+1} + \dfrac{5}{x+2} = \dfrac{12}{x+3}$.
18. $\dfrac{8-x}{2} - \dfrac{2x-11}{x-3} = \dfrac{x-2}{6}$.
19. $\dfrac{2x-1}{2x+1} + \dfrac{2x+1}{2x-1} = 3$.
20. $5x - \dfrac{3(x-1)}{x-3} = 2x + \dfrac{3(x-2)}{2}$.
21. $\dfrac{7x+1}{6\frac{1}{2} - 3x} = \dfrac{\dfrac{80}{3}\left(x - \dfrac{1}{2}\right)}{x - \dfrac{2}{3}}$.

22. $\dfrac{3x+8}{x-4} - \dfrac{5(12-x)}{2x+3} = 11.$ 23. $\dfrac{x}{x-1} + \dfrac{x+2}{x+1} = \dfrac{8x-13}{4(x-2)}.$

24. $2ax^2 + (a-2)x - 1 = 0.$ 25. $.03x^2 - 2.7x = 30.$

26. $\dfrac{(x-1)(x-2)}{2} = (x - 2\tfrac{2}{3})(x - 1\tfrac{3}{4}).$

27. $\dfrac{c}{a-c}\left(x + \dfrac{1}{x}\right) = 1 + \dfrac{a+c}{(a-c)x} + \dfrac{b}{a-c}\left(1 + \dfrac{1}{x}\right).$

28. $\dfrac{a^2}{b+x} + \dfrac{a^2}{b-x} = c.$ 29. $\dfrac{a^2}{b+x} - \dfrac{a^2}{b-x} = c.$

30. $nx + \dfrac{b}{x} = na + \dfrac{b}{a}.$ 31. $abx^2 - (a+b)cx + c^2 = 0.$

32. $\dfrac{2x(a+x)}{3a+2x} = \dfrac{3a}{4}.$ 33. $\left(\dfrac{a+x}{a-x}\right)^2 = 1 + \dfrac{cx}{b}.$

34. $\dfrac{a(x-b)(x-c)}{(a-b)(a-c)} + \dfrac{b(x-c)(x-a)}{(b-c)(b-a)} = x.$

35. $x^6 + (x+1)^6 + 1 = 2(x^2 + x + 1)^3.$

36. $\dfrac{x}{x^2 - 2x - 15} - \dfrac{7.5}{x^2 + 2x - 35} = \dfrac{1}{x^2 + 10x + 21}.$

37. $\dfrac{1}{x+a} + \dfrac{1}{x+2a} + \dfrac{1}{x+3a} = \dfrac{3}{x}.$ 38. $\dfrac{12+2x}{x+3} + \dfrac{4x-3}{1+2x} = \dfrac{4x-1}{x-1}.$

39. $\dfrac{2x^2 - x - 1}{x-2} + \dfrac{2x^2 - 3x - 8}{x-3} = \dfrac{8x^2 - 9}{2x - 3}.$

40. $\dfrac{1+a}{1+ax} = \dfrac{b+x}{1+bx}.$ 41. $\dfrac{x^3 + a^3}{x+a} + \dfrac{x^3 - a^3}{x-a} = 4a^2.$

42. $\dfrac{1}{x+a-b} + \dfrac{1}{x+a+b} + \dfrac{1}{x-a-b} + \dfrac{1}{x-a+b} = 0.$

43. $\dfrac{x-b}{x-a} + \dfrac{x-a}{x-b} = \dfrac{x+b-2a}{x+a-2b} + \dfrac{x+a-2b}{x+b-2a}.$

44. $\dfrac{1}{(a+b)(x-c)} + \dfrac{1}{(b+c)(x-a)} + \dfrac{1}{(c+a)(x-b)} = 0.$

45. $\dfrac{1}{x+a+b} = \dfrac{1}{x} + \dfrac{1}{a} + \dfrac{1}{b}$. 46. $\left(x - \dfrac{ab}{x}\right)^2 = \dfrac{a}{2}(a+b)\left(1 + \dfrac{a^2}{x^2}\right)$.

47. $\dfrac{(x+b)(x+c)}{(x-b)(x-c)} + \dfrac{(x+c)(x+a)}{(x-c)(x-a)} + \dfrac{(x+a)(x+b)}{(x-a)(x-b)} = 3$.

48. $\dfrac{x-a}{x-b} + \dfrac{x-a-b}{x-b-c} = \dfrac{x+a}{x+b} + \dfrac{x+a+b}{x+b+c}$.

49. $\dfrac{ax-bc}{x+b} + \dfrac{bx-ca}{x+c} + \dfrac{cx-ab}{x+a} = 0$ when $a+b+c=0$.

50. $(4a^2 - 9cd)x^2 + (4a^2c^2 + 4abd^2)x + (ac^2 + bd^2)^2 = 0$.

51. Find the value of $\dfrac{x+a}{x-a} + \dfrac{x+b}{x-b}$

when x is a root of the equation $x^2 + x(a+b) = 3ab$.

285. Various artifices are employed in quadratics, as in other equations, to lessen labor. A familiar and useful one is the substitution of *one* symbol for a number of symbols.

Ex. 1.—Solve $\dfrac{x+2}{x-2} - \dfrac{x-2}{x+2} = \dfrac{5}{6}$.

Here the second fraction is the *inverse* or *reciprocal* of the first.

Let $$y = \dfrac{x+2}{x-2},$$

then the equation becomes $y - \dfrac{1}{y} = \dfrac{5}{6}$.

Clearing of fractions, $y^2 - 1 = \dfrac{5}{6} y$

or $y^2 - \dfrac{5}{6} y - 1 = 0$.

Factoring, $\left(y + \dfrac{2}{3}\right)\left(y - \dfrac{3}{2}\right) = 0$;

$\therefore y = \dfrac{3}{2}$ or $-\dfrac{2}{3}$.

QUADRATIC EQUATIONS OF ONE UNKNOWN. 257

Substituting for y its value $\dfrac{x+2}{x-2}$ we have two equations to solve, viz.:—

$$\frac{x+2}{x-2} = \frac{3}{2} \text{ and } \frac{x+2}{x-2} = -\frac{2}{3}.$$

From these equations we get $x = 10$ or $-\dfrac{2}{5}$.

Ex. 2.—Solve

$$\frac{1}{5} \cdot \frac{(x+1)(x-3)}{(x+2)(x-4)} + \frac{1}{9} \cdot \frac{(x+3)(x-5)}{(x+4)(x-6)} - \frac{2}{13} \cdot \frac{(x+5)(x-7)}{(x+6)(x-8)} = \frac{92}{585}.$$

It is easily seen that each numerator and denominator contains $x^2 - 2x$. Put $(x-1)^2 = y$, then the equation becomes

$$\frac{1}{5} \cdot \frac{y-4}{y-9} + \frac{1}{9} \cdot \frac{y-16}{y-25} - \frac{2}{13} \cdot \frac{y-36}{y-49} = \frac{92}{585}. \qquad (1)$$

But

$$\frac{1}{5} + \frac{1}{9} - \frac{2}{13} = \frac{92}{585}. \qquad (2)$$

Subtracting (2) from (1),

$$\frac{1}{5} \cdot \frac{5}{y-9} + \frac{1}{9} \cdot \frac{9}{y-25} - \frac{2}{13} \cdot \frac{13}{y-49} = 0$$

or

$$\frac{1}{y-9} + \frac{1}{y-25} + \frac{2}{y-49} = 0.$$

From this equation $y = 19$;

$$\therefore (x-1)^2 = 19;$$

$$\therefore x = 1 \pm \sqrt{19}.$$

EXERCISE XCI.

Solve according to given examples,

1. $\dfrac{x}{x-1} = \dfrac{3}{2} + \dfrac{x-1}{x}.$

2. $\dfrac{x-6}{x-12} - \dfrac{x-12}{x-6} = \dfrac{5}{6}.$

3. $\dfrac{2x-3}{3x-5} + \dfrac{3x-5}{2x-3} = \dfrac{5}{2}.$

4. $\dfrac{x+4}{x-4} + \dfrac{x-4}{x+4} = \dfrac{10}{3}.$

5. $\dfrac{x}{x+1} + \dfrac{x+1}{x} = \dfrac{13}{6}.$

6. $\dfrac{3x-2}{2x-5} - \dfrac{2x-5}{3x-2} = \dfrac{8}{3}.$

7. $\dfrac{x+1}{x-1} - \dfrac{x-1}{x+1} = 2a.$

8. $\dfrac{x+16}{x-4} + \dfrac{x-4}{x+16} = \dfrac{37}{6}.$

9. $x + 2 - \dfrac{6}{x+2} = 1.$

10. $x^2 + x + 1 = \dfrac{42}{x^2+x}.$

11. $(2x^2 - 3x)^2 - 2(2x^2 - 3x) = 15.$

12. $(x^2 - x)^2 - 8(x^2 - x) + 12 = 0.$

13. $\left(\dfrac{x+1}{x-1}\right)^2 + \dfrac{7}{12}\left(\dfrac{x+1}{x-1}\right) = 1.$

14. $\left(x - \dfrac{1}{x}\right)^2 + \dfrac{5}{6}\left(x - \dfrac{1}{x}\right) = 1.$

15. $x^4 + 2x^2 - 24 = 0.$

16. $x^6 + 19x^3 - 216 = 0.$

17. $(x^2 + x - 2)^2 - 13(x^2 + x - 2) + 36 = 0.$

18. $\dfrac{1}{x^2 + 11x - 8} + \dfrac{1}{x^2 + 2x - 8} + \dfrac{1}{x^2 - 13x - 8} = 0.$

19. $\dfrac{5}{x^2 - 7x + 10} + \dfrac{5}{x^2 - 13x + 40} = x^2 - 10x + 19.$

20. $\dfrac{1}{3} \cdot \dfrac{x^2 - 6x - 1}{x^2 - 6x - 4} + \dfrac{1}{5} \cdot \dfrac{x^2 - 6x - 4}{x^2 - 6x - 9} - \dfrac{2}{9} \cdot \dfrac{x^2 - 6x - 7}{x^2 - 6x - 16}$

$= \dfrac{14}{45} + \dfrac{4}{x^2 - 6x - 9}.$

CHAPTER XVIII.

QUADRATICS INVOLVING SURDS.

286. The methods of solving surd quadratics are in the main the same as those for solving simple equations containing surds. There are, however, two kinds of equations frequently occurring which deserve some notice.

Ex. 1.—Solve $\sqrt[3]{1-2x} + \sqrt[3]{1+2x} = \sqrt[3]{4}$.

We know that $(a+b)^3 = a^3 + b^3 + 3ab(a+b)$. Applying this formula to the given equation we obtain, by cubing both sides,

$$1 - 2x + 1 + 2x + 3\sqrt[3]{(1-2x)(1+2x)}\{\sqrt[3]{1-2x} + \sqrt[3]{1+2x}\} = 4.$$

But $\sqrt[3]{1-2x} + \sqrt[3]{1+2x} = \sqrt[3]{4}$;

$\therefore 1 - 2x + 1 + 2x + 3\sqrt[3]{(1-2x)(1+2x)}\sqrt[3]{4} = 4$

or $2 + 3\sqrt[3]{4(1-4x^2)} = 4.$

Transposing, $3\sqrt[3]{4(1-4x^2)} = 2.$

Dividing both sides by 3, $\sqrt[3]{4-16x^2} = \dfrac{2}{3}.$

Cubing both sides, $4 - 16x^2 = \dfrac{8}{27};$

from which we obtain $x = \pm \dfrac{5}{6\sqrt{3}}.$

QUADRATICS INVOLVING SURDS.

Ex. 2.—Solve $x^2 - 2x + 6\sqrt{x^2 - 2x + 5} = 11$.

If we add 5 to both sides of the equation it will take the form of a quadratic,

$$\therefore (x^2 - 2x + 5) + 6\sqrt{x^2 - 2x + 5} = 16.$$

Let $y = \sqrt{x^2 - 2x + 5}$, then $x^2 - 2x + 5 = y^2$;

$$\therefore y^2 + 6y = 16$$

or $\quad\quad\quad\quad\quad\quad y^2 + 6y - 16 = 0.$

Factoring, $\quad\quad\quad (y + 8)(y - 2) = 0;$

$$\therefore y = -8 \text{ or } 2,$$

that is, $\quad\quad\quad \sqrt{x^2 - 2x + 5} = -8 \text{ or } 2.$

Squaring both sides, $\quad x^2 - 2x + 5 = 64 \text{ or } 4;$

$$\therefore x^2 - 2x = 59 \text{ or } -1.$$

We have now two equations to solve, viz.:—

$$x^2 - 2x - 59 = 0 \quad\quad (1)$$

and $\quad\quad\quad\quad x^2 - 2x + 1 = 0. \quad\quad (2)$

From (1), $\quad\quad\quad\quad x = 1 \pm 2\sqrt{15}.$

From (2), $\quad\quad\quad\quad (x - 1)^2 = 0 \text{ or } x = 1.$

Ex. 3.—Solve $\sqrt{3x^2 + 2x + 4} = 6x^2 + 4x - 622.$

$$6x^2 + 4x - 622 = 2(3x^2 + 2x + 4) - 630.$$

Let $y = \sqrt{3x^2 + 2x + 4}$, then $y^2 = 3x^2 + 2x + 4$;

$$\therefore y = 2y^2 - 630$$

or $\quad\quad\quad\quad 2y^2 - y - 630 = 0. \quad\quad (1)$

From (1) two values of y can be found which, when in turn substituted in $y = \sqrt{3x^2 + 2x + 4}$, will give equations from which the different values of x may be obtained.

QUADRATICS INVOLVING SURDS. 261

EXERCISE XCII.

Solve

1. $\sqrt{x+3} + \sqrt{x+8} = 5\sqrt{x}$.

2. $\sqrt{2x+1} - \sqrt{x+4} = \dfrac{1}{3}\sqrt{x-3}$.

3. $\dfrac{\sqrt{7x^2+4} + 2\sqrt{3x-1}}{\sqrt{7x^2+4} - 2\sqrt{3x-1}} = 7$.

4. $\dfrac{\sqrt{x+2a} - \sqrt{x-2a}}{\sqrt{x-2a} + \sqrt{x+2a}} = \dfrac{x}{2a}$.

5. $\sqrt{\dfrac{x}{4}+3} + \sqrt{\dfrac{x}{4}-3} = \sqrt{\dfrac{2x}{3}}$.

6. $\dfrac{\sqrt{1+x} + \sqrt{1-x}}{\sqrt{1+x} - \sqrt{1-x}} = x$.

7. $\dfrac{x+\sqrt{2-x^2}}{x-\sqrt{2-x^2}} = \dfrac{4}{3}$.

8. $\dfrac{x+\sqrt{9-x}}{x-\sqrt{9-x}} = \dfrac{7}{3}$.

9. $\sqrt[3]{x+22} - \sqrt[3]{x+3} = 1$.

10. $\sqrt[3]{25+x} + \sqrt[3]{25-x} = 2$.

11. $\sqrt[3]{1-x} + \sqrt[3]{1+x} = \sqrt[3]{3}$.

12. $\sqrt[3]{3+x} + \sqrt[3]{3-x} = \sqrt[3]{7}$.

13. $\sqrt[3]{x+1} - \sqrt[3]{x-1} = \sqrt[3]{11}$.

14. $\sqrt[3]{a+x} + \sqrt[3]{a-x} = \sqrt[3]{b}$.

15. $\sqrt[3]{1-x} + \sqrt[3]{8+x} = 3$.

16. $x^2 - 3x - 6\sqrt{x^2-3x-3} = -2$.

17. $x^2 + \sqrt{x^2-7} = 19$.

18. $2x^2 - 2\sqrt{2x^2-5x} = 5(x+3)$.

19. $2x^2 - 2x + 2\sqrt{2x^2-7x+6} = 5x - 6$.

20. $3x(3-x) = 11 - 4\sqrt{x^2-3x+5}$.

21. $x^2 - 3x + 7\sqrt{11x-2x^2+2} = \dfrac{5}{2}x + 21$.

22. $5x - 7x^2 - 8\sqrt{7x^2-5x+1} = 8$.

23. $x + \sqrt{x^2-ax+b^2} = a^{-1}x^2 + b$.

24. $9x - 4x^2 + \sqrt{4x^2-9x+11} = 5$.

QUADRATICS INVOLVING SURDS.

25. $\dfrac{\sqrt{x^2+x+3}}{\sqrt{2x^2+5x-2}} = \dfrac{3}{4}$.

26. $\dfrac{\sqrt{3x^2+x+5}}{\sqrt{4x^2-x+1}} = \dfrac{3}{2}$.

27. $\left(\dfrac{8x^2+12x+1}{4x^2+2x-1}\right)^{-\frac{1}{2}} = \dfrac{1}{3}$.

28. $\dfrac{(2-3x-x^2)^{\frac{1}{2}}}{(8-5x-2x^2)^{\frac{1}{2}}} = \dfrac{4}{5}$.

29. $\dfrac{4}{x+\sqrt{4-x^2}} + \dfrac{4}{x-\sqrt{4-x^2}} = \dfrac{12}{7}$.

30. $\dfrac{5}{x+\sqrt{5+x^2}} - \dfrac{5}{x-\sqrt{5+x^2}} = 6$.

31. $\dfrac{1}{1-\sqrt{1-x^2}} - \dfrac{1}{1+\sqrt{1-x^2}} = \dfrac{\sqrt{3}}{x^2}$.

32. $\dfrac{4x-1}{\sqrt{2x+1}} + 3\sqrt{2x+1} = 7\sqrt{x}$.

33. $\dfrac{\sqrt{3x^2+4} - \sqrt{2x^2+1}}{\sqrt{3x^2+4} + \sqrt{2x^2+1}} = \dfrac{1}{7}$.

34. $\dfrac{\sqrt{27x^2+4} + \sqrt{9x^2+5}}{\sqrt{27x^2+4} - \sqrt{9x^2+5}} = 7$.

35. $\dfrac{\sqrt{5x-4} + \sqrt{5-x}}{\sqrt{5x-4} - \sqrt{5-x}} = \dfrac{\sqrt{4x}+1}{\sqrt{4x}-1}$.

36. $\dfrac{\sqrt{ax+b} + \sqrt{ax}}{\sqrt{ax+b} - \sqrt{ax}} = \dfrac{1+\sqrt{ax-b}}{1-\sqrt{ax-b}}$.

37. $\dfrac{\sqrt{a+x} + \sqrt{a-x}}{\sqrt{a+x} - \sqrt{a-x}} = \dfrac{x+2a}{2x}$.

38. $\sqrt{x} + \sqrt{2a-x} = \dfrac{a}{\sqrt{x}}$.

39. $\dfrac{\sqrt{x} + \sqrt{b}}{\sqrt{x} - \sqrt{b}} = \dfrac{\sqrt{a-x} + \sqrt{b-x}}{\sqrt{a-x} - \sqrt{b-x}}$.

40. $b\sqrt{b^3-x^3} - \dfrac{b(x^3 - 7b^3)}{\sqrt{b^3-x^3}} = 0$.

41. $\sqrt{x^2+x+1} = a - \sqrt{x^2-x+1}$.

QUADRATICS INVOLVING SURDS.

42. $\dfrac{\sqrt{4a^2-x^2}+p\sqrt{4a-x}}{\sqrt{4a^2-x^2}-p\sqrt{4a-x}} = \dfrac{\sqrt{a}+p}{\sqrt{a}-p}.$

43. $\dfrac{\sqrt{a+x}+\sqrt{a-x}}{\sqrt{x+b}+\sqrt{x-b}} = \dfrac{\sqrt{a+x}-\sqrt{a-x}}{\sqrt{x+b}-\sqrt{x-b}}.$

44. $\dfrac{\sqrt{x+2a}+\sqrt{x-2a}}{\sqrt{x-2a}-\sqrt{x+2a}} = \dfrac{x}{2a}.$ 45. $\dfrac{a-x}{\sqrt{a-x}}+\dfrac{x-b}{\sqrt{x-b}} = \sqrt{a-b}$

46. $\sqrt{\dfrac{a-x}{b+x}}-\sqrt{\dfrac{b+x}{a-x}} = c.$ 47. $\sqrt{\dfrac{a-x}{b-x}}+\sqrt{\dfrac{b-x}{a-x}} = c.$

48. $\sqrt{\dfrac{a-x}{b-x}}-\sqrt{\dfrac{b-x}{a-x}} = c.$ 49. $\sqrt{\dfrac{x+2a}{x+2b}} = \dfrac{x+a}{x+b}.$

50. $\sqrt{3x^2+9}-\sqrt{3x^2-9} = \sqrt{34}+4.$

51. $\sqrt{2x^2+5}+\sqrt{2x^2-5} = \sqrt{15}+\sqrt{5}.$

52. $\sqrt{3x^2+10}+\sqrt{3x^2-10} = \sqrt{17}+\sqrt{-3}.$

53. $2\sqrt{x-a}+3\sqrt{2x} = \dfrac{7a+5x}{\sqrt{x-a}}.$ 54. $\dfrac{x}{a+x}+\dfrac{a}{\sqrt{a+x}} = \dfrac{b}{x}.$

55. $\dfrac{x}{\sqrt{x}+\sqrt{a-x}}+\dfrac{x}{\sqrt{x}-\sqrt{a-x}} = \dfrac{b}{\sqrt{x}}.$

56. $\sqrt{x+1}-2\sqrt[4]{x+1} = 4.$ 57. $(a+x)^{\frac{2}{3}}-(a-x)^{\frac{2}{3}} = (a^2-x^2)^{\frac{1}{3}}.$

58. $(a+x)^{\frac{2}{3}}-5(a^2-x^2)^{\frac{1}{3}} = -4(a-x)^{\frac{2}{3}}.$

59. $(7+4\sqrt{3})x^2+(2+\sqrt{3})x = 2.$

60. $x^{\frac{1}{2}}+x^{-\frac{1}{2}} = (1+x)^{\frac{1}{2}}+(1+x)^{-\frac{1}{2}}.$

QUADRATICS INVOLVING SURDS.

287. It sometimes happens that the roots of a surd quadratic *apparently* do not both satisfy the equation. For example, let it be required to find the roots of $x + \sqrt{x} = 20$.

Transposing, $\quad\quad\quad\quad 20 - x = \sqrt{x}$.

Squaring, $\quad\quad\quad\quad 400 - 40x + x^2 = x$

or $\quad\quad\quad\quad x^2 - 41x + 400 = 0$

or $\quad\quad\quad\quad (x - 16)(x - 25) = 0;$

$$\therefore x = 16 \text{ or } 25.$$

Now, if for x in the equation we substitute 25 there is an *apparent* inconsistency; for $\sqrt{25} + 25 = 30$, not 20.

To explain this, we must bear in mind that the root of a quantity may be either positive or negative; therefore $\sqrt{25} = \pm 5$. If we substitute the *negative* root, -5, instead of the positive, $+5$, the equation is satisfied.

So, too, in the following example, the apparent inconsistency disappears if we take the *negative* root instead of the positive.

Solve $3x + \sqrt{2x-2} = 7$.

Transposing and squaring, $\quad (3x - 7)^2 = 2x - 2$

or $\quad\quad\quad\quad 9x^2 - 42x + 49 = 2x - 2$

or $\quad\quad\quad\quad 9x^2 - 44x + 51 = 0.$

Factoring, $\quad\quad\quad (9x - 17)(x - 3) = 0;$

$$\therefore x = \frac{17}{9} \text{ or } 3.$$

Here the value $x = 3$, if substituted, gives $9 + \sqrt{4} = 7$. But $\sqrt{4} = \pm 2$; therefore taking *negative* root, $9 - 2 = 7$.

CHAPTER XIX.

SIMULTANEOUS EQUATIONS INVOLVING QUADRATICS.

288. The solution of simultaneous equations of the second degree can be accomplished in a variety of ways. A common but somewhat cumbrous method is that of substitution.

Ex. 1.—Given $\quad x+y=7, \quad (1)$
$\qquad\qquad\qquad xy=12, \quad (2)$ find x and y.
From (1), $\qquad y=7-x.$
Substituting this value of y in (2),
$\qquad xy = x(7-x) = 12$
or $\qquad x^2 - 7x + 12 = 0; \;\therefore\; x = 3$ or 4.
But $\qquad y = 7 - x; \;\therefore\; y = 4$ or 3.

NOTE.—The student should notice that when x and y are symmetrically involved in an equation the value of their roots are interchangeable. In the above example, for instance, $x=3$ or 4 and $y=4$ or 3, *i.e.*, when $x=4$ $y=3$, and when $x=3$ $y=4$.

This equation, however, can be more neatly solved by the following method:—

Ex. 2.—Solve $\qquad x+y=7, \qquad\qquad (1)$
$\qquad\qquad\qquad xy = 12. \qquad\qquad (2)$
Squaring (1), $\qquad x^2 + 2xy + y^2 = 49 \qquad (3)$
Multiplying (2) by 4, $\quad\underline{\quad 4xy \quad\;\; = 48\quad} \qquad (4)$
Subtracting (4) from (3), $\; x^2 - 2xy + y^2 = \;\; 1$
Extracting square root, $\qquad x - y = \pm 1$
but $\qquad\qquad\qquad\qquad x + y = \;\; 7$
Therefore adding and subtracting, $\; x = 3$ or 4
and $\qquad\qquad\qquad\qquad\qquad y = 4$ or 3.

266 SIMULTANEOUS EQUATIONS INVOLVING QUADRATICS.

Ex. 3.—Solve

$$x - y = 10, \quad (1)$$
$$x^2 + y^2 = 178. \quad (2)$$

Squaring (1) and subtracting the result from (2) we obtain

$$2xy = 78; \quad (3)$$

but $\quad x^2 + y^2 = 178. \quad (2)$

Adding (2) to (3), $\quad (x+y)^2 = 256.$

Extracting square root, $\quad x + y = \pm 16. \quad (4)$

From (4) and (1), by adding and subtracting, we get

$$x = 13 \text{ or } -3$$

and $\quad y = 3 \text{ or } -13.$

Ex. 4.—Solve

$$\frac{1}{x} + \frac{1}{y} = \frac{9}{20}, \quad (1)$$
$$\frac{1}{x^2} + \frac{1}{y^2} = \frac{41}{400}. \quad (2)$$

Squaring (1) and subtracting (2) from the result,

$$\frac{2}{xy} = \frac{40}{400}. \quad (3)$$

Subtracting (3) from (2), and extracting square root,

$$\frac{1}{x} - \frac{1}{y} = \pm \frac{1}{20}. \quad (4)$$

From (4) and (1) we find $x = 4$ or 5 and $y = 5$ or 4.

Ex. 5.—Solve

$$x^3 + y^3 = 35, \quad (1)$$
$$x + y = 5. \quad (2)$$

Divide (1) by (2), then $x^2 - xy + y^2 = 7. \quad (3)$

From (3) and (2) we can find x and y by the ordinary methods.

SIMULTANEOUS EQUATIONS INVOLVING QUADRATICS.

EXERCISE XCIII.

Solve

1. $x + y = 40,$
   $xy = 300.$

2. $x + y = 13,$
   $xy = 36.$

3. $x - y = 45,$
   $xy = 250.$

4. $x + y = 29,$
   $xy = 100.$

5. $x + y = 18,$
   $xy = 72.$

6. $x - y = 15,$
   $xy = 54.$

7. $x + y = 12,$
   $x^2 + y^2 = 104.$

8. $x - y = 14,$
   $x^2 + y^2 = 436.$

9. $x + y = 49,$
   $x^2 + y^2 = 1681.$

10. $x - y = 10,$
    $x^2 + y^2 = 178.$

11. $x + 4y = 5,$
    $x^2 + xy = 5.$

12. $x - 3y = 1,$
    $xy + y^2 = 5.$

13. $2x - 3y = 3,$
    $x^2 - y^2 = 27.$

14. $x + 3y = 10,$
    $3y^2 - x^2 = 27.$

15. $x + y + 3 = 9,$
    $(x-1)^2 + (y-1)^2 = 8.$

16. $\dfrac{1}{x} - \dfrac{1}{y} = 3,$
    $\dfrac{1}{x^2} + \dfrac{1}{xy} = 21.$

17. $\dfrac{1}{x} + \dfrac{1}{y} = \dfrac{3}{4},$
    $\dfrac{1}{x^2} + \dfrac{1}{y^2} = \dfrac{5}{16}.$

18. $\dfrac{1}{x} + \dfrac{1}{y} = 5,$
    $\dfrac{1}{x^2} + \dfrac{1}{y^2} = 13.$

19. $\dfrac{1}{x} + \dfrac{1}{y} = \dfrac{1}{12},$
    $\dfrac{1}{x^2} - \dfrac{1}{xy} = \dfrac{7}{144}.$

20. $\dfrac{1}{x} + \dfrac{1}{y} = 7,$
    $\dfrac{1}{x^2} + \dfrac{1}{y^2} = 25.$

21. $xy = 12,$
    $\dfrac{1}{x} + \dfrac{1}{y} = \dfrac{7}{12}.$

22. $4y = 5x + 1,$
    $2xy = 33 - x^2.$

23. $7x^2 - 8xy = 159,$
    $5x + 2y = 7.$

24. $x^2 - 2xy - y^2 = 1,$
    $x + y = 2.$

25. $x^3 + y^3 = 91,$
    $x + y = 7.$

26. $x^3 + y^3 = 341,$
    $x + y = 11.$

27. $x^3 + y^3 = 1008,$
    $x + y = 12.$

28. $x^3 - y^3 = 56,$
    $x - y = 2.$

29. $x^3 - y^3 = 98,$
    $x - y = 2.$

30. $x^3 - y^3 = 279,$
    $x - y = 3.$

268 SIMULTANEOUS EQUATIONS INVOLVING QUADRATICS.

289. Another class of equations frequently occurs, viz., that in which both equations are of the *second* degree and both *homogeneous*.

Ex.—Solve
$$x^2 + xy + 2y^2 = 74, \qquad (1)$$
$$2x^2 + 2xy + y^2 = 73. \qquad (2)$$

Cross multiplying,
$$73(x^2 + xy + 2y^2) = 74(2x^2 + 2xy + y^2).$$

Collecting like terms,
$$75x^2 + 75xy - 72y^2 = 0$$
or
$$25x^2 + 25xy - 24y^2 = 0. \qquad (3)$$

Factoring (3), $(5x + 8y)(5x - 3y) = 0;$

$\therefore 5x + 8y = 0$ or $5x - 3y = 0.$

$$\therefore x = -\frac{8}{5}y \text{ or } x = \frac{3}{5}y.$$

If $x = -\frac{8}{5}y$ then $x^2 + xy + 2y^2 = \left(\frac{8}{5}\right)^2 y^2 - \frac{8}{5}y^2 + 2y^2 = 74$

or
$$\frac{74}{25}y^2 = 74,$$

from which we get $y = \pm 5,$
and therefore $x = \mp 8.$

Two other values for x and y respectively can be found by substituting for x its value $\frac{3}{5}y$.

290. This equation might have been solved by another method, often used when the equation corresponding to (3) is not capable of being easily factored.

From (3), $25x^2 + 25xy - 24y^2 = 0.$

Dividing by y^2, $25 \cdot \frac{x^2}{y^2} + 25 \cdot \frac{x}{y} - 24 = 0$

This is now a quadratic in $\left(\frac{x}{y}\right)$ and can be solved in the ordi-

SIMULTANEOUS EQUATIONS INVOLVING QUADRATICS. 269

nary manner. Solving we find $\dfrac{x}{y} = \dfrac{3}{5}$ or $-\dfrac{8}{5}$. We now proceed as before to find x and y.

The student should observe that any equation of the form $ax^2 + bxy + cy^2 = 0$ can be treated in this fashion and the values of $\dfrac{x}{y}$ readily found.

291. We have used the equation $\left.\begin{array}{r}x^2 + xy + 2y^2 = 74\\ 2x^2 + 2xy + y^2 = 73\end{array}\right\}$ to illustrate two general methods; but this equation can be solved readily by the following simple artifice:—

Adding the equations together

$$3x^2 + 3xy + 3y^2 = 147$$
or $\qquad x^2 + xy + y^2 = 49$
but $\qquad x^2 + xy + 2y^2 = 74$

Subtracting, $\qquad y^2 = 25$
$\qquad y = \pm 5$

Substituting the values of y in one of the original equations the values of x can be found.

292. We now give examples of equations easily solved by simple artifices.

Ex. 1.—Solve $\qquad x^3 - y^3 = 26,$ $\qquad$ (1)
$\qquad\qquad\qquad\qquad x^2 + xy + y^2 = 13,$ $\qquad$ (2)

Dividing (1) by (2), $\qquad x - y = 2.$ $\qquad$ (3)
Squaring (3), $\qquad x^2 - 2xy + y^2 = 4;$ $\qquad$ (4)
but $\qquad x^2 + xy + y^2 = 13.$ $\qquad$ (2)
Subtracting (4) from (2), $\qquad 3xy = 9;$
$\qquad\qquad\qquad\qquad \therefore xy = 3.$ $\qquad$ (5)
Adding (5) to (2), $\qquad x^2 + 2xy + y^2 = 16.$ $\qquad$ (6)
Extracting square root, $\qquad x + y = \pm 4.$ $\qquad$ (7)

From (7) and (3) x and y can easily be found.

SIMULTANEOUS EQUATIONS INVOLVING QUADRATICS.

Ex. 2.—Solve
$$x^2 + xy = 21, \quad (1)$$
$$y^2 + xy = 28. \quad (2)$$

Dividing (1) by (2), $\dfrac{x}{y} = \dfrac{21}{28} = \dfrac{3}{4}$; $\therefore x = \dfrac{3}{4}y$.

This value of x substituted in (1) or (2) will give a quadratic in y.

EXERCISE XCIV.

Solve

1. $2y^2 - 4xy + 3x^2 = 17,$
   $y^2 - x^2 = 16.$

2. $x^2 - 4y^2 - 9 = 0,$
   $xy + 2y^2 - 3 = 0.$

3. $x^2 + xy + 4y^2 = 6,$
   $3x^2 + 8y^2 = 14.$

4. $x^2 - xy - 35 = 0,$
   $xy + y^2 - 18 = 0.$

5. $x^2 - xy + y^2 = 21,$
   $y^2 - 2xy = -15.$

6. $x^2 + xy + 2y^2 = 44,$
   $2x^2 - xy + y^2 = 16.$

7. $x^2 + xy - 15 = 0,$
   $xy - y^2 - 2 = 0.$

8. $2x^2 + 3xy + y^2 = 70,$
   $6x^2 + xy - y^2 = 50.$

9. $x^2 - xy + y^2 = 7,$
   $3x^2 + 13xy + 8y^2 = 162.$

10. $x^2 - xy - y^2 = 5,$
    $2x^2 + 3xy + y^2 = 28.$

11. $x^2 + xy = 84,$
    $xy + y^2 = 60.$

12. $x^2 - xy = 6,$
    $x^2 + y^2 = 61.$

13. $x^2 + xy = 66,$
    $x^2 - y^2 = 11.$

14. $x^2 + 2xy + 3y^2 = 17,$
    $2x^2 + 3xy + 5y^2 = 28.$

15. $x^3 - y^3 = 37,$
    $x^2 + xy + y^2 = 37.$

16. $x^3 + 6xy = 144,$
    $6xy + 36y^2 = 432.$

17. $x^2 + xy + y^2 = 39,$
    $3y^2 - 5xy = 25.$

18. $3x^2 + 4xy = 20,$
    $5xy + 2y^2 = 12.$

19. $x^2 + y^2 = 225$,
    $xy = 108$.

20. $x^2 + 9xy = 340$,
    $7xy - y^2 = 171$.

21. $x^2 - xy = 35$,
    $xy + y^2 = 18$.

22. $x^2 + y^2 = 68$,
    $xy = 16$.

23. $x^2 + xy - 6y^2 = 24$,
    $x^2 + 3xy - 10y^2 = 32$.

24. $x^2 - xy + y^2 = 21$,
    $y^2 - 2xy + 15 = 0$.

25. $\dfrac{x+y}{x-y} + \dfrac{x-y}{x+y} = \dfrac{10}{3}$,
    $x^2 + y^2 = 20$.

26. $\dfrac{x+y}{x-y} + \dfrac{x-y}{x+y} = \dfrac{5}{2}$,
    $x^2 + y^2 = 20$.

27. $y^2 - 4xy + 20x^2 + 3y - 264x = 0$,
    $5y^2 - 38xy + x^2 - 12y + 1056x = 0$.

28. $\dfrac{1}{x} + \dfrac{1}{y} = \dfrac{x+y}{12} = \dfrac{7}{x+y+5}$.

CHAPTER XX.

PROBLEMS RESULTING IN QUADRATIC EQUATIONS.

293. The statement in algebraic language of the conditions of a problem often gives a quadratic equation. Of the *two* values resulting from the solution of the equation it sometimes happens that only *one* will satisfy the conditions of the problem.

Ex. 1.—The sum of the squares of two consecutive numbers is 481. Find the numbers.

Let $x = $ one number
and $x + 1 = $ the other number.
Then $\quad x^2 + (x+1)^2 = 481.$ \hfill (1)

The solution of (1) gives $x = 15$ or -16; therefore the answer will be 15 and 16 or -15 and -16.

If we confine the meaning of consecutive numbers to arithmetical numbers in the common scale it is evident that the values -16 and -15 do not satisfy the conditions of the problem.

The explanation of this and all other kindred problems lies in the fact that as soon as a condition is expressed in algebraic symbols the unknown quantity is treated as an algebraic number. For instance, in the problem we have just solved, as soon as we let x stand for one number we are required to find an algebraic quantity which will satisfy the condition $x^2 + (x+1)^2 = 481$.

The problem now might be worded as follows:—

Find two algebraic numbers, differing by *unity*, which when squared and added together will give 481. Since algebra permits the use of *negative* quantities it is evident -16 and -15 are values satisfying (1).

PROBLEMS RESULTING IN QUADRATIC EQUATIONS. 273

Ex. 2.—A person bought a number of oxen for £80; if he had bought 4 more for the same sum each ox would have cost £1 less. Find the number of oxen and the price of each.

Let $x =$ number of oxen.

Then $\dfrac{80}{x} =$ price of each in £'s.

If the person had bought 4 more the price of each in £'s would have been $\dfrac{80}{x+4}$; hence, to satisfy the given condition,

$$\frac{80}{x+4} = \frac{80}{x} - 1.$$

Simplifying, $\quad x^2 + 4x = 320;$ \hfill (1)

$$\therefore x = 16 \text{ or } -20.$$

Now, we cannot have -20 oxen, therefore the answer required is 16. But the value -20 is one which will satisfy the condition $x^2 + 4x = 320$, and consequently is a root of that equation. Had this problem been worded as follows no inconsistency would be apparent:—

Find an algebraic number such that when it is divided into 80 the resulting quotient will be *one* more than that obtained when 80 is divided by the number increased by 4.

EXERCISE XCV.

1. The sum of the squares of three consecutive numbers is 365. Find the numbers.

2. Three times the product of two consecutive numbers exceeds four times their sum by 8. Find the numbers.

3. The product of three consecutive numbers is equal to three times the middle number. Find the numbers.

4. A man bought a number of apples for 16 cents. Had he bought 4 more for the same money he would have paid $\frac{1}{3}$ of a cent less for each apple. How many did he buy?

5. Find the price of eggs per score when 10 more in 62½ cents' worth lowers the price 31¼ cents per 100.

6. A merchant bought some pieces of silk for $168.75. He sold the silk for $12 a piece and gained as much as one piece cost him. How much did he pay for each piece?

7. A merchant bought some pieces of silk for $900. Had he bought 3 pieces more for the same money he would have paid $15 less for each piece. How many did he buy?

8. The area of a square may be doubled by increasing its length by 6 inches and its breadth by 4 inches. Determine its side.

9. The length of a rectangular field exceeds the breadth by 1 yard, and the area is 3 acres. Find its dimensions.

10. A grass plot 9 yards long and 6 yards broad has a path around it. The area of the path is equal to that of the plot. Find the width of the path.

11. Divide a line 20 inches long into two parts so that the rectangle contained by the whole and one part may be equal to the square on the other part.

12. A vessel which has two pipes can be filled in 2 hours less time by one than by the other, and by both together in 2 hours 55 minutes. How long will it take each pipe alone to fill the vessel?

13. A vessel which has two pipes can be filled in 2 hours less time by one than by the other, and by both together in 1 hour 52 minutes 30 seconds. How long will it take each pipe alone to fill the vessel?

14. A number is expressed by two digits, one of which is the square of the other, and when 54 is added its digits are interchanged. Find the number.

15. A number is composed of two digits, the first of which exceeds the second by 2. The sum of the square of the number and of that which is obtained by reversing the digits is 4034. What is the number?

16. A number is composed of two digits, the first of which exceeds the second by unity, and the number itself falls short of the sum of the squares of its digits by 35. What is the number?

17. Divide 35 into two parts so that the sum of the two fractions formed by dividing each part by the other may be $2\frac{1}{12}$.

18. A boat's crew row $3\frac{1}{2}$ miles down a river and back again in 1 hour 40 minutes. If the current of the river is 2 miles an hour find the rate of rowing in still water.

19. A jockey sold a horse for $144 and gained as much per cent. as the horse cost. What did the horse cost?

20. A merchant expended a sum of money in goods which he sold again for $24, and lost as much per cent. as the goods cost him. How much did he pay for the goods?

21. A broker bought a number of bank shares ($100 each), when they were at a certain rate per cent. *discount*, for $7,500, and afterwards, when they were at the same rate per cent. *premium*, sold all but 60 for $5,000. How many shares did he buy?

22. A person's gross income is £1,000. After deducting a percentage for income tax, and then a percentage less by one than the income tax rate from the remainder, the income is reduced to £912. Required the rate per cent. at which income tax is charged.

23. If the length and breadth of a rectangle were each increased by 1 the area would be 48; if they were each diminished by 1 the area would be 24. Find the length and breadth.

24. The sum of the squares of the two digits of a number is 25, and the product of the digits is 12. Find the number.

25. The sum, product and difference of the squares of two numbers are all equal. Find the numbers. (Let $x+y$ and $x-y$ represent the numbers.)

26. The sum of two numbers divided by their difference gives the same quotient as if the greater number were divided by the less. Find the quotient.

276   PROBLEMS RESULTING IN QUADRATIC EQUATIONS.

27. The difference of two numbers is $\frac{3}{8}$ of the greater, and the sum of the squares is 356.   What are the numbers?

28. The difference between the hypotenuse and two sides of a right-angled triangle is 3 and 6 respectively.   Find the sides.

29. Find two numbers whose sum is nine times their difference, and whose product diminished by the greater number is equal to twelve times the greater number divided by the less.

30. A person has $13,000, which he divides into two parts, and placing each at interest receives an equal income.   If he placed the first sum at the rate of interest of the second he would receive $360 income, and if he placed the second sum at the rate of the first he would receive $490 income.   What are the two sums and what the rates of interest?

31. A and B have each a quantity of flour, A having 4 barrels more than B. They sell their flour to each other at different prices per barrel, and the account between them is settled by B giving to A £7 16s. B's quantity sold at A's price would have amounted to £28, and A's quantity at B's price to £34.   Find how much was sold by each and the rates per barrel.

32. From a sheet of paper 14 inches long a border of uniform width is cut away all around it, and the area is thereby reduced $\frac{5}{8}$; but had the sheet been 3 inches narrower, and a border of the same width been cut away, the area would have been reduced $\frac{4}{7}$. What was the breadth of the paper?

33. The hypotenuse of a right angle is 20, and the area of the triangle is 96.   Find the sides of the triangle.

34. The fore wheel of a carriage turns in a mile 132 times more than the hind wheel; but if the circumferences were each increased by 2 feet it would turn only 88 times more.   Find the circumference of each.

35. The numerator and denominator of one fraction are each greater by 1 than those of another, and the sum of the two fractions is $1\frac{5}{12}$.   If the numerators were interchanged the sum of the fraction would be $1\frac{1}{2}$.   Find the fractions.

CHAPTER XXI.

THEORY OF QUADRATICS.

294. *A quadratic equation cannot have more than two roots.*

Let the quadratic be reduced to the form $x^2 + px + q = 0$. Then, since $x^2 + px + q$ is of only *two* dimensions, it cannot have more than *two* factors, each of *one* dimension.

If $x - a$ and $x - b$ be the factors of $x^2 + px + q$ then $x^2 + px + q = (x - a)(x - b) = 0$.

It is evident that $x^2 + px + q$ will vanish for $x = a$ and $x = b$; therefore a and b are roots of the equation. Also, a and b are the only roots; for no values other than a and b will make $(x - a)(x - b) = 0$. Hence the quadratic has only two roots, viz., a and b.

295. The following proof is given in many text-books, but it is defective, as it proves that a quadratic cannot have three *unequal* roots, but does not prove that it cannot have three roots, two or more of which are equal.

If possible let α, β and γ be the roots of the quadratic equation $ax^2 + bx + c = 0$.

Then, since α is a root, $\quad a\alpha^2 + b\alpha + c = 0,$ $\quad$ (1)

and since β is a root, $\quad a\beta^2 + b\beta + c = 0;$ $\quad$ (2)

similarly, since γ is a root, $a\gamma^2 + b\gamma + c = 0.$ $\quad$ (3)

Subtracting (2) from (1) and (3) from (1) we obtain

$$a(\alpha^2 - \beta^2) + b(\alpha - \beta) = 0, \quad (4)$$

$$a(\alpha^2 - \gamma^2) + b(\alpha - \gamma) = 0. \quad (5)$$

Dividing (4) by $(\alpha - \beta)$, which is by hypothesis not zero, and (5) by $(\alpha - \gamma)$, which also is not zero, we obtain

$$a(\alpha + \beta) + b = 0, \qquad (6)$$
$$a(\alpha + \gamma) + b = 0. \qquad (7)$$

Subtracting (7) from (6), $\quad a(\beta - \gamma) = 0.\qquad (8)$

By hypothesis $\beta - \gamma$ is not zero and a is not $= 0$; therefore the result is impossible, and the equation cannot have three *unequal* roots.

NOTE.—The reason why a quadratic equation cannot have more than *two* roots is the fact that it cannot have more than *two* factors, each of *one* dimension. For a similar reason a cubic equation cannot have more than three roots, etc.

296. If a and b are the roots of a quadratic equation then $(x - a)$ and $(x - b)$ are the factors of the corresponding quadratic expression. For if the expression vanishes when $x = a$ then $x - a$ is a factor. Similarly $x - b$ is a factor;

$$\therefore \ x^2 + px + q = (x - a)(x - b)$$
or $\qquad x^2 + px + q = x^2 - x(a + b) + ab.$

Now, since this is an identity, by equating coefficients, we obtain

$$p = -(a + b) \quad \text{or} \quad a + b = -p, \qquad (1)$$
and $\quad q = ab \qquad \text{or} \qquad ab = q. \qquad (2)$

Hence the important principle:—

The sum of the roots of a quadratic equation of the form $x^2 + px + q = 0$ is equal to the coefficient of x with its sign changed; and the product of the roots equals the absolute term, that is, the term independent of x.

297. This result may be obtained directly as follows:—

Solving $x^2 + px + q = 0$ we find the values of x to be

$$-\frac{p}{2} + \frac{\sqrt{p^2 - 4q}}{2} \quad \text{and} \quad -\frac{p}{2} - \frac{\sqrt{p^2 - 4q}}{2}.$$

THEORY OF QUADRATICS.

If a and b are the roots then

$$a = -\frac{p}{2} + \frac{\sqrt{p^2 - 4q}}{2}, \qquad (1)$$

$$b = -\frac{p}{2} - \frac{\sqrt{p^2 - 4q}}{2}. \qquad (2)$$

Adding (1) and (2),

$$a + b = -p. \qquad (3)$$

Multiplying (1) and (2) together,

$$ab = \left(-\frac{p}{2}\right)^2 - \frac{p^2 - 4q}{4}$$

$$= \frac{p^2}{4} - \frac{p^2 - 4q}{4} = \frac{4q}{4} = q;$$

$$\therefore ab = q. \qquad (4)$$

298. From the fact that if a and b are roots then $(x-a)$ and $(x-b)$ are factors of the corresponding quadratic expression we can at any time find the equation when its roots are given.

Let a and b be given roots; to find the equation of which they are the roots.

If a and b are roots then $(x-a)$ and $(x-b)$ are factors;

$$\therefore (x-a)(x-b) = 0$$
or $$x^2 - x(a+b) + ab = 0$$

is the required equation.

Ex. 1.—Given 2 and 3 to be roots of an equation, find the equation.

Since 2 and 3 are roots $x-2$ and $x-3$ are factors;

$$\therefore (x-2)(x-3) = 0$$
or $$x^2 - 5x + 6 = 0$$

is the required equation.

Ex. 2.—Given 3 and -4 to be roots, find the equation. Here the factors are $x-3$ and $x-(-4)$ or $x+4$; therefore equation is $(x-3)(x+4) = 0$ or $x^2 + x - 12 = 0$.

280 THEORY OF QUADRATICS.

Ex. 3.—Let $\sqrt{2}$ and $-\sqrt{2}$ be roots, their factors are $x-\sqrt{2}$ and $x+\sqrt{2}$; therefore equation is

$$(x-\sqrt{2})(x+\sqrt{2})=0$$

or
$$x^2-2=0.$$

Hence, to find the equation of which the roots are given, subtract each root in turn from x and multiply the different remainders together. The result equated to zero will be the equation required.

299. *To find the condition that roots of $ax^2+bx+c=0$ may be real and equal, real and unequal, or imaginary.*

Solving $ax^2+bx+c=0$ we find

$$x=-\frac{b}{2a}\pm\frac{\sqrt{b^2-4ac}}{2a}.$$

Let
$$x_1=-\frac{b}{2a}+\frac{\sqrt{b^2-4ac}}{2a}$$

and
$$x_2=-\frac{b}{2a}-\frac{\sqrt{b^2-4ac}}{2a},$$

then x_1 and x_2 are the roots of $ax^2+bx+c=0$.

(*a*) Fixing attention upon the expression $\sqrt{b^2-4ac}$, we see that if $b^2=4ac$ the quantity $\sqrt{b^2-4ac}=0$, and x_1 and x_2 each become $-\frac{b}{2a}$; therefore x_1 and x_2 are *equal* when $b^2=4ac$.

(*b*) But if $b^2>4ac$ then b^2-4ac is a *positive* quantity and $\sqrt{b^2-4ac}$ is a *real* quantity.

Let
$$\pm\sqrt{b^2-4ac}=\pm d,$$

then
$$x_1=-\frac{b}{2a}+d$$

and
$$x_2=-\frac{b}{2a}-d,$$

therefore x_1 and x_2 differ by $2d$, and x_1 and x_2 are *real* and *unequal*.

THEORY OF QUADRATICS. 281

(c) If, however, $b^2 < 4ac$, $b^2 - 4ac$ is *negative*, and $\sqrt{b^2 - 4ac}$ is the square root of a *negative* quantity and therefore *imaginary;* hence x_1 and x_2 are both *imaginary*, as each contains the *imaginary* quantity $\sqrt{b^2 - 4ac}$.

Hence in any quadratic equation the roots are real and equal, real and unequal, or imaginary, according as the square of the coefficient of x is equal to, greater than, or less than, four times the product of the coefficient of x^2 and the absolute term.

Ex. 1.—Let $2x^2 - 5x + 8 = 0$; to find the character of the roots. Here $(5)^2 < 4 \times 2 \times 8$; therefore roots are imaginary.

Ex. 2.—Examine $2x^2 - 8x + 8 = 0$. Since $(8)^2 = 4 \times 2 \times 8$, therefore roots are equal.

Ex. 3.—Examine $3x^2 - 12x + 9 = 0$. Since $(12)^2 > 4 \times 3 \times 9$, therefore roots are real and unequal.

Ex. 4.—What value of n will make $x^2 - 8x + n = 0$ have equal roots? The condition of equal roots is

$$(8)^2 = 4n$$
or $\qquad 64 = 4n; \; \therefore \; n = 16.$

Ex. 5.—What value of a will make $4x^2 + ax + 16 = 0$ have equal roots? To have equal roots,

$$a^2 = 4 \times 4 \times 16$$
or $\qquad a^2 = 256;$
$$\therefore a = \pm 16.$$

Ex. 6.—What value of b will make $bx^2 + 9x + 20 = 0$ have equal roots? Condition required is

$$81 = 4b \times 20$$
or $\qquad 81 = 80b;$
$$\therefore b = \frac{81}{80}.$$

300. The equation $ax^2 + bx + c = 0$ requires examination, (1) when the value of a approaches *zero*, or as is generally said $= 0$; (2) when $a = 0$ and $b = 0$.

THEORY OF QUADRATICS.

(1) Let $ax^2 + bx + c = 0$ have the coefficient of x^2, zero.

Divide by x^2, then $\quad a + \dfrac{b}{x} + \dfrac{c}{x^2} = 0$.

Let $\dfrac{1}{x} = y$, then $\quad a + by + cy^2 = 0$.

But $a = 0$, therefore $\quad by + cy^2 = 0$

or $\quad y(b + cy) = 0$;

$$\therefore y = 0 \text{ or } -\dfrac{b}{c}.$$

But $y = \dfrac{1}{x}$, therefore $\quad \dfrac{1}{x} = 0 \text{ or } -\dfrac{b}{c}$.

Let $\dfrac{1}{x} = -\dfrac{b}{c}$, then $x = -\dfrac{c}{b}$, one of the roots required.

Let $\dfrac{1}{x} = 0$; then, since the value of a fraction diminishes as the denominator increases, the value of x in $\dfrac{1}{x} = 0$ must be very great; for the quotient obtained by dividing 1 by x is very small, or zero. In such examples as this x is said to be infinitely great, or *infinity*, the symbol for which is ∞. Therefore the roots of $ax^2 + bx + c = 0$ are $-\dfrac{c}{b}$ and ∞ when $a = 0$.

(2) Let $a = 0$ and also $b = 0$.

$ax^2 + bx + c = 0$ when divided by x^2 becomes

$$a + \dfrac{b}{x} + \dfrac{c}{x^2} = 0.$$

As before, put $\quad \dfrac{1}{x} = y;$

$$\therefore a + \dfrac{b}{x} + \dfrac{c}{x^2} = a + by + cy^2 = 0.$$

But $a = 0$ and $b = 0$, therefore $a + by + cy^2 = cy^2 = 0$; hence both values of $y = 0$. But $y = \dfrac{1}{x} = 0$, therefore $x = \textit{infinity}$. So that when $a = 0$ and $b = 0$ both values of x are *infinity*.

NOTE.—The student must bear in mind that when we say $a = 0$ we mean that the coefficient of x^2 becomes indefinitely small, so that it differs from zero by a quantity less than any given quantity.

THEORY OF QUADRATICS.

301. We now proceed to give solutions of some problems usually found in connection with the Theory of Quadratics. It is impossible to give examples of all possible forms, but we may state that the most of the problems to be found in text-books on Algebra require for their solution nothing more than a clear conception of the fact that the sum of the roots of a quadratic, in its simplest form, equals the coefficient of x with its sign changed, and the product of the roots equals the absolute term.

Ex. 1.—If a and b are the roots of $x^2 + mx + n = 0$ find the equation whose roots are $\dfrac{1}{a}$ and $\dfrac{1}{b}$.

If $\dfrac{1}{a}$ and $\dfrac{1}{b}$ are the roots of an equation then

$$\left(x - \frac{1}{a}\right)\left(x - \frac{1}{b}\right) = 0 \text{ is the equation,}$$

or $\qquad x^2 - x\left(\dfrac{1}{a} + \dfrac{1}{b}\right) + \dfrac{1}{ab} = 0,$

or $\qquad x^2 - x\left(\dfrac{a+b}{ab}\right) + \dfrac{1}{ab} = 0.$

But $\qquad a + b = -m$ and $ab = n$;

$$\therefore \frac{a+b}{ab} = -\frac{m}{n} \text{ and } \frac{1}{ab} = \frac{1}{n}.$$

$\therefore x^2 - x\left(\dfrac{a+b}{ab}\right) + \dfrac{1}{ab} = 0$ becomes

$$x^2 + \frac{mx}{n} + \frac{1}{n} = 0$$

or $\qquad nx^2 + mx + 1 = 0.$

Otherwise,

Divide $\qquad x^2 + mx + n = 0$ by x^2,

then $\qquad 1 + \dfrac{m}{x} + \dfrac{n}{x^2} = 0.$

If $y = \dfrac{1}{x}$ then the equation becomes $1 + my + ny^2 = 0$, and the values of y will be $\dfrac{1}{a}$ and $\dfrac{1}{b}$, since the values of y are the reciprocals of those of x. Therefore $1 + my + ny^2 = 0$ is the equation whose roots are $\dfrac{1}{a}$ and $\dfrac{1}{b}$.

Ex. 2.—If m and n are the roots of $ax^2 + bx + c = 0$ find the values of

(a) $m^2 + mn + n^2$, (b) $m^2 - n^2$, (c) $m^2 - mn + n^2$,
(d) $m^4 + m^2n^2 + n^4$, (e) $m^3 + n^3$.

(a) Dividing $ax^2 + bx + c = 0$ by a,

$$x^2 + \frac{b}{a}x + \frac{c}{a} = 0;$$

$$\therefore m + n = -\frac{b}{a} \qquad (1)$$

and $\qquad mn = \dfrac{c}{a}.$ \qquad (2)

Squaring (1), $\qquad m^2 + 2mn + n^2 = \dfrac{b^2}{a^2}$ \qquad (3)

and $\qquad mn = \dfrac{c}{a}.$ \qquad (2)

Subtracting, $\qquad m^2 + mn + n^2 = \dfrac{b^2}{a^2} - \dfrac{c}{a} = \dfrac{b^2 - ac}{a^2};$

$$\therefore m^2 + mn + n^2 = \frac{b^2 - ac}{a^2}.$$

(b) To find the value of $m^2 - n^2$.

Since $\qquad m + n = -\dfrac{b}{a}$ \qquad (1)

and $\qquad mn = \dfrac{c}{a},$ \qquad (2)

squaring (1) and subtracting $4mn$,

$$m^2 - 2mn + n^2 = \frac{b^2}{a^2} - \frac{4c}{a} = \frac{b^2 - 4ac}{a^2}.$$

THEORY OF QUADRATICS.

Extracting square root, $\quad m - n = \pm \dfrac{\sqrt{b^2 - 4ac}}{a}.\quad$ (3)

But $\quad m + n = -\dfrac{b}{a}.\quad$ (4)

Multiplying together (3) and (4),

$$m^2 - n^2 = \mp \dfrac{b\sqrt{b^2 - 4ac}}{a^2}.$$

(c) To find the value of $m^2 - mn + n^2$.

$$(m + n)^2 = m^2 + 2mn + n^2 = \dfrac{b^2}{a^2}$$

and $\quad 3mn = \dfrac{3c}{a}.\quad$ From (2)

Subtracting, $\quad m^2 - mn + n^2 = \dfrac{b^2 - 3ac}{a^2}.$

(d) Since $\quad m^2 + mn + n^2 = \dfrac{b^2 - ac}{a^2}$

and $\quad m^2 - mn + n^2 = \dfrac{b^2 - 3ac}{a^2},$

$$\therefore m^4 + m^2 n^2 + n^4 = (m^2 - mn + n^2)(m^2 + mn + n^2)$$
$$= \dfrac{(b^2 - 3ac)(b^2 - ac)}{a^4}.$$

(e) To find the value of $m^3 + n^3$.

$$m^3 + n^3 = (m + n)(m^2 - mn + n^2).$$

But $\quad m + n = -\dfrac{b}{a}$

and $\quad m^2 - mn + n^2 = \dfrac{b^2 - 3ac}{a^2};$

$$\therefore m^3 + n^3 = \dfrac{-b(b^2 - 3ac)}{a^3}.$$

Another Method.

$$m^3 + n^3 = (m + n)^3 - 3mn(m + n) = \left(-\dfrac{b}{a}\right)^3 - 3\left(\dfrac{c}{a}\right)\left(-\dfrac{b}{a}\right)$$
$$= \dfrac{3abc - b^3}{a^3}.$$

THEORY OF QUADRATICS.

Ex. 3.—If one-third the sum of the squares of the roots of the equation $ax^2 + bx + c = 0$ is equal to the product show that $b^2 = 5ac$.

Let m and n be roots of the equation, then

$$m + n = -\frac{b}{a}$$

and

$$mn = \frac{c}{a};$$

also, by hypothesis, $\quad \frac{1}{3}(m^2 + n^2) = mn.$ \hfill (1)

We have now to express $m^2 + n^2$ and mn in terms of a, b and c.

Since $\quad m^2 + n^2 = (m+n)^2 - 2mn = \dfrac{b^2}{a^2} - \dfrac{2c}{a} = \dfrac{b^2 - 2ac}{a^2};$

$$\therefore \ \frac{1}{3}(m^2 + n^2) = \frac{1}{3}\left(\frac{b^2 - 2ac}{a^2}\right);$$

also, $\quad mn = \dfrac{c}{a}.$

Substituting these values of $\frac{1}{3}(m^2 + n^2)$ and mn in (1) we obtain

$$\frac{1}{3}\left(\frac{b^2 - 2ac}{a^2}\right) = \frac{c}{a},$$

or $\quad b^2 - 2ac = 3ac,$

or $\quad b^2 = 5ac.$

Ex. 4.—Find the roots of the equation $\dfrac{a}{x+a} + \dfrac{b}{x+b} + \dfrac{c}{x+c} = 0$ when $a + b + c = 0$.

Clearing of fractions,

$$a(x+b)(x+c) + b(x+a)(x+c) + c(x+a)(x+b) = 0.$$

Multiplying out, and collecting coefficients of x^2 and x,

$$x^2(a+b+c) + x(2ab + 2ac + 2bc) + 3abc = 0.$$

Since $a + b + c$, the coefficient of x^2, $= 0$, one root of the quadratic is *infinity* and the other root is $\dfrac{-3abc}{2(ab + bc + ca)}.$

THEORY OF QUADRATICS.

Ex. 5.—If the equation $x^2 + px + q = 0$ have equal roots show that $ax^2 + p(a+b)x + q(a+2b) = 0$ has one of them, and find the other.

If the roots of $\quad x^2 + px + q = 0$ have equal roots

then $\qquad p^2 = 4q$ or $q = \dfrac{p^2}{4}$;

$\therefore ax^2 + p(a+b)x + q(a+2b) = ax^2 + p(a+b)x + \dfrac{p^2}{4}(a+2b) = 0.$

But $ax^2 + p(a+b)x + \dfrac{p^2}{4}(a+2b) = \left(x + \dfrac{p}{2}\right)\left\{ax + \dfrac{p}{2}(a+2b)\right\};$

$\therefore \left(x + \dfrac{p}{2}\right)\left\{ax + \dfrac{p}{2}(a+2b)\right\} = 0;$

$\therefore x = -\dfrac{p}{2}$ or $-\dfrac{p(a+2b)}{2a}.$

Since $x^2 + px + q = 0$ has equal roots, and $-p = $ sum of roots, therefore $-\dfrac{p}{2} = $ one of the equal roots.

Ex. 6.—If $x^2 - 9x + a = 0,$ (1)

$x^2 - 14x + 3a = 0,$ (2) have a common root solve both of them.

If (1) and (2) have a common root they have a common factor, which must be a factor of their difference, and therefore must be $x - \dfrac{2}{5}a$. Therefore $x = \dfrac{2}{5}a$ is common root.

Substituting this value of x in (1),

$$\left(\dfrac{2}{5}a\right)^2 - 9\left(\dfrac{2}{5}a\right) + a = 0,$$

or $\qquad\qquad 4a^2 - 90a + 25a = 0,$

or $\qquad\qquad 4a^2 = 65a.$

Therefore $a = \dfrac{65}{4}$ and $\dfrac{2}{5}a = \dfrac{13}{2}$; therefore common root $= \dfrac{13}{2}$. But sum of roots in (1) $= 9$ and in (2) $= 14$; therefore remaining roots are $\dfrac{5}{2}$ and $\dfrac{15}{2}$.

THEORY OF QUADRATICS.

Ex. 7.—If the ratio of the roots of the equation $x^2+px+q=0$ be equal to that of the roots of $x^2+p_1x+q_1=0$ then $p^2q_1=p_1^2q$.

Let a and b be roots of $x^2+px+q=0$
and m and n be roots of $x^2+p_1x+q_1=0$,

then
$$\frac{a}{b}=\frac{m}{n}.$$

But $\quad a+b=-p$ and $\quad ab=q$,
also $\quad m+n=-p_1$ and $\quad mn=q_1$;

therefore $\quad \dfrac{(a+b)^2}{ab}=\dfrac{p^2}{q}$ and $\quad \dfrac{(m+n)^2}{mn}=\dfrac{p_1^2}{q_1}$.

But $\quad \dfrac{(a+b)^2}{ab}=\dfrac{a}{b}+\dfrac{b}{a}+2$ and $\quad \dfrac{(m+n)^2}{mn}=\dfrac{m}{n}+\dfrac{n}{m}+2$,

and, by hypothesis, $\quad \dfrac{a}{b}=\dfrac{m}{n}$;

$$\therefore \frac{(a+b)^2}{ab}=\frac{(m+n)^2}{mn},$$

$$\therefore \frac{p^2}{q}=\frac{p_1^2}{q_1}$$

or $\quad p^2q_1=p_1^2q$.

EXERCISE XCVI.

1. If a and b be the roots of $x^2+px+q=0$ prove $a^3+b^3=3pq-p^3$.

2. If a and b be the roots of $px^2+qx+r=0$ prove that the equation whose roots are $\dfrac{a}{b}$ and $\dfrac{b}{a}$ will be $prx^2+(2pr-q^2)x+pr=0$.

3. If m and n be the roots of $ax^2+x+b=0$ show that

$$\left(1+\frac{m}{n}\right)\left(1+\frac{n}{m}\right)=\frac{1}{ab}.$$

4. Find the sum of the roots of the equation

$$\frac{a_1}{x-a}+\frac{b_1}{x-b}+\frac{c_1}{x-c}=0.$$

THEORY OF QUADRATICS.

5. If α and β be the roots of $ax^2 + bx + c = 0$ form the equation whose roots are $\dfrac{\alpha^2}{\beta^2}$ and $\dfrac{\beta^2}{\alpha^2}$; also form the equation whose roots are $\dfrac{\alpha^3}{\beta^3}$ and $\dfrac{\beta^3}{\alpha^3}$.

6. If α and β be the roots of $x^2 + rx + \dfrac{3}{16}r^2 = 0$ form the equation whose roots are $\alpha^3 + \beta^3$ and $\alpha^3 - \beta^3$.

7. If a and b be the roots of the equation $px^2 + qx + q = 0$ show that $\sqrt{\dfrac{a}{b}} + \sqrt{\dfrac{q}{p}} + \sqrt{\dfrac{b}{a}} = 0$.

8. If the roots of $x^2 + px + q = 0$ and $x^2 + qx + p = 0$ differ by the same quantity show that $p + q + 4 = 0$.

9. If a and b be the roots of $x^2 + px + q = 0$, and a and c those of $x^2 + rx + s = 0$, then $(b + c)$ and bc respectively will satisfy the equations $x^2 + (p + r)x + 2(q + s) = 0$ and $x^2 + (q + s - pr)x + qs = 0$.

10. If the roots of $ax^2 + bx + c = 0$ are in the ratio of m to n show that $\dfrac{b^2}{ac} = \dfrac{(m+n)^2}{mn}$.

11. If α and β be the roots of $x^2 - x(1 + a) + \dfrac{1}{2}(1 + a + a^2) = 0$ prove that $\alpha^2 + \beta^2 = a$.

12. Form the equation whose roots are the square of the sum and difference of the roots of $2x^2 + 2(m + n)x + m^2 + n^2 = 0$.

13. If α and β are the roots of $ax^2 + bx + c = 0$ form the equation whose roots are $\alpha^2 + \beta^2$ and $\dfrac{1}{\alpha^2} + \dfrac{1}{\beta^2}$.

14. If α and β be the roots of $ax^2 + bx + c = 0$ prove that the quadratic equation whose roots are $\dfrac{\alpha^4}{\beta}$ and $\dfrac{\beta^4}{\alpha}$ is

$$a^4cx^2 + b(b^4 - 5ab^2c + 5a^2c^2)x + ac^4 = 0.$$

15. Find the relation between the coefficients of $ax^2 + bx + c = 0$ that one root may be double of the other.

16. Form the equation whose roots are $\sqrt{3}$ and $-\sqrt{3}$.

17. Form the equation whose roots are $m + n$ and $m - n$.

THEORY OF QUADRATICS.

18. Find the condition that $ax^2 + bx + c = 0$ and $a_1x^2 + b_1x + c_1 = 0$ may have a common root.

19. If $ax^2 + bx + c = 0$ and $a_1x^2 + b_1x + c_1 = 0$ have respectively two roots, one of which is the reciprocal of the other, prove that $(aa_1 - cc_1)^2 = (ab_1 - bc_1)(a_1b - b_1c)$.

20. For what value of m will the equation $2x^2 + 8x + m = 0$ have equal roots?

21. Show that in every quadratic of the form $ax^2 + bx + a = 0$ the roots are the reciprocal of each other.

22. Show that the roots of $cx^2 + bx + a = 0$ are the reciprocals of those of $ax^2 + bx + c = 0$.

23. If α and β be the roots of the equation $ax^2 + bx + c = 0$, and $\alpha^2 + \beta^2 = 1$, show that $2ac = b^2 - a^2$.

24. If $x^2 - 3x - a = 0$ and $x^2 - 4x - 5 = 0$ have one root in common show that $a^2 - 14a + 40 = 0$.

25. If the roots of the equation $x^2 + px + q = 0$ be respectively any equimultiples of the roots of the equation $x^2 + rx + s = 0$ prove that $sp^2 = qr^2$.

26. If the difference between the roots of the two equations $x^2 + (p-a)x + b^2 = 0$ and $x^2 + (p-b)x + a^2 = 0$ be equal then shall $2p = 5(a + b)$.

27. If a and b be the roots of $x^2 - px + q = 0$, and α and β the roots of $x^2 - qx + p = 0$, show that $\dfrac{1}{a\alpha} + \dfrac{1}{a\beta} + \dfrac{1}{ab} + \dfrac{1}{\beta b} = 1$.

28. Show that the roots of the quadratic equation
$$\frac{1}{3a-x} + \frac{1}{3b-x} + \frac{1}{3c-x} = 0$$
are real if $a^2 + b^2 + c^2$ is greater than $ab + ac + bc$.

29. Find the values of m which will make the equation $3mx^2 + (6m - 12)x + 8 = 0$ have equal roots.

30. If $ab + bc + ca = 0$ find the roots of $\dfrac{a^2}{x-a} + \dfrac{b^2}{x-b} + \dfrac{c^2}{x-c} = 0$.

APPENDIX.

SPECIAL FORMS OF SIMPLE EQUATIONS.

1. By methods already given any simple equation involving but one unknown quantity, x, may always be reduced to the form $ax = b$ where a represents the sum of the coefficients of all the terms containing x, and b the sum of all the other terms.

I. Let $a = 0$, then $ax = 0$; for if one factor $= 0$ the product $= 0$, and therefore no value of x can satisfy the equation. If, however, a be not zero but a very small quantity, then the equation will be satisfied when x is very great; and by making a small enough x may be made greater than any assignable quantity. In this sense it is customary to say that when $a = 0$ $x = \infty$. Similar remarks apply to equations (3) and (4) of the following Art. when the coefficients of x and y vanish.

II. Let $a = 0$ and also $b = 0$, then $ax = 0$, and the equation is satisfied for any value of x. In this case the given equation is an *identity*, and therefore true for all values of x. (Art. 108.)

2. The following simple example will illustrate the meaning of the preceding cases:—

A and B commence business with capitals of $\$a$ and $\$b$ respectively; A gains $\$m$ and B $\$n$ per annum. In how many years will they have an equal amount of money?

292 SPECIAL FORMS OF SIMPLE EQUATIONS.

Let $x =$ the number of years,

then $\quad a + mx = A$'s money at the end of x years

and $\quad b + nx = B$'s money at the end of x years.

Then $\quad a + mx = b + nx;$

$$\therefore (m - n)x = b - a \text{ or } x = \frac{b - a}{m - n}.$$

Now, if $m - n = 0$ no value of x can satisfy the equation unless also $b - a = 0$, and then *any* value of x will satisfy it. Referring to the original problem we see that these results are correct; for if $m = n$ and a is not $= b$ then evidently the two men can *never* have the same amount of money; but if $m = n$ and also $a = b$ then they have *always* the same amount.

3. Simultaneous equations of the first degree involving two unknown quantities may always be reduced to three terms each, and may therefore be expressed in the form

$$ax + by = c, \qquad (1)$$
$$a'x + b'y = c'. \qquad (2)$$

Eliminating y and x successively in the usual way we get

$$(ab' - a'b)x = b'c - bc', \qquad (3)$$
$$(ab' - a'b)y = c'a - ca'. \qquad (4)$$

I. Let $ab' - a'b = 0$, then no values of x and y can satisfy equations (3) and (4), and consequently no values of x and y can satisfy the given equations. In this case the given equations are said to be *inconsistent* or *contradictory*, and no solution is possible.

II. Let $ab' - a'b = 0$ and also $b'c - bc' = 0$, then any value of x will satisfy equation (3); and since $ab' - a'b = 0$, $\frac{a'}{a} = \frac{b'}{b}$; and since $b'c - bc' = 0$, $\frac{b'}{b} = \frac{c'}{c}$, and therefore $\frac{c'}{c} = \frac{a'}{a}$; therefore $c'a - ca' = 0$, and therefore any value of y will satisfy equation (4). In this

SPECIAL FORMS OF SIMPLE EQUATIONS. 293

case the given equations are *not independent*, and the values of x and y are *indeterminate*.

Referring to the original equations we find that in Case I. we have $\dfrac{a'}{a} = \dfrac{b'}{b}$, but $\dfrac{b'}{b}$ not $= \dfrac{c'}{c}$; therefore $a'x + b'y$ is a multiple of $ax + by$, but c' is not the same multiple of c, which is evidently *impossible*.

In Case II. we have $\dfrac{a'}{a} = \dfrac{b'}{b} = \dfrac{c'}{c}$, and therefore $a'x + b'y$ is a multiple of $ax + by$, and c' also the same multiple of c. The second equation is therefore merely a repetition of the first, and an indefinite number of solutions may be given. (Art. 180.)

4. Thus far we have supposed that no one of the quantities a, b, c, a', b', c' is zero, and we have seen that the values of x and y, as determined from (3) and (4), are either both impossible or both indeterminate; but without the preceding condition this result is not necessarily true. For let a and a' each be zero and $b'c - bc'$ not zero, then the value of x is impossible and the value of y indeterminate. Again, let a and a' each be zero and $b'c - bc'$ also zero, then the values of x and y from (3) and (4) *appear* indeterminate. Referring to the original equations we have $by = c$, $b'y = c'$; and since $b'c - bc' = 0$ one equation is a multiple of the other. The value of x is indeterminate because *the equations do not contain x*, and the value of y *appears* indeterminate in (4) because both the original equations were multiplied by the zero multipliers, a' and a; the real value of y is either $\dfrac{b}{c}$ or $\dfrac{b'}{c'}$.

5. When three simultaneous equations are given involving three unknown quantities various peculiarities may arise the discussion of which in detail would be tedious. What has been given is sufficient to enable the student to understand any particular case which may present itself. The most important cases are illustrated by numerical examples in the following exercise.

EXERCISE XCVII.

Solve

1. $(a-b)(x-c) + (b-c)(x-a) + (c-a)(x-b) = 0$.
2. $(a-b)(x-c)^2 + (b-c)(x-a)^2 + (c-a)(x-b)^2 = 0$.
3. $(a-b)(x-c)^3 + (b-c)(x-a)^3 + (c-a)(x-b)^3 = 0$.
4. $(2x-a-b)^3 - (x-a)^3 - (x-b)^3 = 3(x-a)(x-b)(2x-a-b)$.
5. $m(x-a) + n(y-b) = (m+n)(a+b)$.
   $m(x-b) + n(y-a) = (m-n)(a-b)$.

6. $x - y = a$,
   $y - z = b$,
   $z - x = c$.

7. $x + y + z = 20$,
   $x + 2y + 3z = 40$,
   $2x + 3y + 4z = 60$.

8. $x + 2y + 3z = 10$,
   $3x + 4y + 6z = 23$,
   $x + 6y + 9z = 24$,

9. $3x + y - z = 15$,
   $7x + 2y - 2z = 35$,
   $11x + 3y - 3z = 70$.

10. $(a-b)x + (b-c)y + (c-a)z = m$,
    $(b-c)x + (c-a)y + (a-b)z = n$,
    $(c-a)x + (a-b)y + (b-c)z = p$.

11. $x + y = 5$,
    $x^3 + y^3 + 15xy = 125$.

Expose the fallacy in the following:—

12. Let $a = b$, then $ab = b^2$ and $a^2 - ab = a^2 - b^2$;
    $\therefore a(a-b) = (a+b)(a-b)$;
    $\therefore a = a + b = 2a$ since $a = b$;
    $\therefore a = 2a$ or $1 = 2$.

13. Solve
    $axy = c(bx + ay)$,   (1)
    $bxy = c(ax + by)$.   (2)

Multiply (1) by a, (2) by b, and subtract, we get
$(a^2 - b^2)xy = c(a^2 - b^2)y$ or $x = c$.

Substitute this value for x in (1) and we get
$acy = c(bc + ay)$ or $bc^2 = 0$.

ANSWERS.

EXERCISE I. (Page 13.)

1. 5. 2. 31. 3. 7. 4. 38. 5. 49.
6. 90. 7. 183. 8. 198. 9. 2. 10. 87.
11. 21 12. 1. 22. 76, 536.

EXERCISE II. (Page 14.)

1. $a+b$. 2. a^2+b^2. 3. $(a+b)^2$. 4. $6ab$. 5. $\dfrac{a^3+b^3}{a+b}$.

6. $\sqrt{a^2+b^2}$. 7. $(a+b)^2 = a^2+b^2+2ab$. 8. $\dfrac{a^3-b^3}{a-b} = a^2+b^2+ab$.

9. $100x$, $\dfrac{x}{100}$. 10. $12x+y$, $36x+12y$. 11. $\$7x$. 12. $x-5a-7b$.

13. xy miles. 14. $\dfrac{a}{x}$ hrs. 15. $\dfrac{(nq+mp)x}{100}$. 16. mx. 17. xy.

18. xyz, $2xy+2yz+2zx$, $4x+4y+4z$. 19. $\dfrac{xyz}{2592}$. 20. $\dfrac{x-y}{m}$.

21. $x+y+10$. 22. $x-10$. 23. $x-y$. 24. $\dfrac{x}{y}$. 25. $xy+r$.

26. 5, 98, 18, 7, 3. 27. $\dfrac{2(x+3)(a+b)-d}{m+n}$. 28. $\dfrac{ax+by}{a+b}$ cts.

29. $(x+y)(a+b)(p+q)$. 30. $\dfrac{(x+y)^4}{100}$. 31. $\dfrac{100m-ax-by}{c}$.

32. $\left(\dfrac{200x}{m}+\dfrac{300y}{n}\right)\div p$. 33. $\dfrac{x-ab-cd}{y}$. 34. $\dfrac{x-kr}{h-k-\dfrac{m}{60}}$.

35. $\dfrac{m}{x+y}$ hrs., $\dfrac{mx}{x+y}$, $\dfrac{my}{x+y}$.

ANSWERS

EXERCISE III. (Page 19.)

1. 99, 1980.　　　2. 40 lbs.　　　3. 108, 27.
4. 9, or 441 sq. in; 27, or 9261 cu. in.　　5. 5 miles *west*.
6. -8.　7. Debts.　8. -100.

EXERCISE IV. (Page 22.)

1. $+15$.　2. -15.　3. $+9$.　4. -10.　5. -24.　6. -15
7. $-5ab$.　8. $+9m^2$.　9. $10m^2 - 13n^2$.　10. $-3a + b + 7c + d$.
11. $6x - 9z$.　　12. $7x - 13y + 2z$.　　13. $3a^2 - 5ab + b^2$
14. -44, $+1885$, -19, $+1910$.
15. $+20 + (-15) = +5$, $20 + 15 = 35$, *direction*.
16. $1000 + 3000 + (-2000) + (-2500) = -500 = \$500$ *debt*.

EXERCISE V. (Page 23.)

1. $4a + 7b + 7c$.　　2. $3a + 3b + 3c$.　　3. $9a - 13b - 4c$.
4. $-ab - 5bc + 8ac$.　5. $10x^2 - 7xy - 4y^2$.　6. 0.
7. $6(x+y)$.　　8. $9(a^2 + b)$.　　9. $-a(x - y)$.
10. $x + y + z - 2(m + n)^2$.　11. $13(2a - 3b) - 2c$.　12. 0.
13. $9a - 4b - 17c - 12d + 12e$.
14. $6a^3 - 2a^2b - 3ab^2 - 22ac^2 - 7b^2 - c^3$.
15. $12a^3 + 2a^2b + 12ab^2 + b^3 - 4a^2$.　16. $2a^2b^2c$.
17. $3x + 7y + 11z$.　18. $5x + 13y + 6z$.　19. $6a + 6b + 6c$.　20. 0

EXERCISE VI. (Page 25.)

1. $(a + m)x + (b + n)y$.　　2. $(2a - 7)x + (10 - 3b)y$.
3. $(m - b + 1)x + (n - a - 1)y$.　4. $(5c - 10d)x + (a + 7b)y$.
5. $(2a + 2b)y$.　　6. $3ax$.　　7. $nx + y$.
8. $-nx$.　　9. 0.　　10. $(a + b + c)x$.

EXERCISE VII. (Page 27.)

1. $+8$.　　2. -8.　　3. -30.　　4. $+30$.
5. $+41$.　　6. -75.　　7. $6a$.　　8. $-6a$.
9. $-8m$.　　10. $3x$.　　11. $-8ab$.　　12. $8a^2x$.
13. $-3ay$.　　14. $-3aby$.　　15. $17m^2x^3$.　16. $-12xy$.
17. $6x^2$.　　18. $9y^2$.　　19. $m - n$.　　20. 0.
21. $476 - (-753) = 1229$.　　22. $2000 - (-1500) = 3500$.

ANSWERS.

EXERCISE VIII. (PAGE 28.)

1. $a - 3b - 2c$.
2. $-2a - b - 2c$.
3. $2a - 2b - x$.
4. $5x^3 - 5x^2y - 11xy^2$.
5. $2x^3 - 6x^2y + 6xy^2 - 2y^3$.
6. $-x^3 - 2x^2 - x - 4$.
7. $7 - 10x + 10x^2 - 6x^3$.
8. $2a^2 - 2b^2 - 2ac + 2bc$.
9. $x^2 - y^2 - 3xy - a + b$.
10. $a^3b + ab^3$.
11. $5x^2 + 3y^2 - 11xy + 5yz - 7xz$.
12. $2a^2bc + 2abc^2 - 2abc$.
13. $(a - c)x^2 - (b - d)y^2$.
14. $(a - l)x^2 + (2b - 2m)xy + (c - n)y^2$.
15. $(c - b)x^2 + (a - c)xy + (b - a)y^2$.
16. $(a - b) + 8(x + y)$.
17. $2(a + b) - 2(c + d) + 10(x + y) + m - n$.
18. $2bx + 2cy + 2az$.
19. $-x^3 + 15x^2y + 7y^3$.

EXERCISE IX. (PAGE 30.)

1. 0.
2. 0.
3. $2a - b$.
4. $4a - b - c$.
5. $2c - 5d$.
6. $5x - 3a$.
7. $6x - 8y$.
8. $2a - 2b$.
9. $2a - b - d$.
10. $3a$.
11. $-2x - y$.
12. $7x - 6y$.
13. a.
14. $7a - 7b$.
15. $10c - b$.
16. $(a - b) + (c - d) - (e - f)$, $(a - b + c) - (d + e - f)$.
17. $a - \{b - (c - d - e) - f\}$.
18. $a - \{(b - c) + (d + e) - f\}$.
19. $a - [b - \{c - (d + \overline{e - f})\}]$.
20. $-a + b - c$.
21. 17.

EXERCISE X. (PAGE 34.)

1. 15.
2. -15.
3. -42.
4. 99.
5. 196.
6. 15.
7. 135.
8. 0.
9. 162.
10. -2.
11. -450.
12. -21.
13. 4816.
14. $-35a^5$.
15. $-80a^2bc$.
16. $-36a^5x^3y^2$.
17. $24m^4n^4p$.
18. $-30x^5y^5z^5$.
19. $60(abxy)^3$.
20. $0, 0, 0$.
21. $3b - c$.

EXERCISE XI. (PAGE 35.)

1. $6x^3 - 9x^2 + 12x$.
2. $15x^4y - 6x^2y^3$.
3. $-2a^3b^2 + 2a^2b^3 - 2ab^4$.
4. $-8m^3n^2p + 20m^2n^3p$.
5. $-35a^3bxy^2 + 25a^2b^2y^3 - 55ab^3y^3 + 30a^2by^4$.
6. $-24x^3y + 40xy^3 + 56x^2y^2 + 32x^2y + 24xy^2 - 16xy$.
7. $13x^3 - 27x^2 + 44x$.
8. $4a^2 - 11ab + 15b^2$.
9. $8y^2 + 6xy - 24x^2$.
10. $2x^4 + 26x^3 - 12x^2$.
11. $a^2 + b^2 + c^2$.
12. 0.

13. $2ad + 2bc$. 14. $2my - 2nx$. 15. 0. 16. $9b$.
17. $2ax$. 18. $2bx + 2by$. 19. $2bx + 2cy$. 20. 0.

EXERCISE XII. (Page 37.)

1. $x^3 - x^2 - 2x + 8$.
2. $2x^3 - 8x^2 + 13x - 10$.
3. $8x^3 - 6x^2 - 23x + 21$.
4. $-8x^3 + 18x^2 + x - 15$.
5. $x^4 + x^2 + 1$. 6. $x^4 + 4$. 7. $a^3 - b^3$. 8. $a^3 + b^3$.
9. $a^4 - 4a^2 + 12a - 9$. 10. $4a^4 - 13a^2b^2 + 9b^4$.
11. $6x^4 - 23x^3 + 25x^2 - 16x + 15$.
12. $2x^6 - 4x^5 + 5x^4 + 6x^3 - 9x^2 + 3x + 7$. 13. $1 - 4x^2 + 16x^3 - x^4 - 12x^5$.
14. $x^7 + 5x^6 - 3x^5 - 9x^4 + 10x^3 - x^2 - 6x + 3$.
15. $2x^5 - 7x^4 - 5x^3 - 9x^2 - 15x + 18$.
16. $x^9 - x^6 - 2x^5 - x^4 + 2x^3 + x^2 - x - 1$.
17. $x^8 + 2x^6 + 3x^4 + 2x^2 + 1$. 18. $x^3 + 151x - 264$.
19. $x^5 - 41x - 120$. 20. $a^3 + b^3 + c^3 - 3abc$.
21. $2a^2b^2 + 2b^2c^2 + 2a^2c^2 - a^4 - b^4 - c^4$. 22. $x^3 + y^3 + 3xy - 1$.
23. $x^6 + 10x - 33$. 24. $343x^9 - 512y^3$.
25. $x^8 - 4x^6y^2 + 6x^4y^4 - 4x^2y^6 + y^8$. 26. $x^3 - 8y^3 + z^3 + 6xyz$.
27. $x^3 + (a + b + c)x^2 + (ab + bc + ca)x + abc$.
28. $x^3 - (a + b + c)x^2 + (ab + bc + ca)x - abc$.
29. $x^3 - 9x^2 + 26x - 24$. 30. $x^4 - 10x^2 + 9$.
31. $x^6 - a^6$. 32. $x^8 + x^4y^4 + y^8$. 33. $729m^6 - a^6$.

EXERCISE XIII. (Page 38.)

1. $x^2 + 2xy + y^2$. 2. $x^2 - 2xy + y^2$. 3. $x^2 - y^2$.
4. $4x^2 + 4xy + y^2$. 5. $x^2 - 4xy + 4y^2$. 6. $x^2 - 4y^2$.
7. $4x^2 + 12xy + 9y^2$. 8. $4x^2 - 12xy + 9y^2$. 9. $4x^2 - 9y^2$.
10. $x^4 + 2x^2y^2 + y^4$. 11. $x^4 - 2x^2y^2 + y^4$. 12. $x^4 - y^4$.
13. $a^4 + 2a^2bc + b^2c^2$. 14. $a^4 - 2a^2bc + b^2c^2$. 15. $a^4 - b^2c^2$.
16. $a^2x^2 + 2abxy + b^2y^2$. 17. $a^2x^2 - 2abxy + b^2y^2$. 18. $a^2x^2 - b^2y^2$.
19. $1 + 2x^2 + x^4$. 20. $9x^4 - 24x^2 + 16$. 21. $16x^{10} - 1$.
22. $25m^4n^2 + 60m^3n^3 + 36m^2n^4$. 23. $64m^6 - 80m^4p^2 + 25m^2p^4$.
24. $49y^2 - 81x^2$. 25. $x^8 - y^8$. 26. 6399. 27. 9409.
28. 8096. 29. 7000. 30. $10^8 - 1 = 99999999$.
31. $24ab - 120b^2$. 32. $2x^3y^3 + 3x^2y^4 - y^6$. 33. 0. 34. $a^2 - d^2$.

ANSWERS.

EXERCISE XIV. (Page 40.)

1. $x^2+y^2+z^2+2xy+2xz+2yz$.
2. $x^2+y^2+z^2+2xy-2xz-2yz$.
3. $x^2+y^2+z^2-2xy+2xz-2yz$.
4. $x^2+y^2+z^2-2xy-2xz+2yz$.
5. $x^2+y^2+z^2+2xy-2xz-2yz$.
6. $x^2+y^2+z^2-2xy-2xz+2yz$.
7. $a^2+b^2+4c^2+2ab+4ac+4bc$.
8. $x^2+4b^2+9c^2-4ab+6ac-12bc$.
9. $4a^2+b^2+9c^2-4ab+12ac-6bc$.
10. $1+2x+3x^2+2x^3+x^4$.
11. $1-2x+3x^2-2x^3+x^4$.
12. $x^4+2x^3y+3x^2y^2+2xy^3+y^4$.
13. $4x^4-12x^3+25x^2-24x+16$.
14. $x^4-2x^3-3x^2+4x+4$.
15. $4x^4-4x^3-15x^2+8x+16$.
16. $a^2+b^2+c^2+d^2-2ab+2ac-2ad-2bc+2bd-2cd$.
17. $a^4+2a^3b-2a^3c-4a^2bc+a^2b^2-2ab^2c+a^2c^2+2abc^2+b^2c^2$.
18. $x^6-2x^5+3x^4-4x^3+3x^2-2x+1$.
19. $4(x^2+y^2+z^2)$.

EXERCISE XV. (Page 41.)

1. $a^2+2ab+b^2-c^2$.
2. $a^2-b^2-c^2+2bc$.
3. $4x^2-4xy+y^2-9z^2$.
4. $9z^2-x^2-4y^2+4xy$.
5. x^4+x^2+1.
6. $a^4+a^2b^2+b^4$.
7. $9a^4+11a^2b^2+4b^4$.
8. x^4+4a^4.
9. $a^2+b^2-c^2-d^2+2ab+2cd$.
10. $c^2+d^2-a^2-b^2+2cd+2ab$.
11. $-y^2z^2-3z^4$.
12. 0.
13. $2a^2b^2+2b^2c^2+2a^2c^2-a^4-b^4-c^4$.

EXERCISE XVI. (Page 42.)

1. $x^2+7x+12$.
2. $x^2+7x+10$.
3. x^2+7x+6.
4. $x^2+21x+108$.
5. $x^2+21x+90$.
6. $x^2+20x+99$.
7. $x^2-9x+14$.
8. $x^2-13x+30$.
9. $x^2-20x+75$.
10. x^2+x-6.
11. x^2+7x-8.
12. $x^2+15x-100$.
13. $x^2-15x-100$.
14. $x^2-20x-300$.
15. $x^2-20x-21$.
16. x^4-8x^2-20.
17. x^6+10x^3-56.
18. x^5+20x^4-125.
19. $x^2-4xy-5y^2$.
20. $x^4-9x^2yz+14y^2z^2$.
21. $x^2+(a-b)x-ab$.
22. $4x^2-34xy+70y^2$.
23. $25x^2-60x+20$.
24. $9a^4-3(b-c)a^2-bc$.
25. $a^2x^2+a(b+c)x+bc$.

EXERCISE XVII. (Page 43.)

1. x^3+3x^2+3x+1.
2. x^3-3x^2+3x-1.
3. $x^4+4x^3+6x^2+4x+1$.
4. $x^4-4x^3+6x^2-4x+1$

ANSWERS.

5. $x^5 + 5x^4 + 10x^3 + 10x^2 + 5x + 1$.   6. $x^5 - 5x^4 + 10x^3 - 10x^2 + 5x - 1$.
7. $x^3 + 6x^2y + 12xy^2 + 8y^3$.   8. $8x^3 - 12x^2y + 6xy^2 - y^3$.
9. $8a^3 - 36a^2b + 54ab^2 - 27b^3$.   10. $a^4 - 8a^3b + 24a^2b^2 - 32ab^3 + 16b^4$.
11. $16a^4 + 32a^3b + 24a^2b^2 + 8ab^3 + b^4$.
12. $x^5 - 5x^4y + 10x^3y^2 - 10x^2y^3 + 5xy^4 - y^5$.
13. $2a^3 + 6ax^2$.   14. $6a^2x + 2x^3$.   15. $2x^5 + 20x^3y^2 + 10xy^4$.
16. $8x^3y + 8xy^3$.   17. $2a^3 + 2b^3$, 0.   18. $x^2 + y^2 + z^2 - xy - yz - zx$.
19. $2(x^3 + y^3 + z^3 - 3xyz)$, 0.   20. $x^2 + y^2 + z^2 - xy - yz - zx$.

EXERCISE XVIII. (Page 46.)

1. 4.   2. -4.   3. 50.   4. -50.   5. -128.
6. 250.   7. a.   8. $2a^3c^2$.   9. $8a^3b$.   10. $16a^2y$.
11. $-24a^6b^4$.   12. $9x^4y^5$.   13. $-25a^5$.
14. $2a^6b^5c^4d^3$.   15. $2y^2$.   16. $12xy$.
17. $x^2 - 2x - 7$.   18. $-y^3 + ay^2 - by$.   19. $a^2 - 2ab - 3b^2$.
20. $-5a^2 + 6ab + 8b^2$.   21. $x^4 - mpx^2 + p^2$.   22. $-4x^2 + 7xy - 9y^2$.
23. $7x^2z^2 - 9xy$.   24. $4a^5 - 5ab^4 + 6b^5$.   25. $2a^3 + 3a^2b - 4b$.
26. $12x^2 - 11xy + 10y^2$.   27. $2 - 3bc + 5ad^2$.

EXERCISE XIX. (Page 50.)

1. $x + 5$.   2. $x - 4$.   3. $x + 7$.   4. $x - 9$.
5. $x^2 + 7x + 12$.   6. $x^2 + x + 1$.   7. $a + b$.   8. $a^2 + ab + b^2$.
9. $a + b$.   10. $a^4 - a^3b + a^2b^2 - ab^3 + b^4$.   11. $x^2 + 3x + 2$.
12. $2x^2 - 3x + 7$.   13. $2x + 3y$.   14. $4x - 1$.   15. $x^3 - 3x^2 + 3x + 1$.
16. $3x^2 + 4xy + y^2$.   17. $a^3 - 2a^2b + 3ab^2 + 4b^3$.   18. $a^3 - 3a^2x + 2ax^2$.
19. $a^2 - ab + b^2$.   20. $x^4 + x^2 + 1$.   21. $x^5 + y^5$.
22. $x^{12} + x^9y^3 + x^6y^6 + x^3y^9 + y^{12}$.   23. $x^4 + 2x^3 + 3x^2 + 2x + 1$.
24. $x^4 + x^3 + x + 1$.   25. $x^8 + x^7y - x^5y^3 - x^4y^4 - x^3y^5 + xy^7 + y^8$.
26. $2x^3 - 3x^2 + 2x$.   27. $x^2 - 3xy - y^2$.
28. $-13a^2b^2 - 3ab - 1$.   29. $x^4 + 2x^3y + 2x^2y^2 + xy^3$.
30. $x^4 + 3x^3y + 8x^2y^2 - 8y^4$.   31. $27x^2y - 18xy^2 - 9y^3$.
32. $2a^3b + 3a^2b^2 - ab^3 + 4b^4$.   33. $x^5 + 4x^4 - 3x^3 + x^2 - 5$.
34. $x^8 + x^7y - x^5y^3 - x^4y^4 - x^3y^5 + xy^7 + y^8$.

EXERCISE XX. (Page 52.)

1. $x^2 - (a+c)x + ac$, $x^2 - (a+b)x + ab$. 2. $x + a$.
3. $x^2 - (b+d)x + bd$, $x^2 - (a+d)x + ad$. 4. $x^2 + ax - b$.
5. $(1+m)x + (1-n)y$. 6. $a^2 + b^2 + c^2 - ab + bc + ca$.
7. $a + b - c$. 8. $4a^2 + b^2 + c^2 + 2ab + bc - 2ca$. 9. $x + 2y - z$.
10. $2x - 3y + 4z$. 11. $ab + bc + ca$.
12. $ab - bc + ca$. 13. $x - 2y + 3z$.
14. $x^2 + y^2 + z^2 + 2xy - 2yz - 2zx$. 15. $x^2 - ax + a^2$.
16. $x^2 + ax + a^2$. 17. $(x^2 + x + 1)a - (x + 1)$.

EXERCISE XXI. (Page 53.)

1. $x^3 - x^2y + xy^2 - y^3$. 2. $x^4 + x^3y + x^2y^2 + xy^3 + y^4$.
3. $x^5 + x^4y + x^3y^2 + x^2y^3 + xy^4 + y^5$.
4. $x^6 - x^5y + x^4y^2 - x^3y^3 + x^2y^4 - xy^5 + y^6$.
5. $x^2 - x + 1$. 6. $x^5 - x^4 + x^3 - x^2 + x - 1$.
7. $1 + xy + x^2y^2$. 8. $x^4 - x^3yz + x^2y^2z^2 - xy^3z^3 + y^4z^4$.
9. $x^4 - 2x^3 + 4x^2 - 8x + 16$. 10. $9 + 3x + x^2$.
11. $x^4 - 3x^3 + 9x^2 - 27x + 81$. 12. $4x^2 + 6xy + 9y^2$.
13. $x^2 - 2x + 4$. 14. $x^4 + 2x^3 + 4x^2 + 8x + 16$.
15. $8x^3 - 4x^2 + 2x - 1$. 16. $x^3 + 3x^2 + 9x + 27$.
17. $36 + 6x + x^2$. 18. $4x^2 + 14xy + 49y^2$.
19. $9x^2 - 30xy + 100y^2$. 20. $16x^2 + 28xy + 49y^2$.
21. $25x^2 - 40xy + 64y^2$. 22. $x^4 + 5x^2y^2 + y^4$.

EXERCISE XXII. (Page 54.)

1. $3(x^2 - 5)$. 2. $5(2x^2 - 3xy + 4y^2)$.
3. $7p^2(p - 9)$. 4. $11(2m^2 - 3mn - 10n^2)$.
5. $ax(x^2 - bx + 1)$. 6. $27a^3b^6(2 + 4a^3b^2 - 9a^5b^3)$.
7. $35xyz(x + 2y - 3z)$. 8. $(a+b)(x+y)$.
9. $(a-b)(x-y)$. 10. $(c+d)(2a+3b)$. 11. $(3x-y)(a-b)$.
12. $(a+b)(c+1)$. 13. $(x+a)(x+b)$. 14. $(x-a)(x-b)$.
15. $(x-a)(x+b)$. 16. $(x+a)(x-b)$. 17. $(x+b)(a-x)$.
18. $(x+a)(b-x)$. 19. $(ac-bd)(ax-by)$. 20. $(x-y)(ac+bd)$.
21. $(x+y)(ad-bc)$. 22. $(x+1)(x^2+1)$.
23. $(x+1)(x^4-x^2+1)$. 24. $(x-y)(2a+3b-c)$.
25. $(2a^2-3bc)(3b^2-ac)$. 26. $a(x-y)(bx+cy)$.

ANSWERS.

EXERCISE XXIII. (Page 56.)

1. $(x+1)(x+3)$.   2. $(x+5)(x+2)$.   3. $(x+6)(x+1)$.
4. $(x+10)(x+3)$.   5. $(x+6)(x+5)$.   6. $(x-15)(x-2)$.
7. $(x-12)(x-6)$.   8. $(x-20)(x-2)$.   9. $(x-17y)(x-3y)$.
10. $(x^2-18)(x^2-4)$.   11. $(x+6)(x-5)$.   12. $(x+7)(x-6)$.
13. $(x+7)(x-5)$.   14. $(x+11)(x-8)$.   15. $(x+10)(x-1)$.
16. $(x+7)(x-11)$.   17. $(x+2)(x-9)$.   18. $(x^2+1)(x^2-12)$.
19. $(x^3-17)(x^3+4)$.   20. $(x^3-4)(x^3+3)$.   21. $(x^5+20)(x^5-1)$.
22. $(xy+39)(xy-10)$.   23. $(x^2y+20)(x^2y-5)$.   24. $(xyz-20)(xyz+5)$.
25. $3(x-9)(x+8)$.   26. $4(x+3)(x-2)$.
27. $5(x^2-10y^2)(x^2+8y^2)$.   28. $2a(x^3-7a)(x^3+2a)$.
29. $11(x^5+20)(x^5-15)$.   30. $x(1-6x)(1+x)$.
31. $a^2b(1+19x)(1-x)$.   32. $-(x+6)(x-1)$.
33. $-(x+29)(x-1)$.   34. $(x-a)(x-b)$.   35. $(a-b)(b-c)$.

EXERCISE XXIV. (Page 57.)

1. $(x+5)^2$.   2. $(x^2+9)^2$.   3. $(x+10y)^2$.   4. $(ax+2)^2$.
5. $(a^2x+4y^3)^2$.   6. $(m^2-8n^2)^2$.   7. $(x^3-19)^2$.   8. $(lx-my)^2$.
9. $(a^5-7b)^2$.   10. $(9x^2-1)^2$.   11. $(2x+3y)^2$.   12. $(3x-5y)^2$.
13. $(4a^2-3b)^2$.   14. $(a+3b)^2$.   15. $(2a-5b)^2$.   16. $(5ax-7by)^2$.
17. $x^2(3a-7b)^2$.   18. $a^2(2x-5y)^2$.   19. $(6a-4b)^2$.   20. $6(2x+3y)^2$.
21. $3x(ax-3y)^2$.   22. $3y^2(x-3ay)^2$.   23. $(a+b+c)^2$.   24. $(a-b+4)^2$.
25. x^2.   26. $4x^2$.   27. x^2.   28. $(x+y)^2$.
29. $(2a+2b+3c+3d)^2$.   30. $(2a-b+c)^2$.   31. $(a^2-b^2+c^2)^2$.

EXERCISE XXV. (Page 58.)

1. $(x+y)(x-y)$.   2. $(x+4)(x-4)$.   3. $(5+y)(5-y)$.
4. $(2x+3y)(2x-3y)$.   5. $(4x^2+7y)(4x^2-7y)$.
6. $(x^2+4)(x+2)(x-2)$.   7. $(4x^2+9y^2)(2x+3y)(2x-3y)$.
8. $(16x^4+1)(4x^2+1)(2x+1)(2x-1)$.   9. $(xy^2+10z^3)(xy^2-10z^3)$.
10. $5(x^2+2y^4)(x^2-2y^4)$.   11. $3(a^2+3b^2)(a^2-3b^2)$.
12. $2(9xy+11z)(9xy-11z)$.   13. $(a+b+c)(a+b-c)$.
14. $(a-b+c)(a-b-c)$.   15. $(a+b+c)(a-b-c)$.
16. $(a+b-c)(a-b+c)$.   17. $(2a-b)b$.
18. $(a+3b)(a+b)$.   19. $(a+b+c+d)(a+b-c-d)$.
20. $(a-b+c-d)(a-b-c+d)$.   21. $4ab$.

ANSWERS.   303

22. $(x-y+z)(x-y-z)$.
23. $(b+c+a)(b+c-a)$.
24. $(a+b-c)(a-b+c)$.
25. $(ax+by+1)(ax+by-1)$.
26. $(a+2b-3c)(a-2b+3c)$.
27. $(1+4a-5b)(1-4a+5b)$.
28. $(a+b+c-d)(a+b-c+d)$.
29. $(a-b+c+d)(a-b-c-d)$.
30. $(a+b-c-d)(a-b-c+d)$.
31. $(a+b+c+d)(a-b+c-d)$.
32. $12(2x+1)(x-1)$.
33. $(x-1)(x-2)(x-3)(x-4)$.
34. $(x+2)(x-2)(x+4)(x-4)$.
35. $(x+5)^2(x-3)(x-7)$.
36. $3(x+3)(x+5)(x+7)(x-5)$.
37. $(a+d)(a-d)(b+c)(b-c)$.
38. $(x^2+y^2)(x+y)(x-y)(1+n^2)(1+n)(1-n)$.

EXERCISE XXVI. (PAGE 59.)

1. $(x^2+x+1)(x^2-x+1)$.
2. $(x^2+2x+4)(x^2-2x+4)$.
3. $(x^2+3x+9)(x^2-3x+9)$.
4. $(x^2+2xy+2y^2)(x^2-2xy+2y^2)$.
5. $(x^2+xy+3y^2)(x^2-xy+3y^2)$.
6. $(x^2+xy-y^2)(x^2-xy-y^2)$.
7. $(x+y)(x+2y)(x-y)(x-2y)$.
8. $(2x^2+2x+1)(2x^2-2x+1)$.
9. $(2x^2+3xy+3y^2)(2x^2-3xy+3y^2)$.
10. $(x^2+3x+5)(x^2-3x+5)$.
11. $(x+y)(3x+y)(x-y)(3x-y)$.
12. $(x^2+x+1)(x^2-x+1)(x^4-x^2+1)$.
13. $(2x+y)(x+3y)(2x-y)(x-3y)$.
14. $(2x+3y)(x+y)(2x-3y)(x-y)$.
15. $(a^2+3b^2)(b^2+3a^2)$.
16. $(5a^2-2ab+b^2)(a^2-2ab+5b^2)$.
17. $(a+b+c)(a+b-c)(a-b+c)(a-b-c)$.

EXERCISE XXVII. (PAGE 60.)

1. $(a+b)(a^2-ab+b^2)$.
2. $(a+2)(a^2-2a+4)$.
3. $(3+b)(9-3b+b^2)$.
4. $(2x+3y)(4x^2-6xy+9y^2)$.
5. $(3x+4y)(9x^2-12xy+16y^2)$.
6. $(x-y)(x^2+xy+y^2)$.
7. $(x-10)(x^2+10x+100)$.
8. $(9x-8y)(81x^2+72xy+64y^2)$.
9. $(5xy-7z)(25x^2y^2+35xyz+49z^2)$.
10. $(4ab-10c^3)(16a^2b^2+40abc^3+100c^6)$.
11. $(x+y)(x^4-x^3y+x^2y^2-xy^3+y^4)$.
12. $(x+3)(x^4-3x^3+9x^2-27x+81)$.
13. $(3+y)(81-27y+9y^2-3y^3+y^4)$.
14. $(x-y)(x^4+x^3y+x^2y^2+xy^3+y^4)$.
15. $(x-2yz^2)(x^4+2x^3yz^2+4x^2y^2z^4+8xy^3z^6+16y^4z^8)$.
16. $(1-xy^2z^2)(1+xy^2z^2+x^2y^4z^4+x^3y^6z^6+x^4y^8z^8)$.

17. $(a+b)(a-b)(a^2+ab+b^2)(a^2-ab+b^2)$.
18. $(a^2+b^2)(a^4-a^2b^2+b^4)$.   19. $(a+b)(a^2-ab+b^2)(a^6-a^3b^3+b^6)$.
20. $(a-b)(a^2+ab+b^2)(a^6+a^3b^3+b^6)$.
21. $(a+b)(a-b)(a^2+b^2)(a^2+ab+b^2)(a^2-ab+b^2)(a^4-a^2b^2+b^4)$
$(a^6+a^3b^3+b^6)(a^6-a^3b^3+b^6)(a^{12}-a^6b^6+b^{12})$.
22. $(a+b)(a-b)(a^4+a^3b+a^2b^2+ab^3+b^4)(a^4-a^3b+a^2b^2-ab^3+b^4)$.
23. $(a^2+b^2)(a^8-a^6b^2+a^4b^4-a^2b^6+b^8)$.
24. $(a^2+b^2)(a^{12}-a^{10}b^2+a^8b^4-a^6b^6+a^4b^8-a^2b^{10}+b^{12})$.
25. $(a-b)(a+b)(a^2+b^2)(a^4+b^4)(a^8+b^8)$.
26. $b(3a^2+3ab+b^2)$.   27. $(2a+b)(a^2+ab+b^2)$.
28. $(2a-b)(a^2-ab+b^2)$.   29. $(a+b)(a^2-4ab+7b^2)$.
30. $(a-b)(a^2+4ab+7b^2)$.   31. $2a(a^2+3b^2)$.
32. $2b(3a^2+b^2)$.   33. $9(a-b)(a^2-ab+b^2)$.
34. $(a+b)(7a^2-13ab+7b^2)$.   35. $(a+b)^3$.   36. $(a-b)^3$.
37. $(a-b-c)(a^2+b^2+c^2-2ab+ac-bc)$.
38. $(x+1)(x-2)(x^4-2x^3+3x^2-2x+4)$.

EXERCISE XXVIII. (Page 61.)

1. $(x+2y)(2x+y)$.   2. $(x+y)(2x+3y)$.
3. $(2x+y)(x+3y)$.   4. $(3x-4y)(5x-2y)$.
5. $(3x+y)(4x-3y)$.   6. $(14x-y)(x+6y)$.
7. $(17x^2+2)(2x^2+1)$.   8. $(6x^2-y^2)(x^2-2y^2)$.
9. $(4x+9)(2x+1)$.   10. $(3x-5y)(4x+6y)$.
11. $a(6x+y)(x-6y)$.   12. $a^2(5x-11y)(2x+3y)$.
13. $(x+ay)(ax+y)$.   14. $(2x+y)(x+2y)(2x-y)(x-2y)$.

EXERCISE XXIX. (Page 62.)

1. $(2x+y-3z)(x+2y-2z)$.   2. $(2x-y+z)(x-2y+3z)$.
3. $(6x-y+z)(x-6y-z)$.   4. $(3x-2y+3)(2x-3y+2)$.
5. $(x+y+3)(x+y+2)$.   6. $(2x-3y-2)(2x-3y-1)$.
7. $(x-6y-5z)(x+2y-3z)$.   8. $(9x+8y-20)(8x-y-1)$.
9. $(15x^2+8y^2+5z^2)(x^2-2y^2+3z^2)$.
10. $(a+b-c)(a^2+b^2+c^2-ab+bc+ca)$.
11. $(a-b-c)(a^2+b^2+c^2+ab-bc+ca)$.
12. $(a+b+2c)(a^2+b^2+4c^2-ab-2bc-2ca)$.
13. $(2a+b-3c)(4a^2+b^2+9c^2-2ab+3bc+6ca)$.

ANSWERS.   305

EXERCISE XXX. (Page 63.)

1. $(a+b+c)(a+b-c)(a-b+c)(-a+b+c)$.
2. $(a+b+c-d)(a+b-c+d)(a-b+c+d)(-a+b+c+d)$.
3. $(a-1)(a-2)(a-3)(a-4)$.   4. $(a+1)(a-2)(a-7)(a-10)$.
5. $(3x+2y)(3x-2y)(x+y)(y-x)$.
6. $\{(a-b)x-(a+b)y\}\{(a+b)x+(a-b)y\}$.
7. $(a^2+ab+b^2)(a^2-ab+b^2)(a^2-b^2+1)$.
8. $(a^4-a^2b^2+b^4)(a^2+b^2+1)$.   9. $(a+b)(a-b)^2(a^2+ab+b^2)$.
10. $(a+b)^2(a-b)(a^2-ab+b^2)$.
11. $(x^2+y^2+z^2+2xy+3xz+3yz)(x^2+y^2+z^2+2xy-3xz-3yz)$.
12. $16ab(a^2-ab+b^2)$.   13. $(ax+by)(ax-by)(bx+ay)(bx-ay)$.
14. $(x+y)(x-y)^3$.   15. $4xy(x+y)(x-y)$.
16. $2(a^2+ab+b^2)(a-b+1)$.   17. $(x+a)(x-a)^2(x^2+a^2)$.
18. $\{(a-b)x+(a+b)y\}\{(a+b)x-(a-b)y\}$.
19. $(3x+1)(3x-1)(9x+1)(x+1)$.
20. $(ax-b)(x^2-ax-b)$.   21. $(x+a)(x-a)(x-2a)$.
22. $(a^2+b^2)(c^2+d^2)$.   23. $(x+1)(x-1)(x+p)(x-p)$.
24. $(a+b+c)(a+b-c)(a-b+c)(-a+b+c)$.
25. $(a+b+c)(a^2+b^2+c^2+2ab-bc-ca)$.   26. $(a+b)(a^2+b^2+1)$.
27. $(x-y-z)(x^2+xy+y^2)$.   28. $(x^2+xy+y^2)(x^2-xy+y^2+1)$.
29. $(x^2+xy+y^2)(x^2-xy+y^2+x-y)$.   30. $(a^2-ab+b^2)(a+b+1)$.
31. $(a+b+c+d)(a-b-c+d)(a+b-c-d)(a-b+c-d)$.
32. $(x+y)(x^2+xy+y^2)(x^2-xy+y^2)$.   33. $(a-b)(c-a)(c-b)$.
34. $(a-b)(b+1)b$.   35. $(x-y)(y-z)(z-x)$.
36. $(x-y)(y-z)(z-x)(x+y+z)$.   37. $(x^2+y^2)(a^2+b^2+c^2)$.
38. $(c-a)(a^2+b^2+c^2+ab+bc+ca)$.
39. $(a-b)(b-c)(c-a)(a^2+b^2+c^2+ab+bc+ca)$.
40. $(3x^4+7x^2y^2+3y^4)(x^4+5x^2y^2+y^4)$.
41. $(a+2b+c)(a+2b-c)(a-2b+c)(a-2b-c)$.
42. $(a+2b+2c)(a+2b-2c)(a-2b+2c)(-a+2b+2c)$.
43. $\{(a-b)x-(a+b)y\}\{(a+b)x+(a-b)y\}$
        $\{(a-b)x+(a+b)y\}\{(a+b)x-(a-b)y\}$.
44. $16(mx+ny)(mx-ny)(my+nx)(my-nx)$.
45. $(2x+1)^2(x^2+x+6)(2x+3)(2x-1)(x+2)(x-1)$.

EXERCISE XXXI. (PAGE 68.)

1. $x=2$. 2. $x=2$. 3. $x=1$. 4. $x=6$. 5. $x=3$.
6. $x=-4$. 7. $x=2$. 8. $x=7$. 9. $x=8$. 10. $x=0$.
11. $x=6$. 12. $x=4$. 13. $x=10$. 14. $x=-3\frac{1}{2}$. 15. $x=\frac{4}{3}$.
16. $x=-34$. 17. $x=3$. 18. $x=15$. 19. $x=3$. 20. $x=2\frac{7}{10}$.
21. $x=\frac{61}{18}$. 22. $x=2$. 23. $x=a+b$. 24. $x=2a$. 25. $x=-\frac{a}{9}$.
26. $x=0$. 27. $x=a+b$. 28. $x=c-a-b$. 29. $x=1$.
30. $x=-1$. 31. $x=\frac{a-b}{a-2b}$. 32. $x=\frac{cd-ab}{a+b-c-d}$.
33. $x=\frac{ab}{2(a+b)}$. 34. $x=-\frac{a+b+c}{3}$. 35. $x=\frac{a^2b}{a^2+b^2}$.
36. $x=\frac{2ac-ab-bc}{a+c-2b}$. 37. $x=-1$. 38. $x=-1$.

EXERCISE XXXII. (PAGE 70.)

1. $x=a, b$. 2. $x=-a, b$. 3. $x=a, -b$. 4. $x=-a, -b$.
5. $x=b, -c$. 6. $x=-b, -c$. 7. $x=2, 6$. 8. $x=2, 2$.
9. $x=a, b, c$. 10. $x=1, 2, 3$. 11. $x=3, \frac{1}{3}$.
12. $x=a-b, -(a-b)$. 13. $x=-a, 2a$. 14. $x=0, 3b, -b$.
15. $x=\frac{4}{3}, -\frac{3}{2}$. 16. $x=\frac{7}{3}, -\frac{4}{3}$. 17. $x=3, -3$.
18. $x=\frac{7}{5}, -\frac{7}{5}$. 19. $x=1, 1, 1$. 20. $x=1, 3, 5$.
21. $x=1, -1, -1$.

EXERCISE XXXIII. (PAGE 72.)

1. 55, 45. 2. $\frac{22}{13}$. 3. 68. 4. 18, 31.
5. 15 at 11s. 6d.; 5 at 7s. 6d. 6. 538, 441. 7. 37, 30, 20.
8. 9. 9. $400, $200, $100. 10. $300.
11. 19, 9. 12. 40, 20. 13. 35, 25. 14. 8.
15. 4. 16. $177\frac{7}{47}$ miles from Toronto, at end of $7\frac{4}{47}$ hrs.
17. 10 cows, 15 horses. 18. 50 lbs. 19. 25 gals., 15 gals., 5 gals.

ANSWERS. 307

20. $130, $150, $130, $90.   21. A, $350; B, $450; C, $720.
22. 13, 21.   23. 49 gals.   24. 152 men, 76 women, 38 ch.
25. 50, 40.   26. 42, 18.   27. 40, 120.   28. 80, 128.
29. 1.71\frac{3}{7}$, 3.42\frac{6}{7}$.   30. 28 of first, 2 of second.   31. 80.
32. 40.   33. 60 gals. of first, 40 of second.   34. $31\frac{1}{2}$ gals.
35. Each son $1,000, each daughter $500, and his wife $4,000.
36. $20.   37. $700.   38. $500, $1,000, $1,500.   39. 27 days.
40. Length of first field, 180 yds.; breadth, 90 yds. Length of second field, 230 yds.; breadth, 100 yds.

EXERCISE XXXIV. (Page 77.)

1. a^2b^2.   2. $6ab^2m$.   3. $7ab$.   4. $17xy^2$.
5. $4x^2y^2z^2$.   6. $19x^5y^6z^6$.   7. $5x^2y^2$.   8. $7m$.
9. $3m + 5nx$.   10. $x^4 + x^2y^2$.   11. $a+y$.   12. $ab(a+b)$.
13. $x-3$.   14. $x-7$.   15. $x+1$.   16. $x-17$.
17. x^2+y^2.   18. x^2+xy+y^2.   19. x^3-y^3.   20. x^2+y^2.
21. $x^8+x^4y^4+y^8$.   22. $x+7$.   23. $2x+3y$.   24. a^2+ab+b^2.
25. x^2+2x+4.   26. $x+1$.   27. $25x^2+20xy+16y^2$.
28. $x(x+1)$.   29. $2(a-b)$.   30. $c(a-b)$.   31. $a+b+c+d$.
32. $x-y$.   33. $a-b$.   34. $x-y$.   35. $5x^2-1$.

EXERCISE XXXV. (Page 82.)

1. $x+3$.   2. x^2+3x+4.   3. x^2+2x+3.   4. $x+4$.
5. $3x-2$.   6. $2x^2-3x+4$.   7. $x-7$.   8. $3x-7$.
9. $x-1$.   10. x^3-1.   11. $x-3$.   12. x^2+2x+3.
13. $(x+1)^3$.   14. x^2-2x-3.   15. $3x^2+2x+1$.   16. x^2-4x+1.
17. x^4-x^2+1.   18. $a(2a-3x)$.   19. $ax-by$.   20. $x-2$.
21. x^2-1.   22. $x-p$.   23. $x+p$.   24. $ax-b$.
25. x^2+mx+n.   26. $(x+2m)(x-3)$.   27. x^2-x+1.
28. ax^2+bx+c.   29. $a+b+c$.   30. $xy+ab$.

EXERCISE XXXVI. (Page 84.)

1. $6a^2b^2c$.   2. $30a^3b^3cd^2$.   3. $1800p^4q^4r^4$.   4. $210l^5m^4n^3p$.
5. $ab(a^2-b^2)$.   6. a^3-b^3.   7. $(x^3-1)(x+1)$.   8. $x^2y^2(x^2-y^2)$.
9. $60(x^2-y^2)^2$.   10. $15xy(x^6-1)$.   11. a^6-b^6.   12. $a^{12}-b^{12}$.

13. $(x^2-4)(x^2-9)(x^2-16)$. 14. $(x-a)(x-b)(x-c)$.
15. $(x^2-4a^2)^3$. 16. $120(a^4-b^4)^3$. 17. $72(a^2-b^2)^3(a^2+ab+b^2)(a^2+b^2)$.
18. $(x^2-1)(x^2-4)(x^2-9)$. 19. $4a^2b^2-(a^2+b^2-c^2)^2$.
20. $120x^2y^2(x-y)(x+y)^2$. 21. $x(x^6-1)$. 22. $a(x^4-a^4)(x-a)$.

EXERCISE XXXVII. (Page 86.)

1. $(x-1)(x-2)(x-3)(x-4)$. 2. $(x-1)(x-4)(x^2-5x-1)$.
3. $(2x+3)(3x-4)(x^2+3x-1)$. 4. $(x-1)^2(x^2+x+1)(x^2-3x+1)$.
5. $(9x^2-4y^2)(4x^2-9y^2)(x+y)$. 6. $(x-a)^3(x+2a)(x+4a)$.
7. $(x^3+1)(x^2-x-6)$. 8. $(x^2-1)(x^2-9)(x+7)$.
9. $(x^3-1)(2x^2+x-1)(x+3)$. 10. x^6-64.
11. $(x+1)(x+2)(x+3)(x+4)$. 12. $ax(a-x)(a^2-ax-6x^2)$.
13. $(x+1)(x^2+1)(x-1)^3$.

EXERCISE XXXVIII. (Page 88.)

1. $\dfrac{a}{2b}$. 2. $\dfrac{2b}{3c}$. 3. $\dfrac{2ln}{3}$. 4. $\dfrac{2ab^3}{3c}$. 5. $\dfrac{5xy}{6bc}$.

6. $\dfrac{1}{5b^5}$. 7. $\dfrac{5x}{z}$. 8. $\dfrac{4r}{3p}$. 9. $\dfrac{1}{ab}$. 10. $3xy^2$.

11. $\dfrac{a}{a+b}$. 12. $\dfrac{2a}{3a^2-4b^2}$. 13. $\dfrac{1}{3x-5y}$. 14. $\dfrac{m}{ab}$. 15. $\dfrac{xy}{ab}$.

16. $\dfrac{y^2}{2x}$. 17. $\dfrac{xy}{2(x+y)}$. 18. $\dfrac{a-2b}{2a}$. 19. $\dfrac{3}{2a^2-3b^2}$. 20. $\dfrac{a}{c(x+1)}$.

21. $\dfrac{1}{2(x-y)}$. 22. $\dfrac{x^2+a^2}{x^3}$. 23. $\dfrac{x^4+a^2x^2+a^4}{x^2+a^2}$. 24. $\dfrac{1}{x^2+a^2}$.

25. $\dfrac{a+b}{a-b}$. 26. $x^2-2ax+2a^2$. 27. $\dfrac{x-1}{x+1}$. 28. $\dfrac{a+b-c}{a-b+c}$.

29. $\dfrac{a-b+c}{a+b-c}$. 30. $\dfrac{x-b}{x+b}$. 31. $\dfrac{cx+d}{ax+b}$. 32. $\dfrac{1}{x^2+1}$. 33. $\dfrac{x^2+1}{x}$.

34. $\dfrac{1}{b+c-a}$. 35. $\dfrac{a}{b}$.

ANSWERS.   309

EXERCISE XXXIX. (PAGE 90.)

1. $\dfrac{x-4}{x+3}$.   2. $\dfrac{2x+3}{x-4}$.   3. $\dfrac{1}{x^2-1}$.   4. $\dfrac{x}{x^2-4}$.

5. $\dfrac{x^2+2x+3}{x^2+2x}$.   6. $\dfrac{3a^2+9a+5}{a(5a^2-2a-6)}$.   7. $\dfrac{x-2}{x+4}$.   8. $\dfrac{x^2+x-6}{x^2+4x}$.

9. $\dfrac{x-2}{2x-3}$.   10. $\dfrac{3x-1}{2x-1}$.   11. $\dfrac{x-5}{x+5}$.   12. $\dfrac{2x+3}{3x-4}$.

13. $\dfrac{4x+3y}{4(3x^2+y^2)}$.   14. $\dfrac{x-1}{x+1}$.   15. $\dfrac{1}{x^2-2x+2}$.   16. $\dfrac{3ax^2+1}{4a^2x^4+2ax^2-1}$.

17. $\dfrac{2a}{a^2+2b^2}$.   18. $\dfrac{x+5}{x-3}$.   19. $\dfrac{x+y}{x-y}$.   20. $\dfrac{ax+by}{ax-by}$.

21. $\dfrac{x+y-z}{z-x+y}$.   22. $\dfrac{a+b+c}{2}$.   23. $\dfrac{x^2-ax+b^2}{x^2+ax-b^2}$.   24. $\dfrac{x+c}{a+b-x}$.

25. $\dfrac{b^2-3ac}{c^2-4ab}$.   26. $\dfrac{2z^2-3y^2}{3z^2-2y^2}$.   27. $\dfrac{x-b}{bc-ac-ax}$.   28. a.

29. $\dfrac{(a^2+a+1)x^2+(a+1)x+1}{(a-1)^2x^2-(a+1)x-1}$.   30. $\dfrac{2(1-ac)}{1+ac}$.   31. m.   32. 1.

33. $\dfrac{a^2+c^2}{(a+d)(b+c)}$.   34. $\dfrac{b(y+z)+a(x-z)}{b(y+z)-a(x-z)}$.   35. $\dfrac{a}{b}$.

EXERCISE XL. (PAGE 92.)

1. $5x-3+\dfrac{3}{x+2}$.   2. $a-2x+\dfrac{3x^2}{a+x}$.

3. $3(x^2+xy+y^2)+\dfrac{5}{x-y}$.   4. $x^2+xy+y^2+\dfrac{2y^3}{x-y}$.

5. $2x^2+x-1-\dfrac{2}{x^2-x+1}$.   6. $3a-1+\dfrac{1}{3a+1}$.

7. $x-1-\dfrac{7x-4}{3x^2-4x+3}$.   8. $x-1+\dfrac{5x+4}{5x^2+4x-1}$.

9. $x-1-\dfrac{x-2}{x^2+x+1}$.   10. $\dfrac{1}{1-x}$.   11. $\dfrac{1}{1-x}$.

12. $\dfrac{a^2}{a+b}$.   13. $\dfrac{2y}{x+y}$.   14. $\dfrac{2a^2-ab-b^2}{a+b}$.   15. $\dfrac{4ab}{a+b}$.

16. $\dfrac{2(a^2+b^2)}{a-b}$.   17. $\dfrac{x^2}{x-3}$.   18. $\dfrac{x^2-2xy-y^2}{x+y}$.   19. $\dfrac{x^3}{x-y}$.

20. $\dfrac{x^3}{x+y}$.   21. $\dfrac{1}{1-x+x^2}$.   22. $\dfrac{x^3+xy+y^2}{x+a}$.   23. $\dfrac{x+2}{x-2}$.

24. $\dfrac{1+y}{1-y}$.   25. $\dfrac{a}{1-b}$.   26. $\dfrac{a}{1+b}$.

27. $\dfrac{x^2}{3a}+x-\dfrac{5a}{3x}+\dfrac{2a^2}{x^2}$.   28. $\dfrac{a}{z}+\dfrac{y}{x}+\dfrac{x}{y}+\dfrac{z}{a}$.

29. $\dfrac{1}{d}+\dfrac{1}{a}+\dfrac{1}{b}+\dfrac{1}{c}$.   30. $\dfrac{3x^2}{20}-\dfrac{2x}{5}+\dfrac{3}{4}-\dfrac{3}{2x}$.

31. $\dfrac{1}{(a-b)(m+n)}-\dfrac{1}{(a+b)(m-n)}$.   32. $\dfrac{a}{x-2y}+\dfrac{b}{x+y}$.

EXERCISE XLI. (Page 94.)

1. $\dfrac{20x}{30}, \dfrac{24x}{30}, \dfrac{21x}{30}$.   2. $\dfrac{24x-9}{12x^2}, \dfrac{4x-6x^2}{12x^2}$.

3. $\dfrac{a^2}{abc}, \dfrac{b^2}{abc}, \dfrac{c^2}{abc}$.   4. $\dfrac{2+2x}{1-x^2}, \dfrac{3}{1-x^2}$.

5. $\dfrac{a^2+ab}{a^2-b^2}, \dfrac{ab-b^2}{a^2-b^2}$.   6. $\dfrac{x^2}{xy(x-y)}, \dfrac{y^2}{xy(x-y)}$.

7. $\dfrac{(a+b)^2}{a^2-b^2}, \dfrac{(a-b)^2}{a^2-b^2}, \dfrac{ab}{a^2-b^2}$.   8. $\dfrac{75(x+2)}{30(x^2-4)}, \dfrac{70(x-2)}{30(x^2-4)}, \dfrac{48}{30(x^2-4)}$.

9. $\dfrac{x^3-16}{(x-2)(x-3)(x-4)}, \dfrac{x^2-4}{(x-2)(x-3)(x-4)}$.

10. $\dfrac{x(x+1)(x+2)}{x(1-x)(1+x)}, \dfrac{(x+1)(x+3)}{x(1-x)(1+x)}, \dfrac{x+4}{x(1-x)(1+x)}$.

11. $\dfrac{b^2-c^2}{(a-b)(a-c)(b-c)}, \dfrac{c^2-a^2}{(a-b)(a-c)(b-c)}$.

12. $\dfrac{c^2(a^2-b^2)}{abc(a-b)(a-c)(b-c)}, \dfrac{a^2(b^2-c^2)}{abc(a-b)(a-c)(b-c)}$.

ANSWERS.

EXERCISE XLII. (PAGE 95.)

1. $\dfrac{17x}{14}$.
2. $\dfrac{69a+10b}{48}$.
3. $\dfrac{a+x}{2}$.
4. $\dfrac{23x-10y}{30}$.
5. $\dfrac{4x}{7}$.
6. $\dfrac{48x-21}{50}$.
7. 0.
8. $\dfrac{2}{x}$.
9. 1.
10. 0.
11. $\dfrac{(x+y)^2}{x^2y^2}$.
12. $\dfrac{(a+b+c)^2}{abc}$.
13. $\dfrac{a+b}{a-b}$.
14. 2.
15. $\dfrac{3}{x-7}$.
16. $\dfrac{x+y}{xy}$.
17. $\dfrac{x+y}{y}$.
18. $\dfrac{2x^3}{x^4+x^2y^2+y^4}$.
19. $\dfrac{a^2+b^2}{ab}$.
20. $\dfrac{2a}{a+b}$.
21. $\dfrac{2(x^4+3x^2y^2+y^4)}{x^4+x^2y^2+y^4}$.
22. $\dfrac{4a^4}{a^4-b^4}$.
23. $\dfrac{2y^2}{x^3-y^3}$.
24. $\dfrac{2(x+1)}{x^2+x+1}$.
25. $\dfrac{4(a^4x^3-b^4y^3)}{a^4x^4-b^4y^4}$.
26. $\dfrac{2}{x+3}$.
27. $\dfrac{1}{(x^3+1)(x^2+1)}$.
28. $\dfrac{8x^4-x^3-24x^2+5x-4}{(x+1)(x-1)(x+3)(x-2)}$.

EXERCISE XLIII. (PAGE 97.)

1. $\dfrac{x}{14}$.
2. $\dfrac{11}{6}$.
3. $\dfrac{2}{x^2-1}$.
4. $\dfrac{a^2+b^2}{a^2-b^2}$.
5. $\dfrac{4xy}{x^2-y^2}$.
6. $\dfrac{a^2+b^2-2ac}{a^2-b^2}$.
7. $\dfrac{a+x}{ax}$.
8. $\dfrac{m+n}{mn}$.
9. $\dfrac{x}{x+y}$.
10. $\dfrac{a+x}{a^2}$.
11. $\dfrac{ax}{a^2-x^2}$.
12. $\dfrac{2x^3}{1+x^2+x^4}$.
13. $\dfrac{2xy}{x^4+x^2y^2+y^4}$.
14. $\dfrac{3}{(x-2)(x-3)(x-5)}$.
15. $\dfrac{7}{(x-5)(x-7)(x-12)}$.
16. $\dfrac{32}{16x^2+8x-15}$.
17. $\dfrac{4}{x+2}$.
18. $\dfrac{a(x-9a)}{(x-4a)(x-6a)}$.
19. $\dfrac{2(a-b)}{(a+b)(a+3b)}$.
20. $\dfrac{4(x+2)}{x^2-16}$.
21. $\dfrac{c}{(b-c)(c-a)}$.
22. $\dfrac{a}{(x-b)(x-c)}$.
23. $\dfrac{2}{a^2-b^2}$.
24. $\dfrac{1}{(2x^2-3x+4)^2}$.
25. 3.

EXERCISE XLIV. (Page 99.)

1. $\dfrac{2}{x+y}$.   2. $\dfrac{4x}{1-x^4}$.   3. $\dfrac{2x}{1-x^4}$.   4. -1.

5. $\dfrac{8y^7}{x^8-y^8}$.   6. $\dfrac{4x^2y^2}{x^4-y^4}$.   7. $\dfrac{x-2}{x^4-1}$.   8. $\dfrac{x^2}{(x^2-1)^2}$.

9. $\dfrac{3x-1}{x^2(x+1)^2}$.   10. $\dfrac{1+2x^2}{x^3(x+1)}$.   11. $\dfrac{1}{x(x^3-1)}$.   12. $\dfrac{-1}{x(x^3+1)}$.

13. 0.   14. $\dfrac{4a}{a+x}$.   15. $\dfrac{1}{x+2}$.   16. 2.

17. $\dfrac{x}{2(x+y)}$.   18. $\dfrac{x}{x-y}$.   19. 1.   20. 0.   21. 0.

22. $\dfrac{84-186x+93x^2-6x^3}{4(3-2x)^2(3+2x)}$.   23. $\dfrac{(x-1)^2}{x^3(x^2+1)^2}$.   24. 0.   25. 0.

26. 0.   27. $\dfrac{x^3}{(x+1)^3(x-1)^3}$.   28. $\dfrac{x^4}{(x^4-1)^2}$.

EXERCISE XLV. (Page 102.)

1. $\dfrac{a^2}{2b^2}$.   2. $\dfrac{1}{4}$.   3. $\dfrac{5bx}{9ay}$.   4. $\dfrac{2x}{y}$.   5. $2ab$.

6. $\dfrac{cxy}{8abz}$.   7. $\dfrac{4}{9}$.   8. $\dfrac{1}{2}$.   9. $\dfrac{3a}{2b}$.   10. $\dfrac{ac}{bd}$.

11. $\dfrac{(a-b)y}{ax}$.   12. $\dfrac{2ax^2(x-y)}{c}$.   13. $\dfrac{6-x}{x-3}$.   14. 1.   15. 1.

16. $\dfrac{a^4x}{b-a}$.   17. $\dfrac{y}{2x+y}$.   18. $\dfrac{a^2}{2x^2}$.   19. $\dfrac{1}{a^2-ab+b^2}$.

20. $\dfrac{x^2+y^2}{x+y}$.   21. -1.   22. 1.   23. $\dfrac{1}{m^2-n^2}$.   24. $\dfrac{a^2+ax-2x^2}{a^2+ax+x^2}$.

25. $\dfrac{x^2}{12+x-6x^2}$.   26. $\dfrac{x^4+x^2y^2+y^4}{x^2-y^2}$.

EXERCISE XLVI. (Page 104.)

1. $\dfrac{a^3}{b^3}-\dfrac{x^3}{y^3}$.   2. $x^2-\dfrac{1}{x^2}$.   3. $x^2+2+\dfrac{1}{x^2}$.

ANSWERS. 313

4. $x^3 - 2 + \dfrac{1}{x^3}$.   5. $x^2 + 5 + \dfrac{6}{x^2}$.   6. $x^2 + 1 - \dfrac{90}{x^2}$.

7. $abx^2 - (a^2 - b^2) - \dfrac{ab}{x^2}$.   8. $x^3 - \dfrac{1}{x^3}$.   9. $x^3 + \dfrac{1}{x^3}$.

10. $\dfrac{x^3}{6} + \dfrac{x^2}{72} + \dfrac{x}{4} + \dfrac{1}{4}$.   11. $\dfrac{x^2}{6} - x - \dfrac{1}{2} + \dfrac{6}{x} - \dfrac{3}{x^2}$.   12. $x^4 + 1 + \dfrac{1}{x^4}$.

13. $\dfrac{a^4}{b^4} - 2 + \dfrac{b^4}{a^4}$.   14. $1 + \dfrac{a^2}{b^2} + \dfrac{a^4}{b^4}$.   15. $\dfrac{1}{a^2} - \dfrac{1}{b^2} + \dfrac{1}{c^2} + \dfrac{2}{ac}$.

16. $\dfrac{a^2}{b^2} - \dfrac{b^2}{c^2} - \dfrac{c^2}{a^2} + \dfrac{2b}{a}$.   17. $a^4 + \dfrac{4}{a^4}$.   18. $\dfrac{x}{y^3} + \dfrac{1}{xy} + \dfrac{y}{x^3}$.

EXERCISE XLVII. (Page 106.)

1. $\dfrac{7bx}{3ay}$.   2. $\dfrac{8xy}{b}$.   3. $\dfrac{1}{12px^2y^3}$.   4. $\dfrac{x}{6}$.

5. $\dfrac{x^2}{y^2}$.   6. $\dfrac{x}{a^2 - x^2}$.   7. $\dfrac{x(a + 2x)}{a^2}$.   8. $\dfrac{b(a + b)}{3(a - b)^2}$.

9. $\dfrac{a^2}{b^2} - 1 + \dfrac{b^2}{a^2}$.   10. $\dfrac{x + y}{y}$.   11. $\dfrac{x^2}{y^2} - 1 + \dfrac{y^2}{x^2}$.   12. $a^4 - 1$.

13. $\dfrac{1}{a^2} + \dfrac{1}{b^2} + \dfrac{1}{c^2} - \dfrac{1}{ab} - \dfrac{1}{bc} - \dfrac{1}{ca}$.   14. $\dfrac{3x^3}{2} - 5x^2 + \dfrac{x}{4} + 9$.

15. $\dfrac{1}{a} - \dfrac{b}{ax} - \dfrac{b^2(x + a)}{a^2x^2}$.   16. $\dfrac{x^2}{a^2} + \dfrac{a^2}{x^2}$.   17. $\dfrac{a + b - c}{a - b + c}$.

18. $\dfrac{xy}{x^2 + y^2}$.   19. 1.   20. $\dfrac{x^2 + b^2}{x - b}$.   21. $3(x + y)$.

22. $\dfrac{ab(x^2 - 1)}{(a^2 - b^2)x}$.   23. $\dfrac{ab}{a + b}$.

EXERCISE XLVIII. (Page 108.)

1. $\dfrac{15 - 4x}{25}$.   2. $\dfrac{6x}{7 - 2x}$.   3. $\dfrac{16x - 21y}{5(x - y)}$.   4. $\dfrac{9(x - 1)}{12 - x}$.

5. $\dfrac{20 - 3x}{2x - 25}$.   6. $\dfrac{1}{3}$.   7. $-\dfrac{(a + b)^2}{2(a - b)^2}$.   8. $\dfrac{1 + x^2}{2x}$.

9. 1.   10. $\dfrac{b^2}{a^2}$.   11. $\dfrac{1}{a - b - c}$.   12. $a^2 + x^2$.

ANSWERS.

13. $\dfrac{xy^2}{(x-y)^2(x+y)}$. 14. $\dfrac{ac-bd}{ac+bd}$. 15. $\dfrac{4}{3x}$.

16. $\dfrac{4a^3x}{x^4-a^4}$. 17. $\dfrac{(a+b+c)^2}{2bc}$. 18. 1. 19. $\dfrac{a-b}{a+b}$.

20. $a(2a-x-y)$. 21. $\dfrac{4m^2+1}{m(4m^2-1)}$. 22. $\dfrac{2}{x}$. 23. $\dfrac{x-y}{x+y}$.

EXERCISE XLIX. (PAGE 109.)

1. $\dfrac{1}{ab}\left(\dfrac{x}{a}-\dfrac{y}{b}\right)^2$. 2. a. 3. $\dfrac{a+b}{ab}$. 4. 0.

5. x^2+y^2. 6. $\dfrac{2(x-y)}{y}$. 7. -1. 8. $xy(a^2+b^2)$.

9. $\dfrac{ab(x^2-y^2)}{xy(a^2-b^2)}$. 10. 4. 11. $\dfrac{2}{mn(m^2+n^2)}$. 12. $\dfrac{2a-b}{a+b}$.

13. $\dfrac{x(x-1)}{x^2-x+1}$. 14. 0. 15. 0. 16. $2\{(ay-bx)^2+(ax-by)^2\}$.

17. 1. 18. $\dfrac{pr^2x^2+q^2x+prs}{qrs}$. 19. $\dfrac{2(a^3-r^3)}{a^3+r^3}$.

20. 0. 21. $\dfrac{a+3b}{b+c}$. 22. 2. 23. $\dfrac{9(b-c)}{c-2a+b}$.

25. $a^2+b^2+c^2$. 27. 0. 28. 1.

EXERCISE LI. (PAGE 121.)

1. $x=5$. 2. $x=8$. 3. $x=9$. 4. $x=\dfrac{3}{2}$. 5. $x=7$.

6. $x=-5$. 7. $x=10$. 8. $x=6$. 9. $x=12$. 10. $x=7$.

11. $x=9$. 12. $x=\dfrac{2}{3}$. 13. $x=\dfrac{9}{16}$. 14. $x=2$. 15. $=\dfrac{2}{5}$.

16. $x=-\dfrac{20}{19}$. 17. $x=\dfrac{8}{11}$. 18. $x=2\tfrac{1}{4}$. 19. $x=-6\tfrac{1}{3}$. 20. $x=31$.

EXERCISE LII. (PAGE 122.)

1. $x=-19$. 2. $x=9$. 3. $x=4$. 4. $x=20$. 5. $x=3$.

6. $x=-2$. 7. $x=11$. 8. $x=6$. 9. $x=8$. 10. $x=-7$.

ANSWERS. 315

EXERCISE LIII. (Page 124.)

1. $x = \dfrac{6}{19}$. 2. $x = 2$. 3. $x = 12\frac{2}{3}$. 4. $x = \dfrac{2737}{166}$. 5. $x = 151$.
6. $x = 3$. 7. $x = 3\frac{3}{4}$. 8. $x = -1$. 9. $x = 2$. 10. $x = 0$.
11. $x = 0$. 12. $x = 3\frac{8}{15}$. 13. $x = \dfrac{5}{6}$. 14. $x = 5$. 15. $x = 3$.

EXERCISE LIV. (Page 127.)

1. $x = 3\frac{1}{2}$. 2. $x = 2$. 3. $x = 3$. 4. $x = 4$. 5. $x = \dfrac{b}{a}$.
6. $x = \dfrac{1}{5}$. (A cu. equation.) 7. $x = -b$. 8. $x = 2$. 9. $x = \dfrac{50}{29}$.
10. $x = \dfrac{3}{2}$. 11. $x = 0$. 12. $x = 12$. 13. $x = 1$. 14. $x = 7$.
15. $x = -2$. 16. $x = 8$. 17. $x = \dfrac{2}{5}$. 18. $x = 2$. 19. $x = 2\frac{1}{2}$.

EXERCISE LV. (Page 129.)

1. $x = \dfrac{ab}{a+b}$. 2. $x = \dfrac{cd-ab}{a+b-c-d}$. 3. $x = \dfrac{a^2-b^2}{4a-b}$.

4. $x = \dfrac{a(a^2-b^2)}{a^2+b^2}$. 5. $x = a+b+c$. 6. $x = \dfrac{b(a-b+c)}{a}$.

7. $x = \dfrac{c-a}{2}$. 8. $x = \dfrac{c^2-ab}{a+b}$. 9. $x = \dfrac{br-cq}{cp-ar}$.

10. $x = \dfrac{ab(a+b-2c)}{a^2+b^2-ac-bc}$. 11. $x = \dfrac{a-b}{2}$. 12. $x = -(a+b+c)$.

13. $x = \dfrac{2ab}{a+b}$. 14. $x = \dfrac{bn-am}{m-n}$. 15. $x = a$.

16. $x = \dfrac{a^2c}{b^2}$. 17. $x = \dfrac{a^2c + b^2a + c^2b - a - b - c}{ab+bc+ca-1}$. 18. $x = \dfrac{a+b+c}{3}$.

19. $x = 0$ or $\sqrt{-(ab+bc+ca)}$. (This is not a simple equation.)

EXERCISE LVI. (Page 130.)

1. $x = \dfrac{1}{2}$. 2. $x = \dfrac{2(a^2+ab+b^2)}{3(a+b)}$. 3. $x = \dfrac{3-2a}{3a-4}$. 4. $x = \dfrac{2ab}{a+b}$.

5. $x = 1$. (Not a simple equation.) 6. $x = 2\frac{1}{2}$. 7. $x = \dfrac{3}{2}$.

8. $x = \dfrac{4}{3}$. 9. $x = -\dfrac{1}{2}$. 10. $x = \dfrac{-ab(a+b-2c)}{a^2+b^2-ac-bc}$. 11. $x = -\dfrac{11}{3}$.

12. $x = \dfrac{5}{3}$. 13. $x = \dfrac{bc+ca-ab}{b-2c}$. 14. $x = \dfrac{5a}{2}$. 15. $x = \dfrac{45}{86}$.

16. $x = \dfrac{5}{4}$. 17. $x = 5\frac{1}{2}$. 18. $x = a+b$. 19. $x = \dfrac{ab}{a+b}$. 20. $x = 9$.

21. $x = -\dfrac{b+c}{2}$. 22. $x = -\dfrac{bc(m-n)+ca(n-p)+ab(p-m)}{a(m-n)+b(n-p)+c(p-m)}$.

23. $x = a+b+c$. 24. $x = a+b+c$. 25. $x = \dfrac{ab+bc+ca}{a+b+c}$.

26. $x = \dfrac{abc+bc^2+a^2c-2a^2b-ab^2}{bc+ac-ab+c^2-a^2-b^2}$.

27. $x = \dfrac{3abc-a^2b-b^2c-c^2a}{a^2+b^2+c^2-ab-bc-ca}$.

28. $x = \dfrac{b^2c^2(b+c)+c^2a^2(c+a)+a^2b^2(a+b)-2abc(ab+bc+ca)}{2abc(a+b+c)-ac(a^2+c^2)-bc(b^2+c^2)-ab(a^2+b^2)}$.

29. $x = \dfrac{m(ac+bd)}{a+b}$.

EXERCISE LVII. (Page 134.)

1. 168. 2. 80. 3. 28 and 32. 4. 36 and 9.
5. 85 gal. wine, 35 gal. water. 6. 28 and 18. 7. $2\frac{2}{3}$ days.
8. 2 days. 9. $1\frac{1}{20}$ days. 10. $26\frac{2}{3}$ days. 11. $10\frac{4}{5}$ days.
12. B and C in $21\frac{9}{11}$ hrs.; A, B and C in $10\frac{19}{23}$ hrs. 13. 10 days.
14. A, $37\frac{1}{2}$ min.; B, 25 min. 15. $4\frac{4}{5}$ hours. 16. 48 minutes.
17. 72 lbs. 18. 300. 19. 1st, $10\frac{10}{11}$ min. after 2; 2nd, $43\frac{7}{11}$ min. after 2; 3rd, $27\frac{3}{11}$ after 2, and 3 o'clock. 20. 1st, $5\frac{5}{11}$ min. after 7; 2nd, $21\frac{9}{11}$ min. and $54\frac{6}{11}$ min. after 7; 3rd, $38\frac{2}{11}$ min. after 7. 21. $5\frac{5}{11}$ min. after 2 o'clock. 22. $2284. 23. $55,500.
24. $3700. 25. 10 gal. from first, 4 gal. from second.

26. $2\frac{1}{2}$ miles per hr. 27. $2\frac{2}{3}$ miles per hr. 28. $\dfrac{mn(t_1-t)}{tt_1(m+1)}$

ANSWERS. 317

29. $2\frac{2}{5}$ hrs. 30. $\dfrac{mb}{a-b}$ hrs. If $a=b$, the rates being equal, one will never overtake the other. If $a < b$ the last loses ground instead of gaining on the first. If direction of motions were reversed the first would overtake the second in $\dfrac{mb}{b-a}$ hrs. (See Appendix.) 31. 30 miles per hr. 32. Distance, 15 miles; speed, 2 miles per hr. 33. $32\frac{8}{11}$ min. after 12.

34. A in $\dfrac{2mnp}{np+pm-mn}$ days, B in $\dfrac{2mnp}{np+mn-pm}$ days, C in $\dfrac{2mnp}{mp+mn-np}$ days; A, B and C together in $\dfrac{2mnp}{mn+np+pm}$ days.

35. $\dfrac{pb(ma+na+mn-mp)}{ma(n-p)}$ boys. 36. $\dfrac{2mn(2m+n)}{4m^2-n^2+4mn}$ days.

37. 6 hrs.; $31\frac{1}{2}$ miles from A. 38. £540; 17d. in the £.

39. $n=\dfrac{1}{30}$. 40. £600. 41. 60 eggs. 42. £180 and £150.

43. 112 lbs. 44. 84. 45. 1,000,000. 46. £450$\frac{4}{12}$, £156$\frac{2}{3}$.

47. £1750, £3472. 48. $26\frac{8}{11}$ min. after 2 o'clock.

49. 98 lbs. of copper. 50. $\dfrac{12(5h+m)}{11}$ min. after h o'clock.

EXERCISE LVIII. (PAGE 144.)

1. $x=4$, 2. $x=5$, 3. $x=2$, 4. $x=3$, 5. $x=2$,
   $y=5$. $y=4$. $y=1$. $y=4$. $y=5$.

6. $x=\dfrac{1}{2}$, 7. $x=6$, 8. $x=7$, 9. $x=3$, 10. $x=10$,
   $y=\dfrac{1}{3}$. $y=9$. $y=5$. $y=4$. $y=7$.

11. $x=\dfrac{1}{3}$, 12. $x=10$, 13. $x=30$, 14. $x=15$, 15. $x=10$,
    $y=\dfrac{14}{15}$. $y=8$. $y=24$. $y=12$. $y=24$.

16. $x=144$, 17. $x=6$, 18. $x=3$, 19. $x=17$, 20. $x=5$,
    $y=216$. $y=8$. $y=2$. $y=3$. $y=2$.

ANSWERS.

21. $x = 5$,
    $y = 5$.

22. $x = \dfrac{263}{37}$,
    $y = \dfrac{90}{37}$.

23. $x = \dfrac{a - b^2}{1 - ab}$,
    $y = \dfrac{b - a^2}{1 - ab}$.

24. $x = -b$,
    $y = a + b$.

25. $x = \dfrac{bc}{a + b}$,
    $y = \dfrac{ac}{a + b}$.

26. $x = \dfrac{ab(a + b)}{a^2 + b^2}$,
    $y = \dfrac{ab(a - b)}{a^2 + b^2}$.

27. $x = \dfrac{macb^2}{b^2 c - a^2}$,
    $y = \dfrac{ma^3 b}{b^2 c - a^2}$.

28. $x = \dfrac{abc(ab + ac - bc)}{a^2 b^2 + a^2 c^2 - b^2 c^2}$,
    $y = \dfrac{abc(ac - ab - bc)}{a^2 b^2 + a^2 c^2 - b^2 c^2}$.

29. $x = \dfrac{b^2 + c^2 - a^2}{2a}$,
    $y = \dfrac{c^2 + a^2 - b^2}{2b}$.

30. $x = \dfrac{c(a - b)}{a^2 + b^2}$,
    $y = \dfrac{c(a + b)}{a^2 + b^2}$.

31. $x = \dfrac{ac}{a^2 + b^2}$,
    $y = \dfrac{bc}{a^2 + b^2}$.

32. $x = 16$,
    $y = 4$.

33. $x = 5$,
    $y = 4$.

34. $x = \dfrac{383}{29}$,
    $y = \dfrac{93}{29}$.

35. $x = 25$,
    $y = 13$.

36. $x = \dfrac{154}{47}$,
    $y = \dfrac{56}{47}$.

37. $x = 8\tfrac{2}{3}$,
    $y = 8$.

38. $x = 2$,
    $y = 6$.

39. $x = \dfrac{2}{5}$,
    $y = 60$.

40. $x = 6$,
    $y = 8$.

41. $x = \dfrac{1}{63}$,
    $y = 18$.

42. $x = 2$,
    $y = 3$.

43. $x = \dfrac{m^2 + mn + n^2}{m + n}$,
    $y = \dfrac{m^2 - mn + n^2}{m - n}$.

44. $x = \dfrac{p + 1}{mp - 1}$,
    $y = \dfrac{m + 1}{mp - 1}$.

45. $x = \dfrac{m + n}{p}$,
    $y = \dfrac{m - n}{p}$.

46. $x = a + b - c$,
    $y = a + c - b$.

47. $x = \dfrac{m + 1}{mn + 1}$,
    $y = \dfrac{m(n + 1)}{mn + 1}$.

48. $x = \dfrac{a + 1}{b}$,
    $y = \dfrac{b + 1}{a}$.

49. $x = \dfrac{2}{m + n}$,
    $y = \dfrac{2}{m - n}$.

ANSWERS. 319

50. $x = \dfrac{m-n}{p-q}$,

$y = \dfrac{n-m}{pn-qm}$.

51. $x = \dfrac{b(d^2-c^2)-c(a^2-b^2)}{bd-ac}$,

$y = \dfrac{d(a^2-b^2)-a(d^2-c^2)}{bd-ac}$.

52. $x = a+b-c$,
$y = a-b+c$.

53. $x = m+n$,
$y = m-n$.

54. $x = (p^2+1)(q^2-1)$,
$y = (p^2-1)(q^2+1)$.

EXERCISE LIX. (PAGE 149.)

1. $x=2$,
$y=2$,
$z=2$.

2. $x=5$,
$y=6$,
$z=8$.

3. $x=1$,
$y=2$,
$z=3$.

4. $x=5$,
$y=6$,
$z=7$.

5. $x=20$,
$y=10$,
$z=5$.

6. $x=2$,
$y=3$,
$z=1$.

7. $x=-4$,
$y=-\dfrac{11}{3}$,
$z=-\dfrac{9}{2}$.

8. $x=9$,
$y=11$,
$z=13$.

9. $x = \dfrac{a+b}{2a}$,

$y = \dfrac{ab+ac+bc-b^2-2c^2}{2(b^2-c^2)}$.

$z = \dfrac{3bc-ab-ac-b^2}{2(b^2-c^2)}$.

10. $x = \dfrac{b^2+c^2-a^2}{2bc}$,

$y = \dfrac{c^2+a^2-b^2}{2ca}$,

$z = \dfrac{a^2+b^2-c^2}{2ab}$.

11. $x=6$,
$y=-12$,
$z=18$.

12. $x=\dfrac{468}{7}$,
$y=-\dfrac{335}{7}$,
$z=\dfrac{16}{7}$.

13. $x=\dfrac{1}{2}$,
$y=\dfrac{1}{3}$,
$z=\dfrac{1}{4}$.

14. $x=1$,
$y=\dfrac{1}{2}$,
$z=\dfrac{1}{3}$.

15. $x=\dfrac{2}{a+b}$,
$y=\dfrac{2}{a+c}$,
$z=\dfrac{2}{b+c}$.

16. $x=1\tfrac{1}{6}$,
$y=-3\tfrac{1}{2}$,
$z=2\tfrac{1}{10}$.

17. $x=\dfrac{7}{3}$,
$y=\dfrac{2}{3}$,
$z=-\dfrac{1}{3}$.

18. $x=1$,
$y=1$,
$z=1$.

19. $x=2$,
$y=3$,
$z=1$.

20. $x=\dfrac{1}{3}$,
$y=\dfrac{1}{2}$,
$z=\dfrac{1}{4}$.

21. $x=5$,
$y=4$,
$z=3$.

22. $x=3$,
$y=-2$,
$z=1$.

23. $x=4$,
$y=5$,
$z=6$.

ANSWERS.

24. $x = 12$, 25. $x = 2$, 26. $x = \dfrac{abc}{ab+bc+ca}$, 27. $x = a$,
    $y = 4$, $y = 3$, $y = \dfrac{abc}{ab+bc+ca}$, $y = b$,
    $z = 24$. $z = -5$. $z = \dfrac{abc}{ab+bc+ca}$. $z = c$.

28. $x = \dfrac{2abc}{bc+ca-ab}$, 29. $x = \dfrac{b+c}{2}$, 30. $x = n+p-m$,
    $z = \dfrac{2abc}{ca+ab-bc}$, $y = \dfrac{c+a}{2}$, $y = m+p-n$,
    $y = \dfrac{2abc}{ab+bc-ca}$. $z = \dfrac{a+b}{2}$. $z = m+n-p$.

31. $x = -11$, 32. $x = \dfrac{1}{5}$, 33. $x = a$,
    $y = -\dfrac{9}{5}$, $y = \dfrac{1}{3}$, $y = b$,
    $z = -\dfrac{7}{5}$. $z = 1$. $z = -c$.

EXERCISE LX. (Page 156.)

1. 8 and 4.   2. 120 and 90.   3. 42 and 12.   4. $\tfrac{4}{5}$.
5. 63.   6. 26.   7. 54.   8. 18 and 8.
9. $4 and $2.   10. 4 and $2\tfrac{2}{3}$.   11. 25 and 55.   12. 5 and 6.
13. 90 cents per kilo. for butter, 75 cents per kilo. for soap.
14. $3.60 and $3.   15. $900 and $400.   16. 2400 and 1800.
17. $12.40 and $1.60.   18. 140, 90 and 130.
19. 854.   20. 640, 720 and 840.
21. $133\tfrac{1}{3}$ lbs. and 100 lbs.   22. A's, 24s., and B's, 16s.
23. 90 min by A, 1 hr. by B, and 3 hrs. by all together.
24. 2 gals. from A, 14 from B.   25. Tea, 5s.; sugar, 4d.
26. $\dfrac{mn(p-1)}{np-m}$ days.   27. A's, £40; B's, £28.
28. 25 from 1st, 75 from 2nd.   29. 65.   30. Man in $21\tfrac{3}{7}$ days, woman in 50 days.   31. 28 barley, 20 rye, 52 wheat.
32. $1\tfrac{1}{2}$.   33. A, £6; B, £18; C, £36; D, £48.
34. £$21\tfrac{7}{16}$, £$8\tfrac{11}{16}$.   35. A, $5\tfrac{5}{8}$ hrs.; B, $6\tfrac{2}{13}$ hrs.; C, $7\tfrac{7}{15}$ hrs.

ANSWERS. 821

36. Rate of pulling, 8 miles an hour; distance, 20 miles.
37. Time down stream, 4 hrs.; time up, 6 hrs.; rate of stream, $\frac{2}{3}$ mile an hr. 38. 20 lbs. cheaper tea, 10 lbs. dearer tea.
39. 40 lbs. and 90 lbs. 40. 20 bush. of rye and 52 bush. of wheat.
41. A in $1\frac{3}{4}$ hrs., B in $3\frac{1}{2}$ hrs., C in 7 hrs.
42. A in 20 days, B in 30 days, C in 60 days.
43. A in $\dfrac{2abc}{ac+bc-ab}$ min., B in $\dfrac{2abc}{bc+ab-ac}$ min., C in $\dfrac{2abc}{ab+ac-bc}$ min. 44. 12 ft. long, 9 ft. broad.
45. 70 yds. long, 38 yds. wide. 46. 14 ft. long, 10 ft. wide.
47. 4 gallons and 10 gallons. 48. 10 gallons and 4 gallons.

EXERCISE LXI. (PAGE 166.)

1. $2ab$, $4a^6b^2c^4$, $5ab^2c$, $4x^2z$, $13a^6b^4c^6$.
2. $\dfrac{5a^2b^3}{11x^4y^5}$, $\dfrac{16x^6}{17y^4}$, $\dfrac{25a}{18b}$, $\dfrac{7ab^2c^3}{8c^4d^2}$. 3. $x+6$, $x-4$, $2ax+b$.
4. x^2+x+1, $2x^2-x-1$. 5. $x^3-5ax+4a^2$.
6. $7x^2+4xy+y^2$. 7. x^2-xy+y^2. 8. $2a^3-3a^2x-ax^2$.
9. $4x^2-2ab+2b^2$. 10. $2x^2-5x+3$. 11. x^3-2x^2+3x-4.
12. $2a^4+4a^2c^2-4c^4$. 13. $2a^3-a^2-3a+2$. 14. $x^3-2x^2y+2xy^2-y^3$.
15. $5x^3-3x^2y-4xy^2+y^3$. 16. $2y^2-3yz+4z^2$.
17. $a+2b+3c$. 18. $a^3+a^2b+ab^2+b^3$. 19. $3-4x+7x^2-10x^3$.
20. $3a-2b+4c$. 21. $5x^2y-3xy^2+2y^3$. 22. $2y^2x-3yx^2+2x^3$.
23. $5x-2y+3z$ 24. $2x^2+y^2-y$.
25. $m^4-2m^3+3m^2-4m+5$. 26. $2ab+a^2+b^2$.

EXERCISE LXII. (PAGE 167.)

1. $x^2 - \dfrac{xy}{2} - y^2$. 2. $x^2 - 2xy + y^2 - \dfrac{y^3}{x}$. 3. $1 + \dfrac{2}{x} + \dfrac{3}{x^2} + \dfrac{4}{x^3}$.
4. $x^2 + \dfrac{x}{2} - \dfrac{1}{3}$. 5. $\dfrac{a}{b} - 1 + \dfrac{b}{a}$. 6. $2a - 3b + \dfrac{b^2}{4}$.
7. $x^2 + 4 + \dfrac{4}{x^2}$. 8. $\dfrac{1}{x} - \dfrac{2}{y} + \dfrac{3}{z}$. 9. $\dfrac{2x}{z} + \dfrac{z}{x} - \dfrac{3y}{z}$.
10. $\dfrac{a}{3} - \dfrac{b}{4} + \dfrac{c}{5} - \dfrac{d}{2}$.

EXERCISE LXIII. (Page 169.)

1. $3a^2 - ab + 5b^2$. 2. $a^2 - 3ab + b^2$. 3. $a^2 + b^2$.
4. $a^2x^2 + 3ax + 1$. 5. $a^2 + ab + ac + bc$. 6. $a^3 + b^3 + c^3 - 3abc$.
7. $(a^2 + b^2)(c^2 + d^2)$. 8. $a^2 - b^2 + c^2 - d^2$. 9. $ab - ac + bc$.
10. $x - 2 - \dfrac{1}{x}$. 11. $x^2 - xy - xz - yz$.

EXERCISE LXIV. (Page 172.)

1. $x + 2y$. 2. $x + 4$. 3. $a - 3$. 4. $x^2 - ax - a^2$.
5. $x^2 + x + 1$. 6. $x^2 - 2x + 1$. 7. $2a^2 - 3b^2$. 8. $a - b + c$.
9. $2x^2 - 3x + 1$. 10. $1 - 3x + 4x^2$. 11. $4x^2 + 4x - 1$. 12. $a^2 + 3ab - 9b^2$.
13. $c^2 - 4bc + 4b^2$. 14. $x^2 - \dfrac{2}{x}$. 15. $\dfrac{a}{2} - \dfrac{2}{3a^2}$. 16. $x + \dfrac{1}{x}$.
17. $\dfrac{a}{b} - 1 - \dfrac{b}{a}$.

EXERCISE LXV. (Page 179.)

1. $x^{\frac{1}{2}}$, $x^{\frac{2}{3}}$, $x^{\frac{5}{2}}$, $a^{\frac{3}{5}}$, $a^{\frac{7}{4}}$, $a^{\frac{7}{3}}$, $a^{\frac{1}{3}}b^{\frac{1}{4}}$.

2. $x^{\frac{1}{3}}y^{\frac{2}{3}}z^{\frac{1}{3}}$, $x^{\frac{2}{5}}y^{\frac{1}{2}}z^{\frac{3}{5}}$, $a^{\frac{4}{7}}b^{\frac{2}{7}}c$, $5ab^{\frac{1}{3}}c^{\frac{3}{7}}x^2$.

3. $\sqrt[3]{a^2}$, $\sqrt[5]{a^3b}$, $4\sqrt[6]{\dfrac{x}{y^5}}$, $3\sqrt[3]{\dfrac{x}{y^2}}$. 4. $\dfrac{1}{a^2}$, $\dfrac{3}{xy^3}$, $\dfrac{6y}{x^3}$, $\dfrac{x^4}{y^5}$, $\dfrac{6xy^3}{ab^2}$.

5. $3xyz^{-2}$, $zx^{-3}y^{-4}$, $ab^{-1}c^{-1}$, $c^6a^{-3}b^2$, $x^{-\frac{1}{2}}y^{\frac{2}{3}}$, $x^{-2}y^{-\frac{1}{3}}$.

6. $a^{\frac{2}{3}}$, $b^{\frac{1}{2}}$, $c^{\frac{3}{4}}$, $d^{\frac{2}{3}}$. 7. $m^{\frac{1}{3}}$, $a^{\frac{1}{2}}$, 1. 8. a, 1, $y^{\frac{1}{2}}$, $x^{\frac{1}{5}}$.

9. $ac^{\frac{1}{4}}d$. 10. $x^{-\frac{1}{3}}y^{\frac{1}{3}}z^{-\frac{1}{3}}$, $x^{\frac{1}{2}}y^{-\frac{1}{4}}$. 11. $a^{-\frac{7}{12}}$, $y^{\frac{1}{3}}x^{-\frac{1}{6}}b^{-\frac{1}{6}}$.

12. $a^{\frac{1}{3}}c^{\frac{1}{3}}n^{-\frac{1}{3}}$. 13. a^{-1}, $c^{-\frac{1}{2}}$, m^{-2}, n^{-1}. 14. $p^{\frac{1}{2}}$, $q^{-\frac{1}{3}}$, $a^{-\frac{1}{3}}$.

15. $\dfrac{27a^3b^{\frac{3}{4}}}{8}$, $\dfrac{64}{27a^6b^{\frac{3}{2}}}$, $\dfrac{a^2}{3^{\frac{1}{4}}}$.

EXERCISE LXVI. (Page 180.)

1. $x + 1$. 2. $a^4b^{-4} - 2 + a^{-4}b^4$.
3. $4x^{-5} - x^{-4} + 3x^{-3} + 2x^{-2} + x^{-1} + 1$. 4. $x^{4p} + x^{2p}y^{2p} + y^{4p}$.
5. $x^{\frac{3}{2}} + x^{\frac{1}{2}}y - xy^{\frac{1}{2}} - y^{\frac{3}{2}}$. 6. $x^4 - x^2(a + b) + ab$.

ANSWERS. 323

7. $x^{mn} - x^a y^{mr-m} - x^{mn-n}y^m + y^{nn}$. 8. $x + y$.
9. $a + a^{\frac{1}{2}}b^{\frac{1}{2}} - b$. 10. $a^n + 1 + a^{-n}$.
11. $x^{\frac{3}{2}} + 2x^{\frac{1}{2}} + 1 + 2x^{-\frac{1}{2}} + x^{-\frac{3}{2}}$. 12. $x^n + 1$.
13. $64x^{\frac{3}{2}} - 729y^{\frac{3}{2}}$. 14. $x^6 - 1$. 16. $2^{n(n-2)}$.
17. $x^{\frac{1}{3}} - y^{\frac{1}{3}} + z^{\frac{1}{3}}$. 18. $x^{\frac{1}{2}} + 2y^{\frac{1}{4}} - z^{\frac{1}{4}}$. 19. $2r^{-2} + 3x^{-1}$.
20. $x + 2 - x^{-1}$. 21. $x^{\frac{1}{2}} - x^{-\frac{1}{2}} + x^{-1}$. 22. $2a^{\frac{1}{2}} - 3b^{\frac{1}{3}} + 4c^{\frac{1}{4}}$.
23. $16x^{\frac{3}{2}} - 16x^{\frac{1}{2}} + 12 - 4x^{-\frac{1}{2}} + x^{-\frac{3}{2}}$. 24. $\dfrac{y}{x^{\frac{1}{2}}} + y^{\frac{1}{4}}x^{\frac{1}{4}} - \dfrac{x}{2y^{\frac{1}{2}}}$.

25. $x - 2 - x^{-1}$. 26. $\dfrac{2a^{\frac{1}{3}}}{c^{\frac{1}{3}}} - 1 + \dfrac{2c^{\frac{1}{3}}}{a^{\frac{1}{3}}}$. 27. $-(ab)^{a+v}$.

28. $\dfrac{x^3 - 1}{x^6 + 1}$. 29. $\dfrac{3a^{-1}x - 4}{a^{-2}x^2 - 11a^{-1}x + 21}$. 30. $\dfrac{x^{\frac{1}{2}} + 1 - y^{\frac{1}{2}}}{x^{\frac{1}{2}} + 2 - 3y^{\frac{1}{2}}}$.

31. $\dfrac{xe^y + ye^x}{xe^x + ye^y}$. 32. $(a^2 - b^2)^3$. 33. $4(a^{\frac{1}{2}}b^{\frac{1}{2}} + 2b)$.

34. x^{pq}. 35. $x^{p(q-1)} - y^{q(p-1)}$. 36. a^{m-1}.

37. $a^2 - b^2$. 38. $\dfrac{2x(x^2 + 3)}{(1+x)^3}$. 39. $1 + x^{\frac{1}{2}}$.

40. $\dfrac{a^5 x^{10} - a^3 x^6 - a^2 x^4 + 1}{a^{\frac{1}{2}}x - 1}$. 42. x^{a+b+c}. 43. $1 + 2^{-\frac{3}{2}}$.

44. $\dfrac{8}{9}$. 45. 11. 46. $\dfrac{2(x+1)}{x^2 + x + 1}$. 47. 1.

EXERCISE LXVII. (Page 187.)

1. $xy^2\sqrt{z}$, $2a\sqrt{2ab}$, $3ay\sqrt[3]{2ax^2}$, $2\sqrt{6}$, $5a^2d\sqrt{5d}$
2. $10\sqrt[3]{a}$, $2xy^2\sqrt[3]{20xy}$, $3m^3n^3\sqrt[3]{4n}$, $7a^5b^5\sqrt[3]{4b}$.
3. $(a-b)\sqrt[3]{a}$, $5(a-b)\sqrt{2}$.
4. $\dfrac{1}{3}\sqrt{5}$, $\dfrac{3}{4}\sqrt{3}$, $\dfrac{5}{4}\sqrt{2}$, $\dfrac{1}{2}\sqrt[3]{4}$, $\dfrac{1}{5}\sqrt[3]{25}$.
5. $\dfrac{1}{a-b}\sqrt{a^2-b^2}$, $\dfrac{1}{a+b}\sqrt[3]{(a^2-b^2)^2}$, $\dfrac{1}{2a}\sqrt[3]{8a^3b}$, $\dfrac{a}{2cy^2}\sqrt{3bcxy}$.
6. $\sqrt{18}$, $\sqrt[3]{56}$, $\sqrt[3]{405}$, $\sqrt[3]{1125}$, $\sqrt[3]{96}$.

7. $\sqrt{9a}$, $\sqrt{48a^2x}$, $\sqrt[3]{\frac{81}{7}}$, $\sqrt{3a^3x}$.

8. $\sqrt{m^2-n^2}$, $\sqrt{\frac{a+b}{a-b}}$, $\sqrt{\frac{x}{x+y}}$.

9. $4ac\sqrt[3]{5a b^2c^2}$, $27y\sqrt[3]{3x^2z}$, $55\sqrt{6}$.

10. $\frac{1}{z}\sqrt[3]{2xy^2z^2}$, $\frac{1}{2b}\sqrt[3]{8ab}$, $\frac{x^2y}{z}$.   11. $abx\sqrt{x}$, $2a^2b^4x\sqrt[3]{b^2}$.

EXERCISE LXVIII. (Page 189.)

1. $\sqrt[6]{125}$, $\sqrt[6]{16}$.   2. $\sqrt[6]{6}$, $\sqrt[6]{49}$.   3. $\sqrt[12]{16}$, $\sqrt[12]{27}$.
4. $\sqrt[12]{256}$, $\sqrt[12]{512}$.   5. $\sqrt[12]{64}$, $\sqrt[12]{625}$, $\sqrt[12]{216}$.
6. $\sqrt[20]{7^5}$, $\sqrt[20]{5^4}$, $\sqrt[20]{(120)^2}$.   7. $\sqrt[30]{3^{15}}$, $\sqrt[30]{6^{10}}$, $\sqrt[30]{10^6}$.
8. $\sqrt[12]{(a^2-ab+b^2)^4}$, $\sqrt[12]{(a+b)^3}$.   9. $\sqrt[20]{\frac{a^5}{b^6}}$, $\sqrt[20]{\frac{b^4}{a^4}}$.

EXERCISE LXIX. (Page 190.)

1. $3\sqrt{7}$ is the greater.
2. $6\sqrt{7}$, $5\sqrt{10}$, $9\sqrt{3}$. (Descending order of magnitude.)
3. $5\sqrt[3]{3}$, $4\sqrt[3]{4}$, $3\sqrt[3]{5}$. (Descending order of magnitude.)
4. $6\sqrt[3]{7}$, $4\sqrt[3]{9}$, $3\sqrt[3]{18}$. (Descending order of magnitude.)

EXERCISE LXX. (Page 194.)

1. $5\sqrt{3}$, $7\sqrt{13}$.   2. $10\sqrt{7}$, $13\sqrt{10}$.   3. $2\sqrt{11}$, $6\sqrt{3}$.
4. $3\sqrt[3]{3}$, $4\sqrt{2}$.   5. $6\sqrt[3]{4}$, $5\sqrt[3]{2}$.   6. $11\sqrt{7}$, $\frac{7}{2}\sqrt[3]{5}$.
7. $-\sqrt{2}$.   8. $12\sqrt{3}$.   9. $129\sqrt{5}$.   10. $-4\sqrt[3]{3}$.
11. $(c+5)\sqrt[3]{c}$.   12. $(a-b+2)\sqrt[3]{b}$.   13. $(a^2+b^2-2ab)\sqrt{x}$.
14. $10ab\sqrt{7ab}$.   15. $(4a^2b^2-24ab+36)\sqrt{2b}$.
16. $-\frac{13}{3}\sqrt[3]{4}$.   17. $\frac{73}{18}+\frac{31}{15}\sqrt{5}$.   18. $2\sqrt{15}$.
19. $\frac{11}{3}\sqrt{6}$.   20. $\frac{2a^2}{a-x}\sqrt{a^2-x^2}$.   21. $(1+a+2x)\sqrt{a+x}$

ANSWERS.

EXERCISE LXXI. (Page 195.)

1. 9, 10, 6. 2. 3, 4, 8. 3. 3, 6, 2. 4. 11, 27.
5. 72, 25. 6. 60, $\frac{3}{4}$. 7. 10. 8. 4.
9. $\sqrt{5}-4$. 10. $10-2\sqrt{2}$. 11. $-15-19\sqrt{3}$.
12. $36-15\sqrt{6}$. 13. $6\sqrt{2}-3\sqrt{15}+8\sqrt{3}-6\sqrt{10}$.
14. $8-8\sqrt[3]{12}+\sqrt[3]{18}$. 15. $\sqrt{21}$. 16. $\sqrt{7}$. 17. $\sqrt{111}$.
18. 5. 19. 4. 20. 4. 21. 5. 22. 30.
23. $86+\frac{47}{2}\sqrt{6}$. 24. x^4-x^2+1. 25. $2ab+2bc+2ca-a^2-b^2-c^2$.
26. $\sqrt[3]{2000}$, $5\sqrt{30}$. 27. $\sqrt{40}$, $125\sqrt[3]{7^5}$.
28. $\sqrt[15]{\frac{1}{144}}$, $\sqrt[20]{\frac{7\times 625}{729}}$. 29. $\sqrt[36]{\frac{2}{2401\times 3}}$, $\sqrt[20]{\frac{5^6}{2^3\times 3^5}}$.

EXERCISE LXXII. (Page 197.)

1. 9, 2, 3. 2. 3, 2, 5. 3. $\frac{7}{6}, \frac{2}{3}, \frac{2}{7}$.
4. $\sqrt{2}+5\sqrt{6}$. 5. $7\sqrt{3}-3\sqrt{5}$. 6. $4\sqrt{35}+8\sqrt{14}$.
7. $2\sqrt{42}-3\sqrt{14}+5\sqrt{10}$. 8. $2\sqrt[3]{25}-6\sqrt[3]{4}+5\sqrt[3]{36}$.
9. $5\sqrt[3]{75}+3\sqrt[3]{18}-4\sqrt[3]{36}$. 10. $a, 2y$.
11. $\sqrt{b}+1$, $\sqrt{a}+\sqrt{b}$, $1+\sqrt{a}$. 12. $a-\sqrt{ab}+b$, $ax\sqrt{x}-by\sqrt{y}$.
13. 3, $\sqrt[4]{\frac{1}{3}}$. 14. $\frac{4}{15}$, $\sqrt[6]{\frac{2}{3}}$. 15. $\frac{1}{x}\sqrt[6]{\frac{4}{x}}$, $\frac{1}{xy^2}\sqrt[4]{\frac{1}{2xy^2}}$.

EXERCISE LXXIII. (Page 201.)

1. $2-\sqrt{3}$. 2. $1+\sqrt{3}$. 3. $5+\sqrt{2}$. 4. $5-\sqrt{2}$.
5. $1+\sqrt{5}$. 6. $1-\sqrt{5}$. 7. $2+\sqrt{6}$. 8. $4-\sqrt{5}$.
9. $\sqrt{a}+\sqrt{b}$. 10. $a+\sqrt{b}$. 11. $\sqrt{a+b}+\sqrt{a-b}$.
12. $\sqrt{a^2+ab+b^2}+\sqrt{a^2-ab+b^2}$. 13. $\frac{2}{3}+\frac{1}{3}\sqrt{3}$.

14. $\dfrac{2}{5} - \dfrac{1}{10}\sqrt{10}.$ 15. $\dfrac{\sqrt{7}+2}{\sqrt{3}}.$ 16. $\dfrac{3}{2} + \dfrac{1}{2}\sqrt{5}.$

17. $b - \sqrt{a^2 - b^2}.$ 18. $(a-b) - 2\sqrt{ab}.$

19. $\sqrt{\dfrac{3a^2+x^2}{2}} + \sqrt{\dfrac{x^2-a^2}{2}}.$ 20. $\sqrt{\dfrac{6b^2+a^2}{2}} + \sqrt{\dfrac{a^2}{2}}.$

EXERCISE LXXIV. (Page 202.)

1. $\sqrt{7}+\sqrt{3},\ \sqrt{11}+\sqrt{5},\ 2\sqrt{3}-\sqrt{2}.$
2. $\sqrt{10}-\sqrt{3},\ 2\sqrt{5}-3\sqrt{2},\ 3\sqrt{5}-2\sqrt{3}.$
3. $3+\sqrt{5},\ 2\sqrt{3}-2\sqrt{2},\ 3-\sqrt{2}.$
4. $\sqrt{7}-\sqrt{2},\ 3\sqrt{7}-2\sqrt{6},\ 2\sqrt{3}+\sqrt{5}.$

EXERCISE LXXV. (Page 204.)

1. $\dfrac{6+2\sqrt{6}}{3},\ \dfrac{8+\sqrt{55}}{3},\ \dfrac{28\sqrt{5}-60\sqrt{2}+15\sqrt{10}-35}{55}.$

2. $\dfrac{84+21\sqrt{6}}{10},\ 20\sqrt{3}-12\sqrt{5},\ \dfrac{24\sqrt{7}-21\sqrt{2}}{2}.$

3. $4-\sqrt{15},\ \dfrac{23-2\sqrt{130}}{3},\ \dfrac{10\sqrt{3}+5\sqrt{6}-6\sqrt{2}-6}{57}.$

4. $\dfrac{11+3\sqrt{14}}{5},\ 4+\sqrt{15},\ \dfrac{12+20\sqrt{3}+9\sqrt{2}+15\sqrt{6}}{-2}.$

5. $\dfrac{\sqrt{2}+\sqrt{6}}{2},\ \sqrt{2}+\sqrt{3},\ \sqrt{5}+2\sqrt{2}-\sqrt{10}-2.$

6. $\dfrac{195+137\sqrt{6}-75\sqrt{3}-187\sqrt{2}}{94},$

$\dfrac{8\sqrt{10}+8\sqrt{15}-22-5\sqrt{30}-10\sqrt{5}+19\sqrt{2}+16\sqrt{3}-10\sqrt{6}}{2}.$

7. $\dfrac{a+\sqrt{a^2-x^2}}{x},\ \dfrac{a^2+\sqrt{a^4-b^4}}{b^2}.$

8. $\dfrac{x+\sqrt{x^2-9a^2}}{3a},\ \dfrac{2a^2+\sqrt{x^2(4a^2-x^2)}}{x^2-2a^2}.$

9. $\dfrac{2^{\frac{2}{3}} - 4.3^{\frac{1}{2}} + 2^{\frac{2}{3}}.3^{\frac{3}{2}} - 6 + 2^{\frac{1}{3}}.3^{\frac{5}{2}} - 3^{\frac{5}{2}}}{-1}$,

$\dfrac{2.3^{\frac{1}{2}}\left(\begin{array}{c} 4^{\frac{11}{3}} - 4^{\frac{10}{3}}.3^{\frac{1}{2}} + 64.3^{\frac{1}{2}} - 4^{\frac{8}{3}}.3^{\frac{3}{2}} + 4^{\frac{7}{3}}.3 - 16.3^{\frac{5}{2}} + 4^{\frac{5}{3}}.3^{\frac{5}{2}} \\ - 4^{\frac{4}{3}}.3^{\frac{7}{2}} + 4 \times 9 - 4^{\frac{2}{3}}.3^{\frac{7}{2}} + 4^{\frac{1}{3}}.3^{\frac{9}{2}} - 3^{\frac{11}{2}} \end{array}\right)}{229}$,

$\dfrac{(2\sqrt{2} - \sqrt{3})(3^{\frac{3}{2}} + 3.5^{\frac{1}{2}} + 3^{\frac{1}{2}}.5^{\frac{1}{2}} + 5^{\frac{3}{2}})}{4}$.

10. $\dfrac{18 - 34\sqrt{5}}{11}$.   11. 4.   12. $\dfrac{57\sqrt{2} - 25\sqrt{3} - 9\sqrt{6}}{3}$.

13. 5.   14. $\dfrac{\sqrt{a+x}}{a+x}$; $\dfrac{13a - 5b + 12\sqrt{a^2 - b^2}}{13b - 5a}$.

EXERCISE LXXVI. (Page 208.)

1. $\dfrac{b}{c}$, $\dfrac{c}{b}$.   2. a^3, $\dfrac{1}{a^2}$.   3. $\sqrt{\dfrac{4a-b}{b}}$, $\sqrt{\dfrac{b}{4a-b}}$.

4. 10.   5. $1 + \sqrt{2}$.   6. .2679492.   7. $\dfrac{5\sqrt{3}}{3} - 2$

8. 1.   9. 2, $\dfrac{1}{2}$.   10. $\dfrac{1}{2}\left(xy + \dfrac{1}{xy}\right)$.   11. c, $\dfrac{1}{c}$.

EXERCISE LXXVII. (Page 210.)

1. $25\sqrt{-1}$, $5\sqrt{-1}$, $9\sqrt{-1}$, $12\sqrt{-1}$, $b^2\sqrt{-1}$.
2. $4\sqrt{-1}$, $8\sqrt{-1}$, $3x^3\sqrt{-1}$, $9m^3\sqrt{-1}$.
3. $-1, 1, -\sqrt{-1}, 1, \sqrt{-1}$.   4. $15\sqrt{-1}$.   5. $16 + 4\sqrt{-1}$.
6. $18 + (3 + 2b)\sqrt{-1}$.   7. $-12, -35, -12yz^3$.
8. $47 - \sqrt{-1}$, $9x^2 + 24 - 15x\sqrt{-1}$.
9. $2ac + 3bd + (3bc - 2ad)\sqrt{-1}$, $mn + b^2 + b(n-m)\sqrt{-1}$.
10. $-a\sqrt{-1}$, $-\sqrt{-x}$, $\sqrt{x}$, $x\sqrt{2}$.   11. 0.   12. 0.   13. $\dfrac{1}{7}$

ANSWERS.

EXERCISE LXXVIII. (Page 215.)

1. $x = 2$.
2. $x = 3$.
3. $x = \frac{49}{4}$.
4. $x = \frac{1}{9}$.
5. $x = \frac{1}{4}$.
6. $x = \frac{4}{9}$.
7. $x = \frac{1}{20}$.
8. $x = -\frac{1}{3}$.
9. $x = 3\frac{1}{2}$.
10. $x = \frac{8}{5}$.
11. $x = 5$.
12. $x = 49$.
13. $x = 25$.
14. $x = 9$.
15. $x = 5$.
16. $x = \frac{13}{9}$.
17. $x = 25$.
18. $x = 9$.
19. $x = 64$.
20. $x = \frac{36}{5}$.
21. $x = 25$.
22. $x = \frac{1}{9}$.
23. $x = 100$.
24. $x = 64$.
25. $x = \frac{3}{4}$.
26. $x = 0$.
27. $x = -\frac{2275}{88}$.
28. $x = \frac{(a-b)^2}{2a-b}$.
29. $x = 2\frac{1}{8}$.
30. $x = 1$.
31. $x = \frac{5}{4}$.
32. $x = 2a$.
33. $x = -2a$.
34. $x = 25$.
35. $x = 4$.
36. $x = 4$.
37. $x = \frac{a^{\frac{1}{2}}}{a^{\frac{1}{2}}+2}$.
38. $x = \frac{\sqrt{3}}{2}$. (Pure quadratic.)
39. $x = \frac{9}{5}$.
40. $x = 3$. (Pure quadratic.)
41. $x = -\frac{5}{3}$.
42. $x = -\frac{8}{3}$.
43. $x = -\frac{7a}{8}$.

EXERCISE LXXIX. (Page 221.)

1. $2(a^2 + b^2 + c^2 + ab + bc + ca)$.
2. $2(a^2 + b^2 + c^2 - ab - bc - ca)$.
3. $a^2 + b^2 + c^2 + 3(ab + bc + ca)$.
4. $ab + bc + ca - a^2 - b^2 - c^2$.
5. $ab + bc + ca - a^2 - b^2 - c^2$.
6. $a^2 + b^2 + c^2 - ab - bc - ca$.
7. $3(a^2 + b^2 + c^2) - 2(ab + bc + ca)$.
8. $3(a^2 + b^2 + c^2) - 2(ab + bc + ca)$.
9. 0.
10. $2(ab + bc + ca)$.
11. 0.
12. 0.
13. 0.
14. 0.
15. 3.
16. 0.
17. 0.
18. 0.
19. x^2.
20. $3x^2 + 2(a+b+c)x + ab + bc + ca$.
21. $3x^2 - 2(a+b+c)x + ab + bc + ca$.
22. $(a+b+c)x^2 - 2(ab+bc+ca)x + 3abc$.

ANSWERS. 329

23. $(a+b+c)x^2 - 3abc$. 24. 0. 25. $c^2(a-b) + a^2(b-c) + b^2(c-a)$.
26. $3\{a^2(b-c) + \&c.\}x - \{c^3(a-b) + \&c.\}$. 27. $ab(a-b) + \&c$.
28. $2(a+b+c)x^2 - \{ab(a+b) + \&c.\}$. 29. 0.
30. $abc\{bc(b-c) + \&c.\}$. 31. bc. 32. $a^2 + b^2 + c^2$.
33. $ab + bc + ca$. 34. $2(ab + bc + ca)$. 35. abc.
36. s^2. 37. c^3. 38. s^3. 39. $8abc$.

EXERCISE LXXX. (Page 224.)

1. 1. 2. 1. 3. 1. 4. 0. 5. $\dfrac{x}{abc}$. 6. 1.
7. 1. 8. x^2. 9. x^2. 10. 1. 11. 0. 12. $\dfrac{1}{3}$.
13. $\dfrac{x}{(x-a)(x-b)(x-c)}$. 14. $\dfrac{x^2}{(x-a)(x-b)(x-c)}$.

EXERCISE LXXXI. (Page 226.)

1. $-(a-b)(b-c)(c-a)(a+b+c)$. 2. $-(a-b)(b-c)(c-a)(a+b+c)$.
3. $a+b+c$. 4. $a+b+c$. 5. $\dfrac{a+b+c}{abc}$.
6. $-(a+b+c)$. 7. $a+b+c$. 8. $a+b+c-x$.
9. $3x - (a+b+c)$. 10. $abc(a+b+c)$. 11. $\dfrac{a+b+c}{(a+b)(b+c)(c+a)}$.
12. 0. (Let $a-b=x$, $b-c=y$, $c-a=z$.) 13. $\dfrac{4(x+1)(a+b+c)}{(a+b)(b+c)(c+a)}$.
14. $a+b+c - (x+y+z)$. 15. $x = a+b+c$.

EXERCISE LXXXII. (Page 228.)

1. $-(a-b)(b-c)(c-a)(a^2 + b^2 + c^2 + ab + bc + ca)$.
2. $-(a-b)(b-c)(c-a)(a^2 + b^2 + c^2 + ab + bc + ca)$.
3. $a^2 + b^2 + c^2 + ab + bc + ca$. 4. $\dfrac{a^2 + b^2 + c^2 + ab + bc + ca}{(a+b)(b+c)(c+a)}$.
5. $\dfrac{a^2 + b^2 + c^2 + ab + bc + ca}{abc}$. 6. $(a+b+c)^2$.
7. $\dfrac{(a+b+c)^3}{(a+b)(b+c)(c+a)}$. 8. $x^3 - 2(a+b+c)x + a^2 + b^2 + c^2$
$+ ab + bc + ca$. 9. $-3x^3 + 3(a+b+c)x - (a^2+b^2+c^2+ab+bc+ca)$.

10. $ab+bc+ca-a^3-b^3-c^3$. 11. $\frac{3}{2}x(a^2+b^2+c^2+ab+bc+ca)$.

12. $\frac{3}{2}x$. 13. $x=\frac{1}{2}(a^2+b^2+c^2+ab+bc+ca)$.

14. $x=\dfrac{a^2+b^2+c^2+ab+bc+ca}{a+b+c}$. 15. $x=a^2+b^2+c^2+ab+bc+ca$.

EXERCISE LXXXIII. (Page 229.)

1. $-(a-b)(b-c)(c-a)(ab+bc+ca)$.
2. $-(a-b)(b-c)(c-a)(ab+bc+ca)$.
3. $ab+bc+ca$. 4. $\dfrac{ab+bc+ca}{abc}$. 5. $\dfrac{ab+bc+ca}{abc}$.

6. 1. 7. $\dfrac{-(a+b+c)x^2+(a+b+c)^2x-(a+b)(b+c)(c+a)}{(x-a)(x-b)(x-c)}$.

8. $\dfrac{a^2+b^2+c^2}{ab+bc+ca}$. 9. $-\dfrac{1}{6}(a+b+c)$. 10. $x=ab+bc+ca$.

11. $x=\dfrac{abc}{a+b+c}$. 12. $x=0$. 13. $x=0,\ \dfrac{a+b+c}{ab+bc+ca}$.

EXERCISE LXXXIV. (Page 231.)

1. $3ab(a+b)$. 2. $5ab(a+b)(a^2+ab+b^2)$.
3. $7ab(a+b)(a^2+ab+b^2)^2$. 4. $-5ab(a-b)(a^2-ab+b^2)$.
5. $-7ab(a-b)(a^2-ab+b^2)^2$. 6. $7ab(a-b)(a^2-ab+b^2)^2$.
7. $3(x-y)(y-z)(z-x)$.
8. $7(x-y)(y-z)(z-x)(x^2+y^2+z^2-xy-yz-zx)^2$.
9. $(a-b)(b-c)(c-a)(a+b+c)$.
10. $(a-b)(b-c)(c-a)(ab+bc+ca)$.
11. $2(x^2+xy+y^2)^2$. 12. $2(x^2+y^2+z^2-xy-yz-zx)^2$.
13. $3(a+b)(b+c)(c+a)$.
14. $5(a+b)(b-c)(a-c)(a^2+b^2+c^2+ab-bc-ca)$. 15. $24xyz$
16. $3(2x+y+z)(x+2y+z)(x+y+2z)$.
17. $-3(a-c)(b-d)(a-b+c-d)$.
18. $-5(a-c)(b-d)(a-b+c-d)(a^2+b^2+c^2+d^2-ab-bc-cd-da)$.
24. $16(b-c)(c-a)(a-b)(x-a)(x-b)(x-c)$.

ANSWERS. 331

EXERCISE LXXXV. (Page 232.)

1. $(a+b)(b+c)(c+a)$. 2. $(a+b+c)(ab+bc+ca)$.
3. $(a+b)(b+c)(c+a)$. 4. $(a+b)(b+c)(c+a)$.
5. $6abc$. 6. $(a+b+c+abc)(1+ab+bc+ca)$.
7. $(a+b+c-abc)(1-ab-bc-ca)$.
8. $2(a+b+c-abc)(1-ab-bc-ca)$.
9. $2(a+b+c)(a^2+b^2+c^2-ab-bc-ca)$.
10. $(a+b+c)(a^2+b^2+c^2-ab-bc-ca)(x+1)(x^2-x+1)$.

EXERCISE LXXXVI. (Page 236.)

1. $a=-6,\ b=9$. 2. $a=4,\ b=2$. 3. $a=20,\ b=85$.
4. $x=10$. 5. $x=4,\ 6$. 6. $m=12$. 7. $x=\dfrac{pq}{2mq-p}$.
8. $x=0,\ -4a,\ \dfrac{4a(a^2+b^2)}{7b^2-a^2},\ \dfrac{2\sqrt{(a^2-b^2)}}{\sqrt{3}},\ \dfrac{7b^2-a^2}{12a}$, &c.
9. $x=\dfrac{c-q(a-p)}{p(a-p)-(b-q)}$. 10. $b^2=ac$. 15. No. 16. Yes.
17. $8c=a(4b-a^2)$.
    $64d=(4b-a^2)^2$.
19. $x=\dfrac{n^2-n+2}{n^2+n}$.
20. $a=-6,\ a=-14;$ 21. $x=a+2c,$
    $b=2\tfrac{2}{3},\ b=0$. $y=b+3c$.

EXERCISE LXXXVII. (Page 242.)

3. $2a^3-6ab(a-b),\ 2b^3+6ab(a-b),\ 2a^3+2b^3$.
6. $a=8,\ \dfrac{x-4}{x-5}$. 7. 2000. 8. 1. 9. 0. 10. 1230, -382.

EXERCISE LXXXIX. (Page 249.)

1. $x=-2,\ -7$. 2. $x=3,\ 5$. 3. $x=4,\ -3$.
4. $x=\dfrac{1}{2},\ \dfrac{1}{3}$. 5. $x=\dfrac{1}{5},\ -\dfrac{1}{6}$. 6. $x=4,\ -5$.
7. $x=\dfrac{1}{4},\ -\dfrac{1}{3}$. 8. $x=14,\ -\dfrac{1}{2}$. 9. $x=1,\ -\dfrac{2}{5}$.

ANSWERS.

10. $x = 2, -\frac{1}{3}$.
11. $x = 1, -a$.
12. $x = 3, -\frac{3}{5}$.
13. $x = \frac{2}{3}, 17$.
14. $x = \frac{1}{10}, \frac{1}{11}$.
15. $x = \frac{1}{13}, \frac{1}{60}$.
16. $x = 1, -4$.
17. $x = 2, -\frac{33}{13}$.
18. $x = 5, -\frac{1}{6}$.
19. $x = 3, -\frac{4}{5}$.
20. $x = 2, \frac{58}{13}$.
21. $x = \frac{a^2}{b}, -\frac{2a^2}{b}$.
22. $x = \frac{a}{2}, \frac{3a}{4}$.
23. $x = 2a, 2b$.
24. $x = 2a, -8$.
25. $x = \frac{32}{3}, 1$.
26. $x = 7, -\frac{77}{2}$.
27. $x = 1, -\frac{5}{4}$.
28. $x = 8, -1$.
29. $x = 4, -5$.
30. $x = a - 2b, a + 2b$.

EXERCISE XC. (Page 254.)

1. $x = \frac{1}{2}, 3$.
2. $x = 17, \frac{2}{3}$.
3. $x = \frac{3}{2}, -\frac{1}{2}$.
4. $x = 3, 11$.
5. $x = 4, -1$.
6. $x = 6, -1$.
7. $x = 5, -\frac{5}{2}$.
8. $x = 2, -2$.
9. $x = 8, \frac{5}{2}$.
10. $x = 10, -2$.
11. $x = \frac{3}{10}, \frac{2}{3}$.
12. $x = \pm\sqrt{6}$.
13. $x = 1, \frac{3}{5}$.
14. $x = 3, -\frac{4}{5}$.
15. $x = 0, 4$.
16. $x = 0, \frac{4}{3}$.
17. $x = 3, -\frac{5}{3}$.
18. $x = 6, \frac{1}{2}$.
19. $x = \pm\frac{\sqrt{5}}{2}$.
20. $x = 4, -1$.
21. $x = 2, \frac{43}{87}$.
22. $x = 6, -6$.
23. $x = 3, \frac{1}{5}$.
24. $x = \frac{1}{a}, -\frac{1}{2}$.
25. $x = 100, -10$.
26. $x = 4, \frac{11}{6}$.
27. $x = -1, \frac{a+b}{c}$.
28. $x = \pm\sqrt{\frac{b(bc - 2a^2)}{c}}$.
29. $x = \frac{a^2 \pm \sqrt{a^4 + b^2c^2}}{c}$.
30. $x = a, \frac{b}{na}$.
31. $x = \frac{c}{a}, \frac{c}{b}$.
32. $x = \frac{a(-1 \pm \sqrt{73})}{8}$.

33. $x = a \pm 2\sqrt{\dfrac{ab}{c}}$. 34. $x = a, b$. 35. $x = 0, -1$.

36. $x = 5, -3\frac{1}{2}$. 37. $x = \dfrac{a}{6}(-11 \pm \sqrt{13})$. 38. $x = 0, 37$.

39. $x = \dfrac{7 \pm \sqrt{85}}{6}$. 40. $x = 1, \dfrac{b-a-1}{a}$. 41. $x = \pm a$.

42. $x = 0, \pm \sqrt{a^2 + b^2}$. 43. $x = \dfrac{a+b}{2}, \dfrac{a+b}{2}$.

44. $x^2\{a^2 + b^2 + c^2 + 3(ab + bc + ca)\} - 3x(a+b)(b+c)(c+a)$
$\qquad + a^2b^2 + b^2c^2 + c^2a^2 + 3abc(a+b+c) = 0$.

45. $x = -a, -b$. 46. $x = \pm\sqrt{a^2 + 2ab}, \pm\sqrt{\dfrac{ab - a^2}{2}}$.

47. $x = 0, \dfrac{ab + bc + ca}{a + b + c}$. 48. $x = 0, \pm\sqrt{\dfrac{(a-b)(b+c)^2 - b^2(c-a)}{2a - b - c}}$.

49. $x = 0, \dfrac{3abc}{a^2 + b^2 + c^2}$. 50. $-\dfrac{ac^2 + bd^2}{2a + 3\sqrt{cd}}, -\dfrac{ac^2 + bd^2}{2a - 3\sqrt{cd}}$.

51. 1.

EXERCISE XCI. (PAGE 258.)

1. $x = 2, \dfrac{1}{3}$. 2. $x = 24, 8\frac{2}{3}$. 3. $x = 1, \dfrac{7}{4}$.

4. $x = 8, -8$. 5. $x = 2, -3$. 6. $x = 1, 4\frac{1}{3}$.

7. $x = \dfrac{1 \pm \sqrt{a^2 + 1}}{a}$. 8. $x = 8, -20$. 9. $x = 1, -4$.

10. $x = 2, -3, \dfrac{-1 \pm 3\sqrt{-3}}{2}$. 11. $x = \dfrac{5}{2}, -1, \dfrac{3 \pm \sqrt{-15}}{4}$.

12. $x = 2, 3, -2, -1$. 13. $x = -7, \dfrac{1}{7}$.

14. $x = \dfrac{1}{2}, -2, \dfrac{1 \pm \sqrt{10}}{3}$. 15. $x = \pm 2, \pm\sqrt{-6}$.

16. $x = 2, -3$. 17. $x = \dfrac{-1 \pm 3\sqrt{5}}{2}, -3, 2$.

18. $x = \pm 8, \pm 1$. 19. $x = 3, 7, 5 \pm \sqrt{11}$.

20. $y = \dfrac{61 \pm \sqrt{1801}}{8}$ where $y = x^2 - 6x$.

EXERCISE XCII. (Page 261.)

1. $x=1, \frac{1}{21}$.
2. $x=3, 12$.
3. $x=\frac{2}{3}, \frac{50}{21}$.
4. $x=2a, -2a$.
5. $x=\pm 9\sqrt{2}$.
6. $x=\pm 1$.
7. $x=\pm \frac{7}{5}$.
8. $x=5, -\frac{45}{4}$.
9. $x=5, -30$.
10. $x=\pm 22\sqrt{2}$.
11. $x=\pm 4 - \frac{\sqrt{5}}{9}$.
12. $x=\pm \frac{10}{63}\sqrt{357}$.
13. $x=\pm 4\sqrt{-\frac{1}{11}}$.
14. $x=\pm\sqrt{a^2 - \frac{(b-2a)^3}{27b}}$.
15. $x=0, -7$.
16. $x=7, 4, -1, -4$.
17. $x=\pm 4, \pm\sqrt{23}$.
18. $x=5, -\frac{5}{2}, \frac{5\pm\sqrt{97}}{4}$.
19. $x=\frac{3}{2}, 2, \frac{7\pm\sqrt{33}}{4}$.
20. $x=\frac{3\pm\sqrt{5}}{2}, \frac{9\pm\sqrt{-83}}{6}$.
21. $x=2, \frac{7}{2}, \frac{11\pm\sqrt{-663}}{4}$.
22. $x=0, \frac{5}{7}, 3, -\frac{16}{7}$.
23. $x=0, a, \frac{a\pm\sqrt{5a^2-8ab}}{2}$.
24. $x=2, \frac{1}{4}, \frac{9\pm\sqrt{-31}}{8}$.
25. $x=2, -16\frac{1}{2}$.
26. $x=1, -\frac{11}{24}$.
27. $x=\frac{1}{2}, -\frac{5}{7}$.
28. $x=3, -3\frac{4}{7}$.
29. $x=3, -\frac{2}{3}$.
30. $x=\pm 2$.
31. $x=\pm\frac{1}{2}$.
32. $x=4, \frac{1}{2}$.
33. $x=\pm 2$.
34. $x=\pm\frac{2}{3}$.
35. $x=4, -\frac{1}{4}$.
36. $x=\frac{1}{2a}(1\pm\sqrt{1+4b^2})$.
37. $x=\pm\frac{2a}{\sqrt{5}}$.
38. $x=\frac{a(2\pm\sqrt{2})}{2}$.
39. $x=b\pm\sqrt{b^2-ab}$.
40. $x=\pm 2b$.
41. $x=\pm\frac{a}{2}\sqrt{\frac{a^2-4}{a^2-1}}$.
42. $x=0, a$.
43. $x=\pm\sqrt{ab}$.
44. $x=\pm\frac{2a}{\sqrt{-3}}$.
45. $x=a, b$.

ANSWERS. 335

46. $x = \dfrac{a-b}{2} \pm \dfrac{(a+b)c}{2\sqrt{c^2+4}}.$

47. $x = \dfrac{a+b}{2} \pm \dfrac{(a-b)c}{2\sqrt{c^2-4}}.$

48. $x = \dfrac{a+b}{2} \pm \dfrac{a-b}{2c}\sqrt{c^2+4}.$

49. $x = -\dfrac{2ab}{a+b}.$

50. $x = \pm \dfrac{5}{\sqrt{3}}.$

51. $x = \pm\sqrt{5}.$

52. $x = \pm\sqrt{\dfrac{7}{3}}.$

53. $x = 9a, \ -a.$

54. $x = \dfrac{k^2 \pm k\sqrt{k^2+4a}}{2}$ where $k = -\dfrac{a}{2} \pm \dfrac{\sqrt{a^2+4b}}{2}.$

55. $x = \dfrac{b \pm \sqrt{b^2-2ab}}{2}.$

56. $x = 55 \pm 24\sqrt{5}.$

57. $x = -\dfrac{a}{2}(1 \mp \sqrt{5}).$

58. $x = 0, \ \dfrac{63}{65}a.$

59. $x = 2 - \sqrt{3}, \ 2(\sqrt{3}-2).$

60. $x = \dfrac{1}{2}(-1 \pm \sqrt{5}).$

EXERCISE XCIII. (PAGE 267.)

1. $x = 10, 30;$
   $y = 30, 10.$
2. $x = 4, 9;$
   $y = 9, 4.$
3. $x = 50, -5;$
   $y = 5, -50.$
4. $x = 25, 4;$
   $y = 4, 25.$
5. $x = 12, 6;$
   $y = 6, 12.$
6. $x = 18, -3;$
   $y = 3, -18.$
7. $x = 10, 2;$
   $y = 2, 10.$
8. $x = 20, -6;$
   $y = 6, -20.$
9. $x = 40, 9;$
   $y = 9, 40.$
10. $x = 13, -3;$
    $y = 3, -13.$
11. $x = -\dfrac{1}{6}(5 \pm \sqrt{265}),$
    $y = \dfrac{35 \pm \sqrt{265}}{24}.$
12. $x = 4, -\dfrac{11}{4};$
    $y = 1, -\dfrac{5}{4}.$
13. $x = 6, -\dfrac{42}{5};$
    $y = 3, -\dfrac{33}{5}.$
14. $x = -\dfrac{10 \pm \sqrt{138}}{2},$
    $y = \dfrac{30 \pm \sqrt{138}}{6}.$

15. $x = 3, 3;$
    $y = 3, 3.$

16. $x = -\dfrac{3 \pm \sqrt{177}}{42},$
    $y = \dfrac{9 \pm \sqrt{177}}{24}.$

17. $x = 4, 2;$
    $y = 2, 4.$

18. $x = \dfrac{1}{2}, \dfrac{1}{3};$
    $y = \dfrac{1}{3}, \dfrac{1}{2}.$

19. $x = -\dfrac{6 \pm 6\sqrt{57}}{7},$
    $y = -3 \pm \sqrt{57}.$

20. $x = \dfrac{1}{3}, \dfrac{1}{4};$
    $y = \dfrac{1}{4}, \dfrac{1}{3}.$

EXERCISE XCIV. (Page 270.)

1. $x = \pm \dfrac{5}{3}, \pm 3;$
   $y = \pm \dfrac{13}{3}, \pm 5.$

2. $x = \pm 5\sqrt{\dfrac{3}{7}},$
   $y = \pm \sqrt{\dfrac{3}{7}}.$

3. $x = \pm 2, \mp \sqrt{\dfrac{2}{5}};$
   $y = \pm \dfrac{1}{2}, \pm 2\sqrt{\dfrac{2}{5}}.$

4. $x = \pm 7, \mp \dfrac{5}{\sqrt{2}};$
   $y = \pm 2, \pm \dfrac{9}{\sqrt{2}}.$

5. $x = \pm 3\sqrt{3}, \pm 4;$
   $y = \pm \sqrt{3}, \pm 5.$

6. $x = \pm 2, \pm \sqrt{2};$
   $y = \pm 4, \pm 3\sqrt{2}.$

7. $x = \pm 3, \pm 5\sqrt{\dfrac{1}{2}};$
   $y = \pm 2, \pm \sqrt{\dfrac{1}{2}}.$

8. $x = \pm 3,$
   $y = \pm 4.$

9. $x = \pm 2, \pm 2\tfrac{15}{16};$
   $y = \pm 3, \pm 2\tfrac{8}{15}.$

10. $x = \pm 3,$
    $y = \pm 1.$

11. $x = \pm 7,$
    $y = \pm 5.$

12. $x = \pm 6, \pm \sqrt{\dfrac{1}{2}};$
    $y = \pm 5, \mp 11\sqrt{\dfrac{1}{2}}.$

13. $x = \pm 6,$
    $y = \pm 5.$

14. $x = \pm 1, \pm \dfrac{1}{\sqrt{2}};$
    $y = \pm 2, \pm \dfrac{3}{\sqrt{2}}.$

15. $x = 4, -3;$
    $y = 3, -4.$

ANSWERS.   337

16. $x = \pm 6,$ 17. $x = \pm 2, \pm \dfrac{46}{7};$ 18. $x = \pm 2, \pm \dfrac{10}{3}\sqrt{-\dfrac{3}{7}};$
    $y = \pm 3.$
     $y = \pm 5, \pm \dfrac{5}{7}.$ $y = \pm 1, \mp 6\sqrt{-\dfrac{3}{7}}.$

19. $x = \pm 12, \pm 9;$ 20. $x = \pm 4, \pm \dfrac{85}{8};$ 21. $x = \pm \dfrac{5}{\sqrt{2}}, \pm 7;$
    $y = \pm 9, \pm 12.$
     $y = \pm 9, \pm \dfrac{19}{8}.$ $y = \pm \dfrac{9}{\sqrt{2}}, \pm 2.$

22. $x = \pm 8, \pm 2;$ 23. $x = \pm 6,$ 24. $x = \pm 3\sqrt{3}, \pm 4;$
    $y = \pm 2, \pm 8.$ $y = \pm 2.$ $y = \pm \sqrt{3}, \pm 5.$

25. $x = \pm 4, \pm 2;$ 26. $x = \pm 3\sqrt{2},$ 27. $x = 0, 15;$
    $y = \pm 2, \pm 4.$ $y = \pm \sqrt{2}.$ $y = 0, 45.$

28. $x = 3, 4, -6 \pm 2\sqrt{6};$
    $y = 4, 3, -6 \mp 2\sqrt{6}.$

EXERCISE XCV. (Page 273.)

1. 10, 11, 12.   2. 3, 4.   3. 1, 2, 3.   4. 12.
5. $31\frac{1}{4}$ cents.   6. $11.25.   7. 12 pieces.   8. 12 inches.
9. 121 and 120 yards.   10. $1\frac{1}{2}$ yards.
11. $30 - 10\sqrt{5}$ and $10\sqrt{5} - 10$.   12. 7 hours and 5 hours.
13. 5 hours and 3 hours.   14. 39.   15. 35.
16. 78.   17. 15, 20.   18. 5 miles per hour.   19. $80
20. $60.   21. 100 shares.   22. 5 per cent.   23. 7, 5.
24. 34, 43.   25. $\dfrac{1}{2}(3 \pm \sqrt{5}), \dfrac{1}{2}(1 \pm \sqrt{5})$.   26. $1 \pm \sqrt{2}$.
27. 16, 10.   28. 9, 12, 15.   29. 5, 4.
30. $6000, $7000; 7 per cent. and 6 per cent.
31. 20 barrels by A, 16 by B. A's price, £1 15s.; B's, £1 14s
32. 12 inches.   33. 12, 16, 20.   34. 8 and 10 ft.   35. $\dfrac{2}{3}, \dfrac{3}{4}.$

23

EXERCISE XCVI. (Page 288.)

4. $\dfrac{a_1(b+c) + b_1(c+a) + c_1(a+b)}{a_1 + b_1 + c_1}$.

5. $x^2 - x\left\{\left(\dfrac{b^2 - 2ac}{ac}\right)^2 - 2\right\} + 1 = 0$, $x^2 - x\left\{\dfrac{(3abc - b^3)^2}{c^3 a^3} - 2\right\} + 1 = 0$

6. $x^2 + \dfrac{r^3}{32}x - \dfrac{91 r^6}{512} = 0$.   12. $x^2 - 4mnx - (m^2 - n^2)^2 = 0$.

13. $x^2 - x\left\{\dfrac{(b^2 - 2ac)(a^2 + c^2)}{a^2 c^2}\right\} + \left(\dfrac{b^2 - 2ac}{ac}\right)^2 = 0$.

15. $2b^2 = 9ac$.   16. $x^2 - 3 = 0$.   17. $x^2 - 2mx + m^2 - n^2 = 0$

18. $(a_1 c - ac_1)^2 = (a_1 b - ab_1)(b_1 c - bc_1)$.   20. $m = 8$.   29. $m = 6, \dfrac{2}{3}$

30. Roots of $x^2(a+b+c)^2 + 3x(abc) + abc(a+b+c) = 0$.

EXERCISE XCVII. (Page 294.)

1. An *identity*, true for any value of x.
2. No value of x can satisfy the equation; it is impossible unless $a = b$ or c.
3. $x = \dfrac{a + b + c}{3}$.
4. True for any value of x.
5. Impossible unless $(m + n)(a + b) = 0$.
6. Impossible unless $a + b + c = 0$.
7. An indefinite number of solutions may be given. One solution is $x = 5$, $y = 10$, $z = 5$.
8. $x = 3$, but y and z are indeterminate and may have any values so that $2y + 3z = 7$. One solution is $x = 3$, $y = 2$, $z = 1$.
9. Impossible.
10. Impossible unless $m + n + p = 0$, and then an indefinite number of solutions may be given.
11. The second equation is simply the cube of the first.

www.ingramcontent.com/pod-product-compliance
Lightning Source LLC
Chambersburg PA
CBHW030007240426

43672CB00007B/855